The Foundations of Signal Integrity

Paul G. Huray

IEEE
IEEE PRESS

WILEY

A John Wiley & Sons, Inc., Publication

Published by John Wiley & Sons, Inc., Hoboken, New Jersey
Published simultaneously in Canada

Library of Congress Cataloging-in-Publication Data:

Huray, Paul G., 1941–
The foundations of signal integrity / Paul G. Huray.
p. cm.
Includes bibliographical references and index.
ISBN 978-0-470-34360-9
1. Signal integrity (Electronics) 2. Electromagnetic interference—Prevention.
3. Electric lines. I. Title.
TK7867.2.H87 2010
621.382′2–dc22

2009018610

Printed in the United States of America

10 9 8 7 6 5 4 3 2 1

Contents

4. Ideal Models vs Real-World Systems 109

5. Complex Permittivity of Propagating Media 145

6. Surface Roughness 216

Preface

This book marries the principles of solid-state physics with the mathematics of time-retarded solutions to Maxwell's equations. It includes the quantum mechanical nature of magnetism in thermal equilibrium with materials to explain how electromagnetic waves propagate in solid materials and across boundaries between dielectrics and insulators. The text uses electromagnetic scattering analysis to show how electromagnetic fields induce electric and magnetic multipoles in "good" conductors and how that process leads to delay, attenuation, and dispersion of signals in transmission lines. The text explains the basis for boundary conditions used with the vector forms of Maxwell's equations to describe analytic problems that can be solved by the first and second Born approximation for real-world applications through successive approximations of

- perfect flat boundaries to boundaries with nanometer deviations,
- perfect electric conductors to materials with finite conductivity, and
- inclusions of multiple impurities in otherwise homogeneous media.

Finally, the text gives examples of how system-level printed circuit board (PCB) geometries can use these principles to numerically simulate solutions for very complex systems.

This book is intended to be a foundation for the discipline of electricity and magnetism upon which measurements, simulations, and "rules-of-thumb" are built through the rigorous application of Maxwell's equations. Assumptions are stated when they are employed, and the set of steps known as the Born approximations is used to show the relative magnitude of neglected terms. In that sense, this is intended to be a book that takes carefully applied theory to practice. It is written in the language of an electrical engineer rather than a mathematician or physicist and is intended to support engineering practice.*

PROBLEMS ADDRESSED

As bit rates of computers have increased into the tens of gigahertz, scientists and engineers have recognized that a less-than-rigorous knowledge of electromagnetic

* Textbooks that support design practices are *Advanced Signal Integrity for High-Speed Digital Designs* by Stephen H. Hall and Howard L. Heck (John Wiley & Sons, 2009); *High-Speed Digital System Design* by Stephen H. Hall, Garrett W. Hall, and James A. McCall (John Wiley & Sons, 2000); and *High-Speed Signal Propagation: Advanced Black Magic* by Howard Johnson and Martin Graham (Prentice Hall, 2003).

field propagation can yield incomplete or even contradictory concepts about attenuation, phase, and dispersion of received electric signals that represent information. Many books present concepts of electricity and magnetism via models of transmitted power in terms of low-frequency harmonic potentials and currents that then yield "rules-of-thumb" that are extended to higher frequencies by modifying the definition of resistance, capacitance, or inductance. Simulation codes often neglect the relatively slow propagation of electromagnetic fields in conductors when solving for propagation of those quantities in a dielectric medium. On physically large circuit boards, the propagation speed of electric signals requires dozens or even hundreds of bits of information to be "on their way" from a transmitter to a receiver, so that timing budgets require picoseconds precision. Some solutions are made by using quasistatic (or other) approximations that are forgotten when applied to situations that violate those assumptions; for numerical simulation software, the assumptions are often not even stated. Most engineering models that are chosen to represent "real-world" transmission lines, vias, or packages make simplifying assumptions that cannot be justified based on the complexity of microscopic examination. Power losses on printed circuit boards are so large at high frequencies that signal-to-noise ratio is unacceptable to preserve targeted bit error rates or to recommend new procedures or processes for fabrication needed for higher speed applications. In short, many intuitive concepts that are learned in undergraduate courses for simple transverse electromagnetic (TEM) field propagations simply do not carry over into the real world of conducting boundaries when employing microwave frequencies is tried.

Most existing texts on signal integrity do **not** provide a **foundational basis** of signal integrity principles based on the propagation of electromagnetic fields but base explanations on traditional circuit theory parameter (resistance, inductance, conductance, capacitance—RLGC) models with plausibility arguments that are comforting to the intuition. However, some of these plausible explanations lead to incorrect pictures of behavior of currents, which cause conundrums for the students. These texts do not explain how electron charge and currents physically distribute themselves in space and time for a complex transmission line that includes "good" conductors and "complex dielectrics." The nonrigorous solutions can also lead students to causal contradictions, conduction electrons that travel faster than the speed of light, and nonsense phrases like currents that "rush-over" imperfections or "crowd" at discontinuous surfaces.

FEATURES OF THE BOOK

Causal electric and magnetic field quantities are color coordinated throughout the book. For example, electric charge density, electric field intensity, electric flux density, scalar electric potential, and vector electric potential, versus current density, magnetic field intensity, magnetic flux density, scalar magnetic potential, and vector magnetic potential are consistently identified, along with the symbols that pertain to those quantities in equations and vector lines that correspond in figures. It is revealing to see that time derivatives of those quantities (e.g., dq/dt) change their causal character and that it is equivalent to state that electric charge *causes* electric field intensity (current *causes* magnetic field intensity) or vice versa. Electric and magnetic field intensity is shown *inside* conductors in the quasistatic approximation, and an analysis of how they move with time is shown to yield dynamic properties that cause them to be conservative (close on themselves).

By using colors, Maxwell's equations are seen to be even more beautifully symmetric than in their black-and-white formats.

RECOGNITION

The author owes a debt of gratitude to Dr. Yinchao Chen of the Electrical Engineering Department at the University of South Carolina, Columbia. Dr. Chen has published articles with the author and has had many discussions on the techniques and meaning of the solutions to Maxwell's equations and their applications. Other USC professors who contributed to the physical and chemical understanding of PCB materials were Michael Myrick of the Chemistry Department and Richard Webb of the Physics Department.

Huray, Chen, and three Signal Integrity engineers (Brian Knotts, Hao Li, and Richard Mellitz) from the Intel Corporation (Columbia, SC) created the first graduate Signal Integrity program in 2003, which has since produced more than 80 practicing Signal Integrity engineers, many of whom read and corrected early drafts of this text.

Huray conducts industrial research on a part-time basis with the Intel Corporation in the area of high-speed electromagnetic signals. In this work, he has had the privilege to work closely with Richard Mellitz and Stephen Hall, on applications of electromagnetism for practical use. It was their penetrating questions that prompted many of the explanations in this text. Another Intel employee, Dan Hua, provided a sequence of exchanged articles on the evaluation of scattering and absorption in the language of vector spherical harmonics; it was through these discussions that the sections on absorption by small good conducting spheres arose. Gary Brist taught the author (and many of his graduate students) about the process of manufacturing PCB stack-ups and stimulated many of the questions that are sprinkled throughout the book. Anusha Moonshiram and Chaitanya Sreerema conducted many of the high-frequency vector network analyzer (VNA) measurements in this text. Femi Oluwafemi conducted many of the numerical simulations on phase analysis to identify time-dependent fields inside good conductors and provided many of the final comparisons to the VNA data. Guy Barnes and Paul Hamilton provided the Fabry–Perot measurements of permittivity. Brandon Gore helped work on magnetic losses, and David Aerne assisted the analysis of spherical composition profiles and near-neighbor interference effects. Peng Ye was a sounding board for arguments about the analytical analysis associated with electromagnetic field dynamics. Kevin Slattery introduced the author to near-field scanning electromagnetic probes and helped direct the work of two USC graduate students, Jason Ramage and Christy Madden Jones, whose work on proof of Snell's law at microwave frequencies and absorption by impurities appears in the text. Intel engineers such as Howard Heck,

Richard Kunze, Ted Ballou, Steve Krooswyk, Matt Hendrick, David Blakenbeckler, and Johnny Gibson passed through USC during the writing of this text to present lectures to the author's Signal Integrity classes and to build richness into the intellectual atmosphere. Mark Fitzmaurice was always ready to help make the Signal Integrity program at USC a success through his support for measurement equipment, student internships, and common sense.

Many USC undergraduate and graduate students contributed to the testing and writing of this book. Steven Pytel worked with the author on scanning electron microscope SEM and analysis measurements at the Oak Ridge National Laboratory in Oak Ridge, TN, and, while working for Intel, was the sounding board for many of the arguments presented here. After receiving his PhD, he became an employee of the Ansoft Corporation, Pittsburg, PA, where he became an applications engineer for Signal Integrity tools. He is primarily responsible for the material in Chapter 8 on numerical simulations. Ken Young helped with editing, Fisayo Adepetun provided assistance with figures, and David London supported Web pages for testing and transmittal of the chapters. Tom McDonough gave lectures to the Signal Integrity classes on the use of Synopsys Corporation, Boston, MA HSPICE software and helped in the analysis of ceramic capacitor fields.

John Fatcheric of the Oak Mitsui Corporation, Camden, SC, assisted the presentation on copper surface production. Bob Helsby, Charles Banyon, and Zol Cendes of the Ansoft Corporation supported the use of forefront numerical solutions to Maxwell's equations. James Rautio of Sonnet Software, Syracuse, NY, assisted on the history of Maxwell and the use of his portrait. Mike Resso of Agilent Corporation, Santa Rosa, CA, supported a joint Intel–Agilent VNA donation. Lee Riedinger, Harry M. Meyer III, Larry Walker, and Marc Garland of the Oak Ridge National Laboratory assisted in making qualitative and quantitative measurements of PCB components by SEM and Auger analysis. José E. Rayas Sánchez of ITESO, Guadalajara, Mexico, James Gover of Kettering University, Flint, MI, and John David Jackson of UC-Berkeley and LBL, Berkeley, CA, provided discussions on Maxwell's interpretations and Signal Integrity of high-speed circuits.

This book is dedicated to the author's lifelong partner:

Susan Lyons Huray

Intent of the Book

The Foundations of Signal Integrity is intended to be a text for a one-semester course in Signal Integrity, under the assumption that the students have a solid foundation in the development and solution techniques of Maxwell's equations. A preliminary text by the author[1] presents that information at a relatively complete level, but it is recognized that students may have had other textbooks for that material. This book presents equations, words, and figures in a consistent, color-coded format so that students can see the relationship between variables of a common type or color. Generally, other textbooks will have used either the symmetric or the asymmetric form of Maxwell's equations as defined below but may have used other symbols for the variables, and they will not generally be color-coded. This section thus presents the form of Maxwell's equations used in *The Foundations of Signal Integrity* with enough introduction that the text may be used by itself.

The Foundations of Signal Integrity concentrates on the solutions to Maxwell's equations in a variety of media and with a variety of boundary conditions. Here, techniques that show how to obtain analytic solutions to Maxwell's equations for ideal materials and boundary conditions are presented. These solutions are then used as a benchmark for the student to solve "real world" problems via computational techniques; first confirming that a computational technique gives the same answer as the analytic solution for an ideal problem.

This information is presented to 21st-century students* in the hope that they will consider mathematical and physical concepts as *integral*. The student is challenged not to accept uncertainty but to be honest within him- or herself in appreciating and understanding the derivations of the electromagnetic giants. After the mathematical solution has been obtained, we hope the student will ask, "What are these equations telling me?" and "How could I use this in some other application?" Perhaps the student will delve even deeper to ask, "What are the physical phenomenon that cause fields to exist, to move, to reflect or to transmit through materials?" With such an armada of knowledge, the student can take these electromagnetic concepts to further applications and to further "stand on the shoulders of giants"†

* One reader from the Physics Web poll that rated Maxwell's equations as the most beautiful equations ever derived recalled how he learned Maxwell's equations during his second year as an undergraduate student. "I still vividly remember the day I was introduced to Maxwell's equations in vector notation," he wrote. That these four equations should describe so much was extraordinary ... For the first time, I understood what people meant when they talked about elegance and beauty in mathematics or physics. It was spine-tingling and a turning point in my undergraduate career."

† The quote "If I have seen farther than others, it is because I have stood on the shoulders of giants" was attributed to Sir Isaac Newton because it appeared in a letter he wrote to Robert Hooke in 1675, but it was also used by an 11th-century monk named John of Salisbury, and there is evidence he may have read it in an older text while studying with Abelard in France.

(perhaps for monetary gain). Sometimes, open-ended questions are asked so that the student questions the giants or questions his or her own set of learned models.

In *Maxwell's Equations*, the justification for using the symmetric form of the equations given in the following table was developed.

Symmetric Form of Maxwell's Equations

Differential form	Integral form	Name
$\vec{\nabla} \times \vec{E} = -\vec{J} - \partial\vec{B}/\partial t$	$\oint_C \vec{E}\cdot d\vec{l} = -I - \iint_S (\partial\vec{B}/\partial t)\cdot d\vec{s}$	Faraday's law
$\vec{\nabla} \times \vec{H} = \vec{J} + \partial\vec{D}/\partial t$	$\oint_C \vec{H}\cdot d\vec{l} = I + \iint_S (\partial\vec{D}/\partial t)\cdot d\vec{s}$	Ampere's law
$\vec{\nabla} \cdot \vec{D} = \rho_V$	$\oiint_S \vec{D}\cdot d\vec{s} = Q$	Gauss's law for electric charge
$\vec{\nabla} \cdot \vec{B} = \rho_V$	$\oiint_S \vec{B}\cdot d\vec{s} = Q$	Gauss's law for magnetic charge

The symmetric form of Maxwell's equations represents the vector field quantities:

$\vec{E}$ = Electric field intensity (Volts/meter).
$\vec{H}$ = Magnetic field intensity (Ampere/meter)
$\vec{D}$ = Electric flux density (Coulombs/meter^2)
$\vec{B}$ = Magnetic flux density (Weber/meter^2 or Tesla)
ρ_V = Electric charge density (Coulomb/meter^3)
ρ_V = Magnetic charge density (Weber/meter^3)
$\vec{J}$ = Magnetic current density (Volts/meter^2)
$\vec{J}$ = Electric current density (Ampere/meter^2)

with the units of the new field quantities in SI units shown in parenthesis.

The equation of continuity was developed for both electric and magnetic charge density by using conservation of charge to write the symmetric forms[2]:

$$\vec{\nabla}\cdot\vec{J} = -\partial\rho_V/\partial t$$

$$\vec{\nabla}\cdot\vec{J} = -\partial\rho_V/\partial t$$

Based on the symmetric equations, we can see that, in a *magnetic charge-free* region of space, $\vec{B}$ is solenoidal ($\vec{\nabla} \cdot \vec{B} = 0$), and, because the divergence of the curl of *any* vector field is identically zero, we can thus assume that $\vec{B}$ may be written in terms of another vector field, $\vec{A}$, called the *magnetic vector potential*:

$$\vec{B} = \vec{\nabla} \times \vec{A}.$$

In a *magnetic current-free* region of space, the symmetric equations are the same as the asymmetric equations most physicists use as Maxwell's equations.

In an *electric charge-free* region of space, $\vec{D}$ is solenoidal ($\vec{\nabla} \cdot \vec{D} = 0$), and we can assume that $\vec{D}$ may be written in terms of another vector field, $\vec{A}$, called the *electric vector potential:*

$$\vec{D} = \vec{\nabla} \times \vec{A}$$

For charge and current density-free space ($\rho_V = 0$, $\rho_V = 0$, $\vec{J} = 0$ and $\vec{J} = 0$), a **unique** definition of the vector fields, $\vec{A}$ and $\vec{A}$, may be specified through additional restrictions ($\vec{\nabla} \times \vec{E} = -\partial\vec{B}/\partial t$) and ($\vec{\nabla} \times \vec{H} = \partial\vec{D}/\partial t$), so we can write

$$\vec{\nabla} \times \vec{E} = -\partial\left(\vec{\nabla} \times \vec{A}\right)/\partial t \quad \text{or} \quad \vec{\nabla} \times \left(\vec{E} + \partial\vec{A}/\partial t\right) = 0$$
$$\vec{\nabla} \times \vec{H} = -\partial\left(\vec{\nabla} \times \vec{A}\right)/\partial t \quad \text{or} \quad \vec{\nabla} \times \left(\vec{H} + \partial\vec{A}/\partial t\right) = 0$$

One can also show that $\vec{\nabla} \times (-\vec{\nabla}V) = 0$ for *any* scalar field.[3] Thus, because the curl of the vector field shown in parentheses above is zero, then that field can be written as the negative gradient of another scalar field that is successively called the *electric scalar potential*, V, and the *magnetic scalar potential*, V, with

$$\vec{E} + \partial\vec{A}/\partial t = -\vec{\nabla}V \quad \text{or} \quad \vec{E} = -\vec{\nabla}V - \partial\vec{A}/\partial t$$
$$\vec{H} + \partial\vec{A}/\partial t = -\vec{\nabla}V \quad \text{or} \quad \vec{H} = -\vec{\nabla}V - \partial\vec{A}/\partial t$$

We can see from the first of these equations that the electric field intensity, $\vec{E}$, can be written in terms of the electric scalar potential, V, and the time derivative of the magnetic vector potential, $\vec{A}$. As long as these scalar and vector potentials are unique, the electric field intensity produced by them will also be unique. *Note:* In the special case of *static* (time independent) fields and potentials, $\partial\vec{A}/\partial t = 0$, and $\partial\vec{A}/\partial t = 0$ the electric and magnetic field intensities reduce to $\vec{E} = -\vec{\nabla}V$ and $\vec{H} = -\vec{\nabla}V$ as Maxwell originally proposed.

For homogeneous media in time-varying fields ($\vec{B} = \mu\vec{H}$ and $\vec{D} = \varepsilon\vec{E}$), the symmetric forms yield $\vec{\nabla} \times \vec{B} = \mu\vec{J} + \mu\varepsilon\, \partial\vec{E}/\partial t$ or $\vec{\nabla} \times (\vec{\nabla} \times \vec{A}) = \mu\vec{J} + \mu\varepsilon\, \partial\vec{E}/\partial t$ or $\vec{\nabla} \times \vec{\nabla} \times \vec{A} = \mu\vec{J} + \mu\varepsilon\, \partial(-\vec{\nabla}V - \partial\vec{A}/\partial t)/\partial t$, and using identity $\vec{\nabla} \times \vec{\nabla} \times \vec{A} = \vec{\nabla}(\vec{\nabla} \cdot \vec{A}) - \vec{\nabla}^2\vec{A}$

$$\vec{\nabla}\left(\vec{\nabla}\cdot\vec{A}\right) - \vec{\nabla}^2\vec{A} = \mu\vec{J} - \vec{\nabla}(\mu\varepsilon\,\partial V/\partial t) - \mu\varepsilon\,\partial^2\vec{A}/\partial t^2 \quad \text{or}$$
$$\vec{\nabla}^2\vec{A} - \mu\varepsilon\partial^2\vec{A}/\partial t^2 = -\mu\vec{J} + \vec{\nabla}\left(\vec{\nabla}\cdot\vec{A} + \mu\varepsilon\,\partial V/\partial t\right).$$

Likewise, the symmetric form $\vec{\nabla} \times \vec{E} = -\vec{J} - \partial\vec{B}/\partial t$ or $\vec{\nabla} \times (\vec{\nabla} \times \vec{A}) = -\varepsilon\vec{J} - \varepsilon\partial\vec{B}/\partial t$ or $\vec{\nabla} \times \vec{\nabla} \times \vec{A} = \varepsilon\vec{J} + \mu\varepsilon\, \partial(-\vec{\nabla}V - \partial\vec{A}/\partial t)/\partial t$ and using identity $\vec{\nabla} \times \vec{\nabla} \times \vec{A} = \vec{\nabla}(\vec{\nabla} \cdot \vec{A}) - \vec{\nabla}^2\vec{A}$

$$\vec{\nabla}\left(\vec{\nabla}\cdot\vec{A}\right) - \vec{\nabla}^2\vec{A} = \varepsilon\vec{J} - \vec{\nabla}(\mu\varepsilon\,\partial V/\partial t) - \mu\varepsilon\,\partial^2\vec{A}/\partial t^2 \quad \text{or}$$
$$\vec{\nabla}^2\vec{A} - \mu\varepsilon\partial^2\vec{A}/\partial t^2 = -\varepsilon\vec{J} + \vec{\nabla}\left(\vec{\nabla}\cdot\vec{A} + \mu\varepsilon\,\partial V/\partial t\right)$$

Now, the definition of a **unique** vector field $\vec{A}$ or $\vec{A}$ requires an additional restriction or gauge. One way to provide this restriction (gauge) is to specify their divergence. Lorenz used the now-called Lorenz gauge to write[4]

$$\vec{\nabla}\cdot\vec{A}+\mu\varepsilon\partial V/\partial t=0$$

$$\vec{\nabla}\cdot\vec{A}+\mu\varepsilon\partial V/\partial t=0$$

From a mathematical solutions perspective, that choice is convenient because it requires $\vec{A}$ and $\vec{A}$ to satisfy second-order, linear, inhomogeneous partial differential equations (PDEs):

$$\vec{\nabla}^2\vec{A}-\mu\varepsilon\partial^2\vec{A}/\partial t^2=-\mu\vec{J}$$

$$\vec{\nabla}^2\vec{A}-\mu\varepsilon\partial^2\vec{A}/\partial t^2=-\varepsilon\vec{J},$$

which are called the inhomogeneous wave equation for the *magnetic vector potential* and the inhomogeneous wave equation for the *electric vector potential.* To solve these equations for $\vec{A}$ or $\vec{A}$, the current density, $\vec{J}$ or $\vec{J}$, is needed.

A corresponding wave equation for the electric scalar potential can be found by using Gauss's law $\vec{\nabla}\cdot\vec{D}=\rho_V$ and $\vec{\nabla}\cdot\vec{E}=\rho_V/\varepsilon \Rightarrow \vec{\nabla}\cdot(\vec{\nabla}V+\partial\vec{A}/\partial t)=-\rho_V/\varepsilon$, which leads to $\vec{\nabla}^2V+\partial(\vec{\nabla}\cdot\vec{A})/\partial t=-\rho_V/\varepsilon$, and, by using the Lorenz gauge ($\vec{\nabla}\cdot\vec{A}+\mu\varepsilon\partial V/\partial t=0$), we see that the *electric scalar potential*, V, also satisfies the inhomogeneous wave equation

$$\vec{\nabla}^2V-\mu\varepsilon\partial^2V/\partial t^2=-\rho_V/\varepsilon$$

This equation needs only ρ_V to solve for the electric scalar potential, V.

Likewise, a corresponding wave equation for the magnetic scalar potential can be found by using Gauss's law $\vec{\nabla}\cdot\vec{B}=\rho_V$ and $\vec{\nabla}\cdot\vec{H}=\mu\rho_V \Rightarrow \vec{\nabla}\cdot(\vec{\nabla}V+\partial\vec{A}/\partial t)=-\mu\rho_V$ or $\vec{\nabla}^2V+\partial(\vec{\nabla}\cdot\vec{A})/\partial t=-\mu\rho_V$, and, by using the Lorenz gauge ($\vec{\nabla}\cdot\vec{A}+\mu\varepsilon\,\partial V/\partial t=0$), we see that the *magnetic scalar potential*, V, also satisfies the inhomogeneous wave equation

$$\vec{\nabla}^2V-\mu\varepsilon\partial^2V/\partial t^2=-\mu\rho_V$$

This equation needs only ρ_V to solve for the magnetic scalar potential, V.

Symmetric Form Conclusion

With a prior knowledge of ρ_V, ρ_V, $\vec{J}$, and $\vec{J}$, we can separate the x, y, and z components of the wave equations and solve for V and V and each component of $\vec{A}$ and $\vec{A}$ independently of the others. All four of these equations are of in the form of the same inhomogeneous wave equation and are independent of one another. Thus, given the electric charge density, the magnetic charge density, the vector electric current density, and the vector magnetic current density, we can solve the inhomogeneous wave equation (subject to boundary conditions specified by a particular application) to find the potentials V, V, $\vec{A}$, and $\vec{A}$ from which we can then find all of the components of the electric field intensity and magnetic field intensity. The inhomogeneous wave equations for V, V, $\vec{A}$, and $\vec{A}$ form a set of four equations equivalent in all respects to the symmetric Maxwell's equations (subject to the restriction of the Lorenz gauge). However, unlike Maxwell's equations, these four inhomogeneous PDEs are independent of one another so they are often easier to solve.

NOTE Using the electric vector potential and the magnetic vector potential results in electric and magnetic fields that originate from $\vec{B} = \vec{\nabla} \times \vec{A}$, $\vec{D} = \vec{\nabla} \times \vec{A}$, $\vec{E} = -\vec{\nabla}V - \partial\vec{A}/\partial t$, and $\vec{H} = -\vec{\nabla}V - \partial\vec{A}/\partial t$. The resulting electric and magnetic field intensity is the vector sum as a result of both potentials: $\vec{E}_{total} = -\vec{\nabla}V - \partial\vec{A}/\partial t + \vec{\nabla} \times \vec{A}/\varepsilon$ and $\vec{H}_{total} = -\vec{\nabla}V - \partial\vec{A}/\partial t + \vec{\nabla} \times \vec{A}/\mu$.

Engineers sometimes use electric vector potential and magnetic vector potential to develop solutions because they are easier to find via the inhomogeneous wave equations with boundary conditions. The solutions can be chosen to have boundary conditions so that one part of the solution yields a transverse electromagnetic (TEM), transverse electric (TE), or transverse magnetic (TM) solution in a particular coordinate system. However, this approximation is poor when considering fields in the microscopic near-field regime so that the two-vector potential technique will not suffice for the analysis of crystal field effects or fields internal to atoms or molecules.

The physics community usually assumes that there is no such thing as magnetic charge density or magnetic current density so that $\rho_V = 0$ and $\vec{J} = 0$. In this formalism, Maxwell's equations are equivalent to their asymmetric form shown below. Because we will often evaluate near-fields, the asymmetric form of Maxwell's equations will be used in this book to find solutions to applied problems in Signal Integrity.

Asymmetric Form of Maxwell's Equations[‡,5]

Differential form	Integral form
$\vec{\nabla} \times \vec{E} = -\dfrac{\partial \vec{B}}{\partial t}$	$\oint_C \vec{E} \cdot d\vec{l} = -\dfrac{d\Phi_B}{dt}$
$\vec{\nabla} \times \vec{H} = \vec{J} + \dfrac{\partial \vec{D}}{\partial t}$	$\oint_C \vec{H} \cdot d\vec{l} = I + \int_S \dfrac{\partial \vec{D}}{\partial t} \cdot d\vec{s}$
$\vec{\nabla} \cdot \vec{D} = \rho_V$	$\oint_C \vec{D} \cdot d\vec{s} = Q$
$\vec{\nabla} \cdot \vec{B} = 0$	$\oint_S \vec{B} \cdot d\vec{s} = 0$

For the *special case* of *source-free* problems (i.e., $\rho_V = 0$ and $\vec{J} = 0$), we can see that both the symmetric and asymmetric forms of Maxwell's equations reduce to:

‡ Oliver Heaviside reformulated Maxwell's equations (originally in quaternion format) to this asymmetric vector form.

Maxwell's Equations for *Source-Free* Problems

Differential form	Integral form	Name of law
$\vec{\nabla} \times \vec{E} = -\partial\vec{B}/\partial t$	$\oint_C \vec{E}\cdot d\vec{l} = -\iint_S (\partial\vec{B}/\partial t)\cdot d\vec{s}$	Faraday's law
$\vec{\nabla} \times \vec{H} = \partial\vec{D}/\partial t$	$\oint_C \vec{H}\cdot d\vec{l} = \iint_S (\partial\vec{D}/\partial t)\cdot d\vec{s}$	Ampere's law
$\vec{\nabla} \cdot \vec{D} = 0$	$\oint_S \vec{D}\cdot d\vec{s} = 0$	Gauss's law
$\vec{\nabla} \cdot \vec{B} = 0$	$\oint_S \vec{B}\cdot d\vec{s} = 0$	No isolated magnetic charge

So if we take the curl of Faraday's law, $\vec{\nabla} \times \vec{\nabla} \times \vec{E} = -\vec{\nabla} \times \partial\vec{B}/\partial t$ or $\vec{\nabla}(\vec{\nabla} \cdot \vec{E}) - \vec{\nabla}^2\vec{E} = -\mu\partial(\vec{\nabla} \times \vec{H})/\partial t$ and substitute Gauss's law ($\vec{\nabla} \cdot \vec{E} = 0$) and Ampere's Law, we see

$$\vec{\nabla}^2\vec{E} - \mu\varepsilon\partial^2\vec{E}/\partial t^2 = 0$$

Likewise, taking the curl of Ampere's law,

$$\vec{\nabla}\times\vec{\nabla}\times\vec{H} = \vec{\nabla}\times\partial\vec{D}/\partial t \text{ or } \vec{\nabla}\left(\vec{\nabla}\cdot\vec{H}\right) - \vec{\nabla}^2\vec{H} = \varepsilon\partial\left(\vec{\nabla}\times\vec{D}\right)/\partial t$$

and using ($\vec{\nabla} \cdot \vec{H} = 0$) with Faraday's law, we see

$$\vec{\nabla}^2\vec{H} - \mu\varepsilon\partial^2\vec{H}/\partial t^2 = 0$$

Asymmetric Form Conclusion

In *source-free space*, V, all of the components of $\vec{A}$, all of the components of $\vec{V}$, and all of the components of $\vec{H}$ satisfy the homogeneous wave equation, and we will label $\mu\varepsilon = 1/u_p^2$ and $\mu_0\varepsilon_0 = 1/c^2$.

TIME-RETARDED SOLUTIONS TO MAXWELL'S EQUATIONS

The solution of the inhomogeneous wave equation is a linear combination of the general solution to the homogeneous equation (with coefficients determined by boundary conditions) plus a particular solution of the inhomogeneous wave equation. For the equations above,

$$\vec{\nabla}^2\psi - \mu\varepsilon\partial^2\psi/\partial t^2 = f(\vec{x},t)$$

$$\text{where}\quad f(\vec{x},t) = -\rho(\vec{x},t)/\varepsilon \quad \text{when } \psi(\vec{x},t) = V(\vec{x},t) \quad \text{and}$$

$$f(\vec{x},t) = -\mu J_i(\vec{x},t) \quad \text{when } \psi(\vec{x},t) = A_i(\vec{x},t)$$

for each of the i components of the *magnetic vector potential* in Cartesian coordinates.

Any technique that provides *a* solution of the inhomogeneous part provides *the* solution because the particular solution is unique. Some authors (e.g., Matthews and Walker) use an informed guess technique, and others (e.g., Jackson) use a formal Green's function technique to obtain an answer. Using the latter Green's function technique for time-varying fields, we can find the solution for an inhomogeneous PDE by first taking its Fourier transform with respect to the variable t.

In 1824, George Green claimed that, if we solve the equation $(\vec{\nabla}^2 - \mu\varepsilon\, \partial^2/\partial t^2)$ $G(\vec{x}, t; \vec{x}', t') = \delta(\vec{x} - \vec{x}')\delta(t - t')$, then (in infinite space with no boundary surfaces) the solution will be

$$\psi(\vec{x}, t) = \int\iiint G(\vec{x}, t; \vec{x}', t') f(\vec{x}', t') d^3x'dt'$$

To solve the differential equation with delta functions on the right-hand side, we can insert the four-dimensional Fourier transform of the Green's function, $g(\vec{k}, \omega)$, on the left-hand side of the equation and the four-dimensional delta function representation on the right-hand side of the equation as follows:

$$G(\vec{x}, t; \vec{x}', t') = \iiint d^3k \int d\omega g(\vec{k}, \omega) e^{j\vec{k}\cdot(\vec{x}-\vec{x}')} e^{-j\omega(t-t')}$$

$$\delta(\vec{x} - \vec{x}')\delta(t - t') = 1/(2\pi)^4 \iiint d^3k \int d\omega e^{j\vec{k}\cdot(\vec{x}-\vec{x}')} e^{-j\omega(t-t')}$$

The result is a simple algebraic equation:

$$g(\vec{k}, \omega) = \left[1/(2\pi)^4\right]\left(k^2 - \mu\varepsilon\omega^2\right)^{-1} = \left[1/(2\pi)^4\right]\left(k^2 - \omega^2/c^2\right)^{-1}$$

and the answer is

$$G(\vec{x}, t; \vec{x}', t') = (-1/4\pi|\vec{x} - \vec{x}'|)\delta((t - t') - |\vec{x} - \vec{x}'|/c)$$

This Green's function is called the *Retarded Green's function* because it exhibits causal behavior associated with the propagation of a wave source to a response location; that is, an effect observed at a point $\vec{x}$ as a result of a source at a point $\vec{x}'$ and time t' will not occur until the wave has had time to propagate the distance $|\vec{x} - \vec{x}'|$, traveling at speed $c = 1/\sqrt{\mu\varepsilon}$.

Finally, we can use the Green's function to find the solution to the inhomogeneous wave equation in the absence of boundary conditions as

$$\psi(\vec{x}, t) = -\int\iiint \frac{\delta((t - t') - |\vec{x} - \vec{x}'|/c)}{4\pi|\vec{x} - \vec{x}'|} f(\vec{x}', t') d^3x'dt'$$

The integration over dt' can be performed to yield the "retarded potential solution"

$$\psi(\vec{x}, t) = -\iiint \frac{[f(\vec{x}', t')]_{retarded}}{4\pi|\vec{x} - \vec{x}'|} d^3x'$$

Conclusion

In this special case, the propagating electric field intensity waves, magnetic field intensity waves, and magnetic vector potential waves are all orthogonal to one another. We call such propagating waves transverse electric (TE^z) and transverse magnetic (TM^z) because they are moving in the z-direction, in phase with the magnetic vector potential. When both TE and TM waves occur in the same propagation (as they do here), the waves are transverse electromagnetic and labeled TEM^z waves.

Relative Magnitudes

We can also use the relationship $k_z = \omega\sqrt{\mu\varepsilon}$ to compare the components of the electric and magnetic field intensity for TEM^z waves as

$$\frac{E^+_{S,x}}{H^+_{S,y}} = \sqrt{\frac{\mu}{\varepsilon}} = Z^+_W = \eta \tag{1.8a}$$

$$-\frac{E^+_{S,y}}{H^+_{S,x}} = \sqrt{\frac{\mu}{\varepsilon}} = Z^+_W = \eta \tag{1.8b}$$

The quantity η is called the ***intrinsic impedance*** of the medium because it is a function only of the permeability and permittivity of the medium. Some texts call this ratio, Z_W, which they call the ***wave impedance***, to remind us that the ratio of an electric field intensity and magnetic field intensity has units of ohms. Thus, this quantity is a measure of the impedance of the medium; the ratio is labeled Z_0 in the case of waves propagating in a vacuum. In air or a vacuum, $\varepsilon = \varepsilon_0 \approx (1/36\pi) \times 10^{-9}$ F/m or (s/Ωm) and $\mu = \mu_0 = 4\pi \times 10^{-7}$H/m or (Ωs/m) so $\eta = Z_0 \approx 120\pi\Omega = 377\Omega$. This is called the ***intrinsic impedance of free space***.

Physical Meaning of the Propagating Wave Equations

Equations 1.6 give us the relative vector directions, phase, and magnitude of $\vec{E}$ and $\vec{H}$ relative to the magnetic vector potential, $\vec{A}$. Without some knowledge of how $\vec{A}$ varies with x and y, we cannot take the partial derivatives. However, the x-direction is just as arbitrary as the z-direction, which we choose to be in the direction of propagation of $\vec{A}$. We can therefore choose the x-direction to be in the direction of the electric field intensity vector, in which case, we write

$$\vec{E}^+(\vec{x}, t) = E_0^+ e^{-j(k_z z - \omega t)}\hat{a}_x \tag{1.9a}$$

$$\vec{H}^+(\vec{x}, t) = \left(E_0^+/\eta\right) e^{-j(k_z z - \omega t)}\hat{a}_y \tag{1.9b}$$

Here, we have chosen the component of $\vec{H}$ to satisfy the ratio condition required by Equation 1.8a.

Assuming the coefficient in 1.9a is a real number, let us now diagram the propagating waves for the real part of the functions 1.9:

$$\mathrm{Re}\left[\vec{E}^{+}(\vec{x},t)\right]=E_0^{+}\cos(k_z z-\omega t)\hat{a}_x \tag{1.10a}$$

$$\mathrm{Re}\left[\vec{H}^{+}(\vec{x},t)\right]=\left(E_0^{+}/\eta\right)\cos(k_z z-\omega t)\hat{a}_y \tag{1.10b}$$

A graph of these functions is shown in Figure 1.1 at time $t = 0$.

In Figure 1.1, we see that, at time $t = 0$, both the electric field intensity and the magnetic field intensity are distributed under a cosine curve envelope in space with a wavelength $\lambda = 2\pi/k_z$ and both envelopes are propagating along the positive z-axis with velocity $u_p = \lambda f = 1/\sqrt{\mu\varepsilon}$.

In this figure, the x-axis direction has been chosen to lie in the direction of the electric field, and Equations 1.7 thus require that the magnetic field must lie in the y-direction. We may use the right-hand rule to see that $\vec{E}\times\vec{H}$ lies in the direction of $\vec{A}$ (the z-direction) at every point in space. Furthermore, the electric field intensity and the magnetic field intensity remain in phase with one another (both are a maximum at the same point in space and both are zero at the same point). For later values of time, both continue to point in their respective x- and y-directions so we say that they are linearly polarized. Finally, we note that the magnitude of the magnetic field envelope $H_0^{+} = E_0^{+}/\eta$, where E_0^{+} is the magnitude of the electric field intensity envelope and $\eta = \sqrt{\mu/\varepsilon}$ is the intrinsic impedance of the medium in which the wave is propagating.

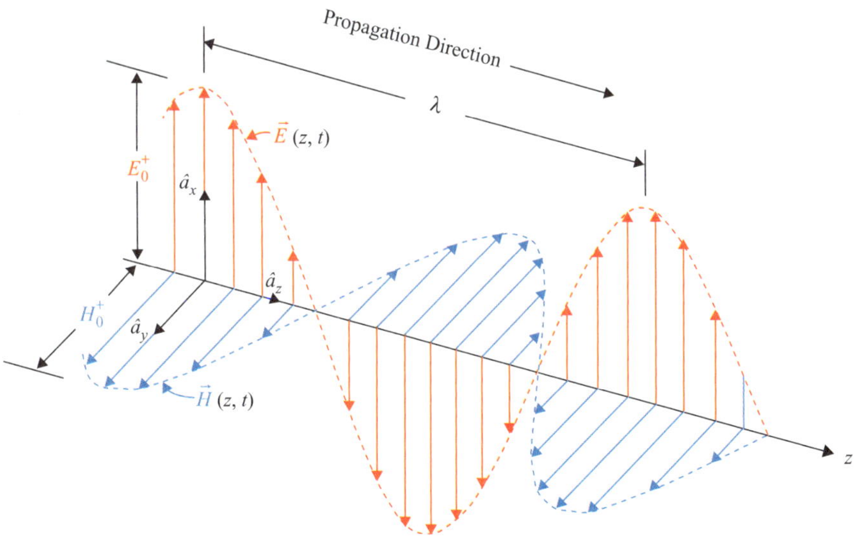

Figure 1.1 Plot of the real parts of the electric and magnetic field intensity as a function of position z, at time $t = 0$ when the x-axis is chosen to lie in the direction of the electric field intensity vector.

NOTE Some texts prefer to graph the magnetic flux density $\vec{B} = \mu\vec{H}$ rather than the magnetic field intensity

$$B_0^+ = \mu\sqrt{\frac{\varepsilon}{\mu}}E_0^+ = \sqrt{\mu\varepsilon}\,E_0^+ = \frac{E_0^+}{u_P} \tag{1.11}$$

because, in the ***special case*** when the propagating medium (e.g., air) has the same permeability and permittivity of free space, $B_0^+ = E_0^+/c$, where c is the speed of light in a vacuum, 2.99792458×10^8 m/s. When the electric field intensity of an electromagnetic wave remains in the same direction as it propagates in a medium, it is said to be linearly polarized. Of course, the relations above show that the magnetic field intensity associated with the wave is also linearly polarized.

1.2 POLARIZED PLANE WAVES

An observer located along the z-axis at a position of maximum electric field (i.e., at position $z = n\lambda$ with n = an integer at $t = 0$) looking back in the $-z$ direction (as shown in Figure 1.2a) would see the electric and magnetic field intensity, as shown in Figure 1.2b.

As a function of time, an observer at $z = n\lambda$ would measure the electric field intensity to be a maximum (in the x-direction) at time $t = 0$, as shown in Figure 1.2b, then observe it to decrease to zero by time $t = (1/4)(\lambda/c)$, then observe it to further decrease to its maximum negative value by time $t = (1/2)(\lambda/c)$, then increase back to zero by $t = (3/4)(\lambda/c)$, then increase back to its maximum positive value by $t = \lambda/c$, and so forth in a cosinusoidal manner with time. The magnetic field intensity

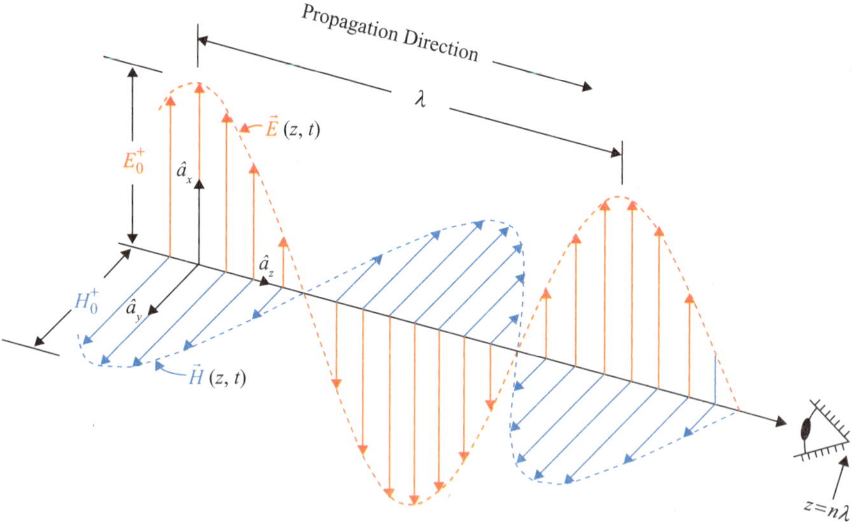

Figure 1.2 (a) Observer at $z = n\lambda$ (n = integer);

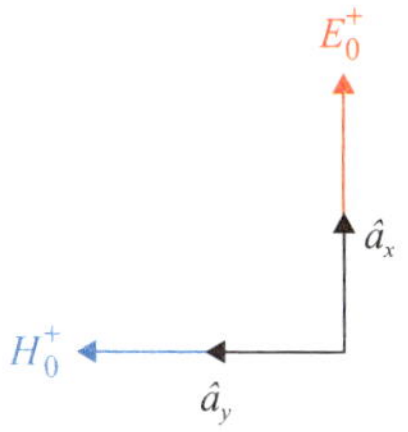

Figure 1.2 (b) electric and magnetic field intensity components observed at time $t = 0$.

would be behaving in a similar manner except it would occur only in the y-direction, and its amplitude would be $H_0^+ = E_0^+/\eta$.

More General Case

If we express the field intensity in the general case (not choosing the x-axis to lie in the direction of the electric field intensity), Equations 1.6a and 1.6b specify their components:

$$\vec{E}_S^+(z) = \frac{-k_z}{\omega\varepsilon\mu}\frac{\partial A_z^+}{\partial x}e^{-jk_z z}\hat{a}_x + \frac{-k_z}{\omega\varepsilon\mu}\frac{\partial A_z^+}{\partial y}e^{-jk_z z}\hat{a}_y$$
$$\vec{E}^+(z,t) = E_{0,x}^+ e^{-j(k_z z-\omega t)}\hat{a}_x + E_{0,y}^+ e^{-j(k_z z-\omega t)}\hat{a}_y \tag{1.12a}$$

$$\vec{H}_S^+(z) = \frac{1}{\mu}\frac{\partial A_z^+}{\partial y}e^{-jk_z z}\hat{a}_x - \frac{1}{\mu}\frac{\partial A_z^+}{\partial x}e^{-jk_z z}\hat{a}_y$$
$$\vec{H}^+(z,t) = H_{0,x}^+ e^{-j(k_z z-\omega t)}\hat{a}_x + H_{0,y}^+ e^{-j(k_z z-\omega t)}\hat{a}_y, \tag{1.12b}$$

where the components of $\vec{E}$ and $\vec{H}$ obey the relations 1.8a and 1.8b, $E_{0,x}^+ = H_{0,y}^+ = \eta$, and $E_{0,y}^+/H_{0,x}^+ = -\eta$. In this case, we can draw the electric field measured by the observer at position $z = n\lambda$ (n = integer) at time $t = 0$ to be that shown in Figure 1.3.

As seen from a point $z = n\lambda$ on the z-axis, the two components of electric field would add vectorally to form a resultant vector $\vec{E}_{0,R}^+$ whose components would vary with time cosinusoidally. Thus, $\vec{E}_{0,R}^+$ would be seen as a linearly polarized field at angle

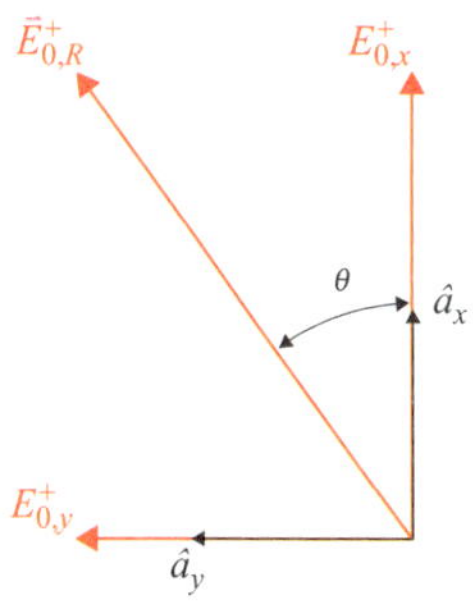

Figure 1.3 Components of the electric field intensity observed at time $t = 0$ (components of the magnetic field intensity are orthogonal to these components but are not shown).

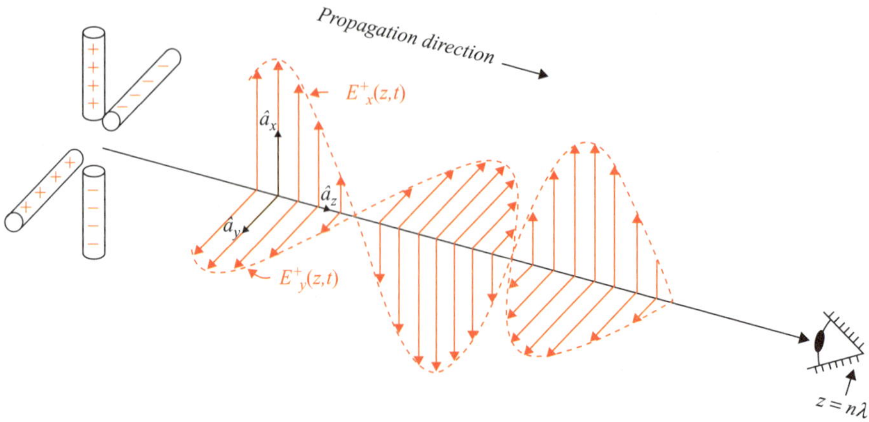

Figure 1.4 Two electric field intensities produced by orthogonal dipole antennas operating at the same frequency and with the same phase.

$$\theta = \tan^{-1}\left(E^+_{0,y}/E^+_{0,x}\right) \tag{1.13}$$

with respect to the x-axis. We would say that the two components of the electric field are in ***space quadrature*** with one another. While both of the measured components change with time in a $\cos\omega t$ manner, the angle θ remains constant so the resultant polarized electric field oscillates in amplitude with the same orientation with respect to the x-axis.

A simple way to picture the resultant of two components is to picture them as originating from two orthogonal sources such as the two dipole antennas shown in Figure 1.4.

Even More General Case

If the two dipole antennas that create the two space quadrature polarized electric field intensities are displaced from one another along the z-axis by an amount $z = \lambda/4$, as shown in Figure 1.5 and are driven at the same frequency and in the same phase, the resulting electric field intensities will be *displaced from one another in phase* by one quarter of a cycle. As seen by the observer at $z = n\lambda$, the second electric field intensity (oriented in the y-direction) will be delayed in time from the first (oriented in the x-direction) by $t = (\pi/2)/\omega$.

The equivalent equation for the observed electric fields at point z is

$$\mathrm{Re}\left[\vec{E}^+_S(z,t)\right] = \mathrm{Re}\left[E^+_{0,x}e^{-j(k_z z-\omega t)}\hat{a}_x + E^+_{0,y}e^{-j\left(k_z z-\omega t-\frac{\pi}{2}\right)}\hat{a}_y\right] \quad \text{or} \tag{1.14}$$

$$\mathrm{Re}\left[\vec{E}^+_S(z,t)\right] = E^+_{0,x}\cos(k_z z-\omega t)\hat{a}_x + E^+_{0,y}\cos\left(k_z z-\omega t+\frac{\pi}{2}\right)\hat{a}_y$$

$$= E^+_{0,x}\cos(k_z z-\omega t)\hat{a}_x - E^+_{0,y}\sin(k_z z-\omega t)\hat{a}_y \tag{1.15}$$

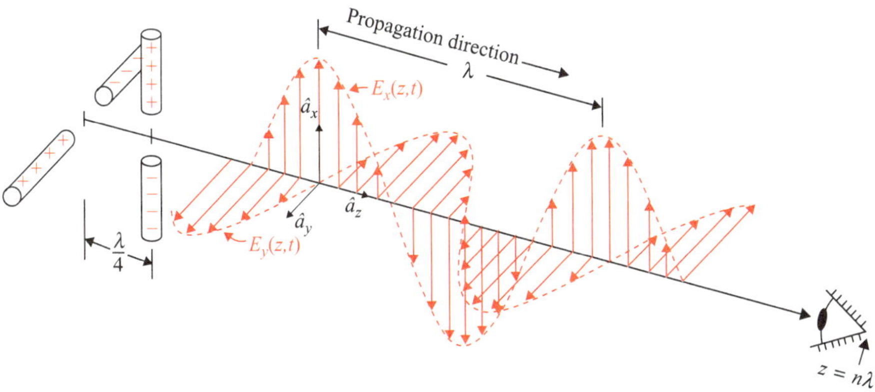

Figure 1.5 Polarized electric field intensities in space and time quadrature.

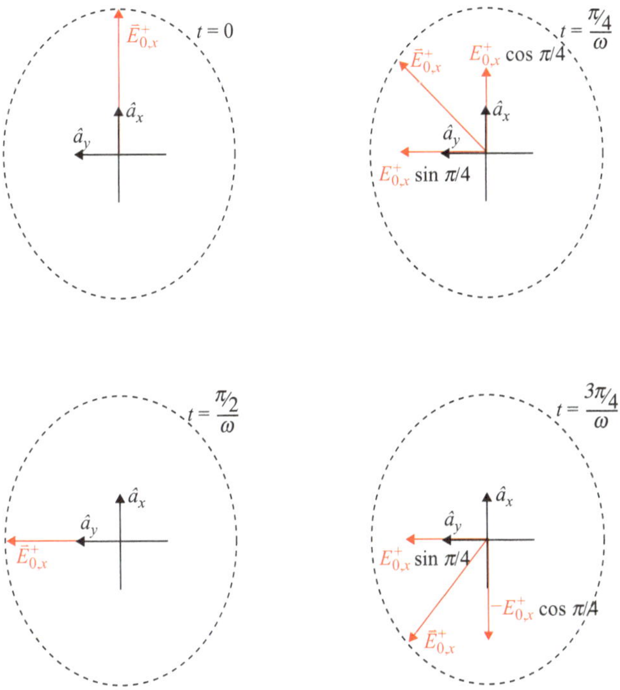

Figure 1.6 Vector sum of the electric field intensities produced by two sources, one of which lags the other by $\pi/2$.

If we plot these terms for $z = n\lambda$ on a graph like that shown in Figure 1.3 for a sequence of times, we get the sequence shown in Figure 1.6.

We can see from the resultant vector in Figure 1.6 that $\vec{E}_R$ rotates in a counter-clockwise manner about the origin, with radial frequency ω, and traces out the path of an ellipse in time as the wave propagates along the z-axis. Such resultant electromagnetic waves are called ***right-hand elliptically polarized*** waves.

Similarly, we can see that, if the component of the electric field intensity in the y-direction leads the component in the x-direction by $t = (\pi/2)/\omega$, the result will be ***left-hand elliptically polarized*** waves.

1.3 DOPPLER SHIFT

Each evening, the news channel brings us the local Doppler radar map of weather in our area, the police track our automobile speed with Doppler laser reflection, the universe is said to be expanding because we can observe and measure the "Red Shift" of stars, scientists use Mössbauer measurements to determine the magnetic flux density at a nucleus, and a trip to a NASCAR event is made more exciting by the change in pitch of a car engine as it zooms past us in the stands. A physician may take a Doppler angiogram movie of a beating heart, or an ultrasound technician may make pictures of a moving fetus in a womb. These events, as well as some troublesome problems such as the change in frequency of a mobile cell phone as measured by a base station, are caused by the motion of a source of waves relative to a receiver.

We can understand the phenomenon of Doppler shift by considering the change in waves produced by a stationary source of electromagnetic waves as it differs from a source in motion with constant velocity, as shown in Figure 1.7.

Suppose a source of electromagnetic waves (such as a quasar) produces TEM^{r} waves, with period Δt_0 between the crests of electric field intensity. These waves move in the z-direction, with velocity c toward an observer at rest with respect to the source, as shown in the top sketch in Figure 1.7. The distance between the crests (the wavelength) is then $\lambda_0 = c\Delta t = c\Delta t_0$.

Now, let us view that same source of electromagnetic waves as it moves away from the observer at velocity, v. Because the source is moving with respect to the observer, there will be a change in the period of the source that follows time dilation, according to the special theory of relativity:

$$\Delta t = \frac{\Delta t_0}{\sqrt{1 - v^2/c^2}} \tag{1.16}$$

Now, the distance between crests of the electric field intensity will be

$$\lambda = c\,\Delta t + v\,\Delta t = \frac{(c+v)\Delta t_0}{\sqrt{1 - v^2/c^2}} = c\,\Delta t_0 \frac{(1+v/c)}{\sqrt{(1+v/c)(1-v/c)}} = \lambda_0 \sqrt{\frac{(1+v/c)}{(1-v/c)}} \tag{1.17}$$

Because the speed of light, c, is the same for all observers,

$$f = \frac{c}{\lambda} = \frac{c}{\lambda_0}\sqrt{\frac{(1-v/c)}{(1+v/c)}} = f_0 \sqrt{\frac{(1-v/c)}{(1+v/c)}} \tag{1.18}$$

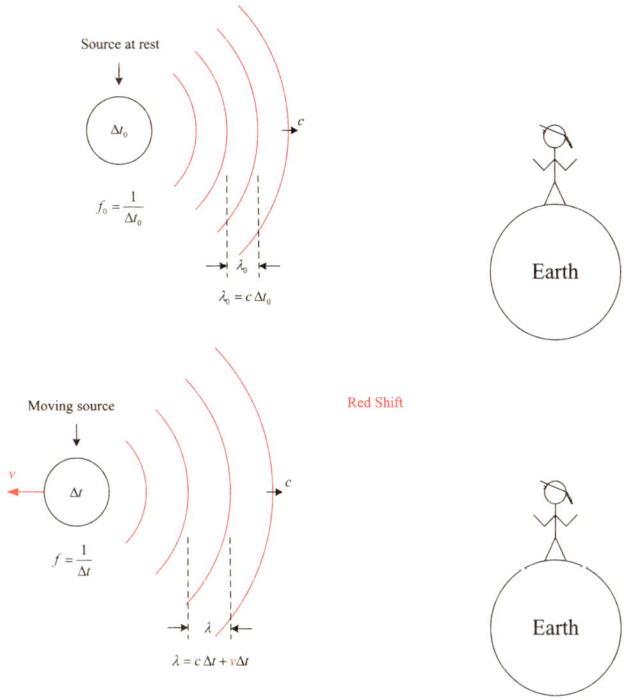

Figure 1.7 Change in the frequency and wavelength of electromagnetic waves from a source at rest versus a source moving at velocity v relative to the observer. Here, the crest (highest intensity) of the transverse electric field waves is shown as outward expanding circles about their source.

Equation 1.18 is the ***Doppler equation*** for the frequency of a moving source relative to a stationary observer. Equation 1.18 is often written in its series form

$$\frac{f}{f_0} = \left[1 - \frac{v}{c} + \frac{1}{2}\frac{v^2}{c^2} - \frac{1}{2}\frac{v^3}{c^3} + \cdots\right] \tag{1.19}$$

because the velocity of the source is normally much less than the speed of light, so we can make a good approximation to the size of the Doppler shift by keeping only the first two terms in Equation 1.19.

However, one theory of quasars is that they were expelled in the "Big Bang" at tremendous velocities; some close to the speed of light. For these sources of electromagnetic waves, we can write an expression for the relative shift in wavelength as

$$Z = \frac{\lambda - \lambda_0}{\lambda_0} = \sqrt{\frac{(1 + v/c)}{(1 - v/c)}} - 1 \tag{1.20}$$

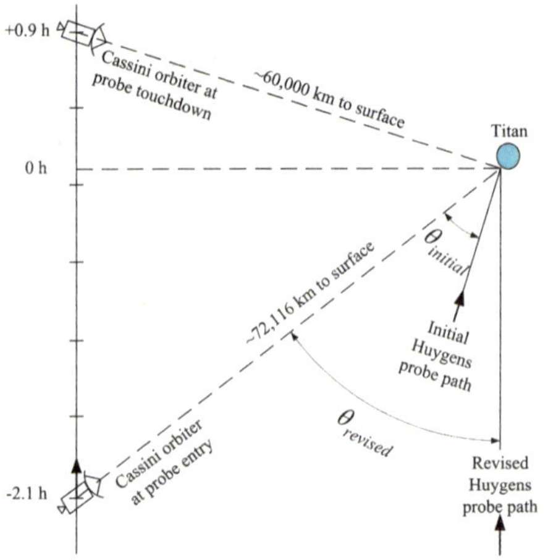

Figure 1.9 Revised position of the Cassini Orbiter relative to the Huygens Probe during entry (and transmission of pictures) to minimize the relative component of velocity, $v\cos\theta$, between transmitter and receiver.

probe by writing the Doppler equation in its series form $f/f_0 = [1 - v/c + (1/2)v^2/c^2 - (1/2)v^3/c^3 + \cdots]$, or $f/f_0 \approx [1 - v/c]$ if the velocity of the source is much less than the speed of light. Thus, the frequency shift $\Delta f \approx f - f_0 = [1 - v/c]\, f_0 - f_0 = -(v/c)f_0$ and for $\Delta f = -38$ kHz, $f_0 \approx (3 \times 10^8$ m/s/5.5 $\times 10^3$ m/s) (38 kHz) = 2.07 GHz.

NASA engineers solved the problem by launching the Huygens probe on the third (rather than the second) orbit about Saturn so that the Cassini receivers were moving nearly perpendicular to the probe decent (thus reducing the relative speed, v, to $v\cos\theta$ between the transmitter and the receiver). This change in relative motion reduced the Doppler shift to the point that the Cassini receivers would not loose lock on the carrier frequency. Figure 1.9 shows the revised location of the Cassini Orbiter as it began to communicate with the Huygens Probe during its descent onto Titan.

PROBLEMS

1.1 With the aid of drawings, explain what happened to the frequency of the signals received by the Cassini orbiter as it moved to an angle $\theta_{revised}$ relative to the path of the falling Huygens probe (assuming it was falling at its terminal velocity). Hint: The effect of time dilation is still valid when perpendicular relative motion is involved.

1.2 Calculate the Doppler shift of a 1.8 GHz cell phone due to its motion in a moving automobile at 70 mph if it is traveling (a) toward or (b) away from a Base Station.

The Mössbauer effect has been used to show that the Doppler shift also works for frequencies of 10^{19} Hz and for velocities as low as 10^{-5} m/s. The following section gives an example of the Mössbauer effect for ^{57}Fe nuclei and shows that the Doppler shift is so precise that it can be used to explain high-Q nuclear linewidths.

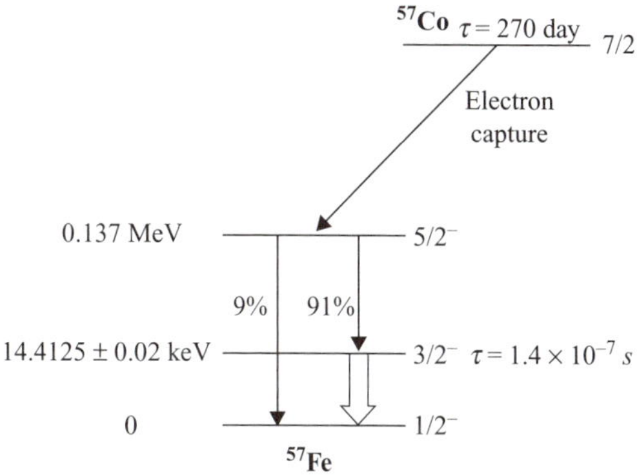

Figure 1.10 Nuclear energy levels of a ^{57}Fe nucleus following its population from the electron capture of a ^{57}Co nucleus (Table of Isotopes).

The Mössbauer Effect Example

When ^{57}Co captures an electron, it populates the 14.4125 keV excited state of ^{57}Fe (with a 98 ns half-life or lifetime $\tau = 1.4 \times 10^{-7}$ s), as shown in Figure 1.10.

The nuclear angular momentum quantum number of the excited state is $\frac{3}{2}$ and that of the ground state is $\frac{1}{2}$. Thus, in the absence of a magnetic field intensity at the ^{57}Fe nucleus, there are mono-energetic γ-rays emitted, with an energy of 14.4125 keV.* Mössbauer showed that these γ-rays are predominantly emitted in a nearly recoilless fashion because the ^{57}Fe nuclei are in a crystal lattice of mass $M = N_A \times m_{Fe}$ absorbs the momentum of the outgoing γ-ray. The frequency of the emitted γ-rays is thus $f_\gamma = 14.4125 \times 10^3$ eV/4.13566727 $\times 10^{-15}$ eV $= 3.484227 \times 10^{18}$ Hz. When γ-rays of this frequency impinge upon ^{57}Fe nuclei in a target material, as shown in Figure 1.11, they are often absorbed by those nuclei and later reemitted in a random direction. Thus, a detector behind the target will see a reduced number of γ-rays when there is absorption (at the resonant frequency).

By moving the source of nuclei (just like the quasar) away from the absorber at a velocity of $v = 0.3$ mm/s, we can Doppler shift their frequencies by a very small amount $v/c = 0.3$ mm/s/3 $\times 10^8$ m/s $= 10^{-12}$ (a vanishingly small amount compared with the frequency shift of a quasar), as shown in Figure 1.11.

As we see in Figure 1.12, this small Doppler shift is sufficient to completely take the γ-rays out of resonance so that the detector sees less absorption (the count rate goes up). This is called a Mössbauer effect absorption spectrum.

If the absorber nuclei experience a magnetic flux density, $\vec{B}$, then the excited and ground states of the absorber split into energy levels according to the Zeeman effect, $U_e = -\vec{\mu}_e \cdot \vec{B}_e$, where $\vec{\mu}_e$ is the nuclear magnetic moment of the nucleus in its excited state and $U_g = -\vec{\mu}_g \cdot \vec{B}_g$, where $\vec{\mu}_g$ is the nuclear magnetic moment of the nucleus in its ground state. The effect of the splitting is shown schematically in

*The energy of the 98 ns ^{57}Fe γ-ray is given here to 6 decimal places (our ability to measure it) but the nucleus knows this energy to about 15 decimal places as is shown below.

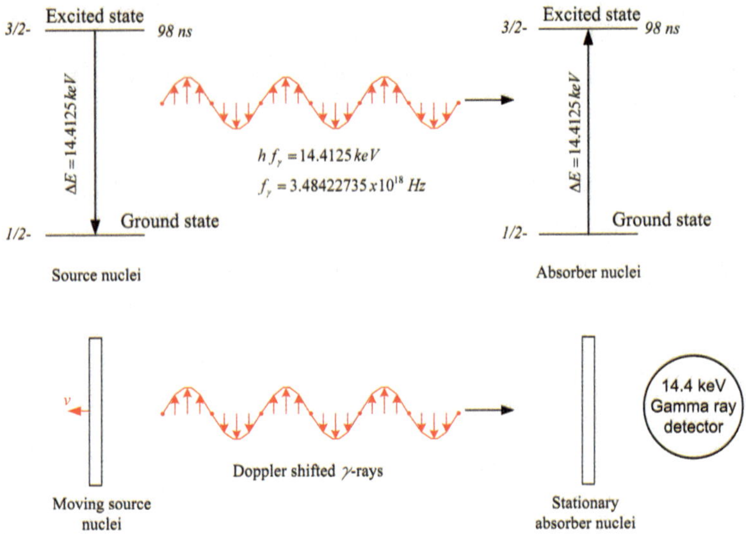

Figure 1.11 Gamma ray emission and absorption scheme for recoilless ^{57}Fe nuclei and a mechanism for shifting their frequency by a Doppler velocity of the emitted nuclei.

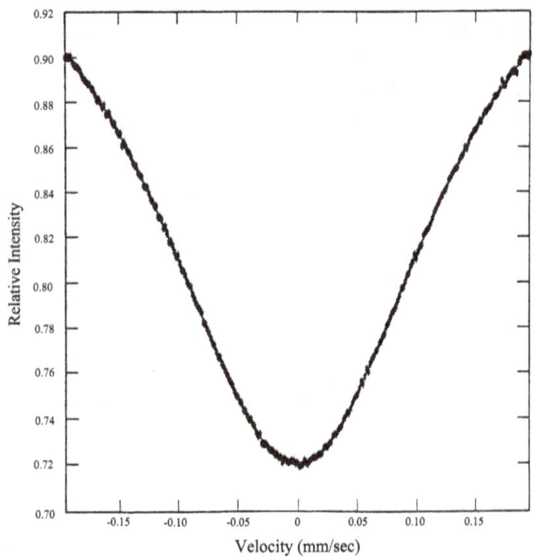

Figure 1.12 14.4125 keV γ-ray counts detected as a function of the Doppler velocity of the source for a non-magnetic source and a non-magnetic absorber.

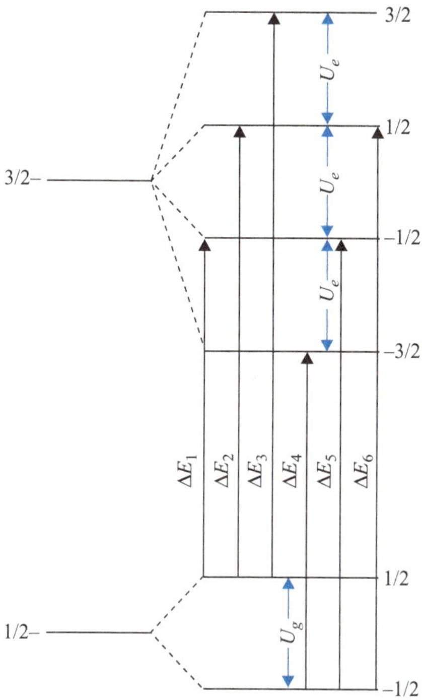

Figure 1.13 Energy splitting of the nuclear states of ^{57}Fe absorber nuclei brought about by Zeeman energy shifts for nuclei with magnetic moment in magnetic flux density.

Figure 1.13. Here, the splitting is *greatly exaggerated* as compared with the energy of the incident γ-rays.

Because the angular momentum of the incident γ-rays is 0 or $\pm 1\ \hbar\omega$ (depending upon whether the photon is linearly, right, or left circularly polarized), the transition between an absorber $-\frac{1}{2}$ ground state to a $+\frac{3}{2}$ excited state (or a $+\frac{1}{2}$ ground state to a $-\frac{3}{2}$ excited state) is not possible; it is said to be a forbidden transition. Thus, there are only six different energies that the incident γ-rays can have that will be absorbed (as shown in the schematic of Figure 1.13). We would thus expect six different Doppler velocities for which Mössbauer absorption will occur. A typical absorption spectrum of a nonmagnetic ^{57}Fe source with a magnetic ^{57}Fe absorber is shown in Figure 1.14.

The Mössbauer absorption spectrum gives us a way to measure the magnetic flux density at absorber nuclei. The energy levels and the distribution of the intensity levels can be strongly dependent on the neighboring atoms to the absorber nuclei. In many cases, a Mössbauer absorption spectrum can give us qualitative and quantitative measures of the atomic structure of an otherwise unknown sample and they can give us the values of magnetic field intensity and electric field gradient at the nuclei of atoms (a subatomic effect we normally ignore in our macroscopic treatment of electromagnetic fields). This effect is the subject of a whole class of experimental studies.

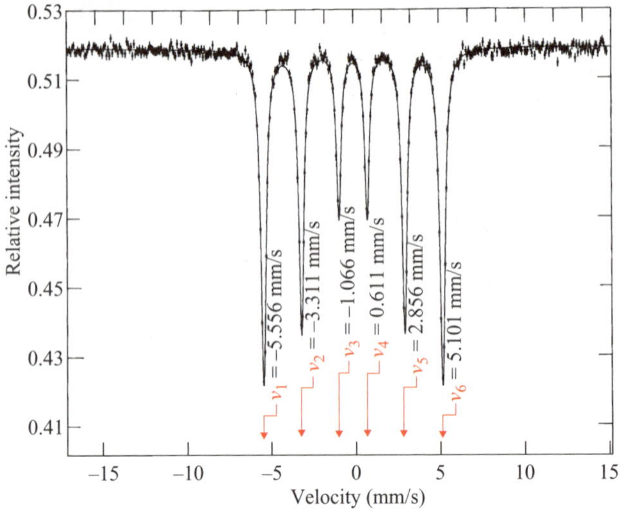

Figure 1.14 Mössbauer spectrum of a nonmagnetic ^{57}Fe source and a magnetic ^{57}Fe absorber.

Using the Doppler shift in the Heisenberg uncertainty principle, $\Delta E_{1/2}\Delta t_{1/2} \geq \hbar/2$ gives $(\Delta v_{1/2}/c)(E_\gamma)\Delta t_{1/2} \geq \hbar/2$ as the uncertainty for the Doppler velocity of the nuclear decay and absorption process. For a 14.4125 keV gamma ray with a half-life of 98 ns, we find $\Delta v_{1/2} \geq 0.07$ mm/s for the source nuclei and the same for the absorber nuclei for an expected uncertainty (Half Width at Half Max [HWHM]) of any of the Mössbauer absorption peaks of 0.14 mm/s, which compares well with the absorption peaks in Figure 1.12 or 1.14.

Conclusion

The uncertainty in the Doppler velocity for a nuclear decay and absorption process is limited only by the Heisenberg uncertainty principle. We that conclude the absorbing nuclei know the resonant energy of the emitted gamma rays at 3.5×10^{18} Hz to a precision of better than 10^{-13}.

Unified Field Theory Application

Gravitational Potential

The precision of the Mössbauer effect was one of its characteristics that permitted a measurement[2] to verify Einstein's principle of equivalence regarding gravitation and space-time. Einstein postulated that, in an enclosed elevator, it would be impossible to distinguish between a force due to a gravitating body like the earth (which caused a weight on a scale) and that due to an upward accelerating elevator. Because acceleration gave the same result as a force, he said that it was equivalent to invoke either a linear space-time with an additional force due to gravity or a curvature in space near a gravitating body with no additional gravitational forces. An experimental test of this equivalence was given by Eddington and Dyson, who observed an eclipse of May 29, 1919, on the islands of Sobral (off Brazil) and Principe (in the Gulf of Guinea). They observed the light from a star that passed behind the sun at the instant of eclipse to continue to be visible as a result of the apparent curvature of space around the sun by an angle of 1.8 s of arc.[3] A simple explanation of the equivalence is that this would be the equivalent deflection of a photon with effective mass $m_{photon} = E_{photon}/c^2 = hf/c^2$ as it passed by the enormous mass of the sun (in this case, f is the frequency of visible light [~10^{15} Hz]).

However, it was 40 years before a test of the time component of space-time curvature could be measured. This part of the principle of equivalence is often stated by the mnemonic that "lower clocks run slower." This equivalence is observed by noting that an emitted photon (with very well-defined frequency) at a height, h, relative to an absorber (with a very well-defined absorption resonant frequency) at a lower point in a gravitational field will experience an equivalent acceleration as a result of its effective mass by the force of gravity. This acceleration will cause the energy of the photon to increase as it "falls" through the distance h by $\Delta E_{photon} = m_{effective}gh = (hf/c^2)gh$. To the absorber nuclei below the source nuclei, the energy of the photon will thus appear to be greater than needed, so a Doppler shift in the same direction as that of the falling photon will be needed to achieve perfect resonance. To an external observer (who does not recognize gravitational forces), the frequency of the absorber nuclei appears to be running according to a slower clock than that which defines the frequency of the source nuclei.

Pound and Rebka placed an emitter of 14.4125 keV Fe^{57} γ-rays at the top of Harvard University's Jefferson laboratory, h = 22.5 m above absorber nuclei at ground level. They also interchanged the location of source and absorber to see the effect of a photon that must "climb" the height h and therefore loose energy. Electronics and clocks today are so much more precise than they were in 1959 that we can observe that "lower clocks run slower" if they are only separated by a height difference of 1 cm! This makes the concept of a tabletop measure of the equivalence principle very realistic.

PROBLEM

1.3 **a.** Compute the shift in energy caused by the falling photon in the Pound and Rebka experiment and determine the Doppler shift required to achieve resonance.

b. Compute the precision needed to observe a Doppler shift of those photons as they fall through a distance of 1 cm.

Electrostatic Potential

Einstein tried in vain to unify gravitation and electromagnetic fields into a single theory in which no external forces would be needed to explain either. He called this

$$k = (\omega\sqrt{\mu\varepsilon}/\sqrt{2})\left[\sqrt{1+(\sigma/\omega\varepsilon)^2}+1\right]^{1/2} - j(\omega\sqrt{\mu\varepsilon}/\sqrt{2})\left[\sqrt{1+(\sigma/\omega\varepsilon)^2}-1\right]^{1/2} \quad (1.29)$$

1. For a **non-conducting medium**, $\sigma = 0$ in Equation 1.29, and k reduces to

$$k = \omega\sqrt{\mu\varepsilon} \quad \text{non-conducting} \quad (1.30a)$$

as in the case of plane waves propagating in a pure dielectric medium.

2. For a **weakly conducting medium** in which $x = (\sigma/\omega\varepsilon) << 1$, we can use a series expansion of the square root terms in Equation 1.29 $[1 + x^2]^{1/2} = [1 + (1/2)x^2 - (1 \cdot 1/2 \cdot 4)x^4 + (1 \cdot 1 \cdot 3/2 \cdot 4 \cdot 6)x^6 - \cdots]$ to see

$$k \approx \omega\sqrt{\mu\varepsilon}\,[1 - j\sigma/2\omega\varepsilon] \quad \text{weakly conducting} \quad (1.30b)$$

3. For a **strongly conducting medium** in which $x = (\sigma/\omega\varepsilon) >> 1$, we can see

$$k \approx \omega\sqrt{\mu\varepsilon}\left[(1-j)/\sqrt{2}\sqrt{\sigma/\omega\varepsilon}\right] \quad \text{strongly conducting} \quad (1.30c)$$

Orienting the Cartesian coordinates such that the z-axis lies in the direction of $\hat{a}_k$,

$$\begin{bmatrix} \vec{E}(\vec{x},t) \\ \vec{H}(\vec{x},t) \end{bmatrix} = \begin{bmatrix} \vec{E}_0 \\ \vec{H}_0 \end{bmatrix} e^{-\alpha z} e^{-j(\beta z - \omega t)} \quad (1.31)$$

The constant α in the loss term in Equation 1.31 depends on the relative value of $x = (\sigma/\omega\varepsilon)$ to 1. This quantity is often referred to as the ***static loss tangent***:

$$\tan\delta_S = \sigma/\omega\varepsilon \quad (1.32)$$

NOTE The phase velocity of electromagnetic waves in each of these media is $u_p = \omega/\beta$. Thus, $u_p = c/\sqrt{\varepsilon_r}$ for a nonconducting medium, but $u_p = (c/\sqrt{\varepsilon_r})\sqrt{2}\sqrt{\omega\varepsilon/\sigma}$ for a strongly conducting medium.

Complex Permittivity

When an electric field is applied to a dielectric material, it orients molecules with electric dipoles in proportion to the size of the electric field. If the applied field oscillates in time (e.g., in an electromagnetic wave), the dipole orientation will try to follow the direction of the applied field. However, the polar molecules being oriented have a mass m that leads to an inertia of the molecule so that it cannot exactly follow the driving frequency in time so that it sometimes lags and can even become completely out of phase with the driving field. Furthermore, the dipoles that oscillate in an external field may lose energy to their neighbors with a damping coefficient, b, through friction. The result is a set of N per unit volume dipoles that are driven, damped, harmonic oscillators. In Chapter 5, we show how they produce a relative complex permittivity:

$$\tilde{\varepsilon}_r = 1 + \frac{N\alpha e^2/\varepsilon_0 m}{(\omega_0^2 - \omega^2) + j\omega b/m} = \varepsilon_r' - j\varepsilon_r'', \tag{1.33}$$

where ω_0 is a resonant frequency of the polar molecules. The tilde over ε_r reminds us that the permittivity can be a complex quantity at high frequencies. At very high frequencies, the model permits a displacement of a negative plasma of electronic charge relative to its positive atomic cores; at high frequencies, the model includes the additional displacement of ionic charge in individual atoms; at lower frequencies, the traditional orientation of polar molecules give rise to additional permittivity; and, at low frequencies, in conductors and semiconductors, the driving electric field can also displace free electric charges relative to holes in the material to give a complex permittivity that takes into account the conductivity of the material in the form $\varepsilon_{r,eff} = (1 - j\sigma/\omega\varepsilon_0)$, as stated in a previous footnote. The additional effects lead to a behavior that is similar to the orientation of the polar molecules, but the resonances are at different frequencies so that

$$\tilde{\varepsilon}_r = 1 + \sum_{i=1}^{n} \frac{N_i \alpha_i e^2/\varepsilon_0 m}{(\omega_i^2 - \omega^2) + j\omega b/m} = \varepsilon_r' - j\varepsilon_r'' \tag{1.34}$$

Some texts prefer to write

$$\tilde{\varepsilon}_r(\omega) = 1 + \tilde{\chi}_e(\omega) \tag{1.35}$$

where $\tilde{\chi}_e$ is the electric susceptibility,

$$\tilde{\chi}_e(\omega) = \chi_e'(\omega) - j\chi_e''(\omega) \tag{1.36}$$

Many scientists and engineers (e.g., Kramers-Kronig, Debye, Clausius-Mosotti) have contributed to this field so it is a subdiscipline in its own right. Real materials have their own individual characteristics that do not fit a single characteristic set of variations, but each typically has a real and an imaginary part that vary with frequency. The loss mechanisms depend on the ratio of the imaginary and real parts so the ***alternating electric loss tangent*** is defined as

$$\tan\delta_a = \varepsilon''/\varepsilon' \tag{1.37}$$

For materials with conductivity and dielectric losses, the ***effective electric loss tangent*** is

$$\tan\delta_e = \tan\delta_S + \tan\delta_a = \sigma/\omega\varepsilon' + \varepsilon''/\varepsilon' \tag{1.38}$$

$$\frac{E_x}{H_y} = \tilde{\eta} = \frac{\omega\mu}{\sqrt{\alpha^2 + \beta^2}} e^{j\phi} \tag{1.46}$$

For **nonconducting materials**, we have previously found (Equation 1.30a) that $\alpha = 0$ and $\beta = \omega\sqrt{\mu\varepsilon}$ so that $\varphi = 0$ and $E_x/H_y = \sqrt{\mu/\varepsilon}$, as we previously found in Equation 1.9.

For **weakly conducting materials**, we have found (1.30b) that $\beta - j\alpha \approx \omega\sqrt{\mu\varepsilon}\,[1 - j(\sigma/2\omega\varepsilon)]$ so that $\varphi = \tan^{-1}(\sigma/2\omega\varepsilon)$ and $E_x/H_y \approx \sqrt{\mu/\varepsilon}$, where the term in the parentheses is small. Thus, there is a small phase shift of the electric field intensity to the magnetic field intensity, but the magnitude is about the same as it was for a nonconductor.

For **strongly conducting materials**, we have found (Equation 1.30c) that $\beta - j\alpha \approx \omega\sqrt{\mu\varepsilon}\left[\sqrt{\sigma/\omega\varepsilon}\,(1-j)/\sqrt{2}\right]$ so that $\varphi = \tan^{-1}1/1 = \pi/4\ rad = 45^\circ$ and $E_x/H_y = \omega\mu/\sqrt{\alpha^2 + \beta^2} = \sqrt{\mu/\varepsilon}/\sqrt{(\sigma/\omega\varepsilon)}$, where the term in the parentheses is much larger than 1. Thus, there is a phase shift of 45° and a decrease in the electric field intensity (relative to the magnetic field intensity) over that of a nonconductor.

The decay and phase shift for a strongly conducting medium are shown in Figure 1.16.

Examples

Describe the character of electromagnetic propagation in copper, seawater, and distilled water if σ and ε_r are given as shown in Table 1.3.

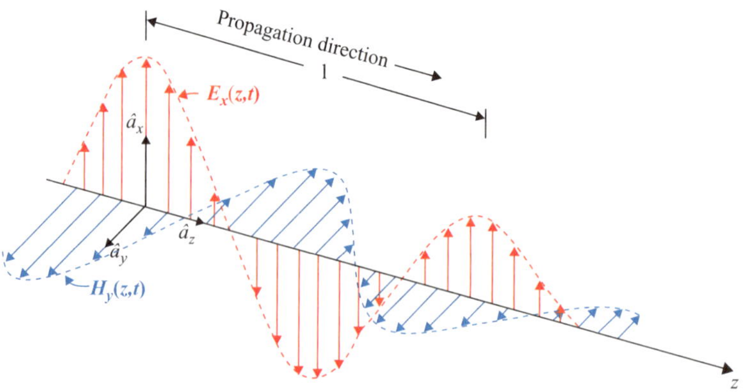

Figure 1.16 Propagation of the electric field intensity and magnetic field intensity of an electromagnetic wave in a strongly conducting material.

Table 1.3 Properties* of selected materials

Material	σ (S/m)	ε_r	$\varepsilon''/\varepsilon'$	μ_r	μ''/μ'
Copper	5.8×10^7	1	0	1	0
Cast iron	1.0×10^6	1	0	60	*
Seawater	4	72	4	1	0
Ferrite (Fe_2O_3)	1.3×10^{-3}	1	0	1000	*
Distilled water	2.0×10^{-4}	80	4×10^{-2}	1	0
Glass	10^{-12}	7	2×10^{-2}	1	0
Resin (FR4)	10^{-15}	4.0	2×10^{-3}	1	0

*Qualifications: Values vary with measurement temperature (typically room temperature), purity, and frequency (typically <10 GHz) of E&M wave. Relative losses of ferromagnetic materials are calculated in Chapter 4.

We have seen from Equation 1.30b and 1.30c that

$$\beta - j\alpha \approx \omega\sqrt{\mu\varepsilon}\,[1 - j\sigma/2\omega\varepsilon] \quad \text{weakly conducting}$$

$$\beta - j\alpha \approx \omega\sqrt{\mu\varepsilon}\left[\sqrt{\sigma/\omega\varepsilon}\,(1-j)/\sqrt{2}\right] \quad \text{strongly conducting}$$

and that $\tan\delta_s = \sigma/\omega\varepsilon$ is the deciding criterion. Choosing the criteria to be $10^{\pm 2}$,

tan δ	α	β	η	φ
<0.01	$\sqrt{\mu/\varepsilon}\,\sigma/2$	$\omega\sqrt{\mu\varepsilon}$	$\sqrt{\mu/\varepsilon}$	$\tan^{-1}\sigma/2\omega\varepsilon$
>100	$\sqrt{\omega\sigma\mu}/\sqrt{2}$	$\sqrt{\omega\sigma\mu}/\sqrt{2}$	$\sqrt{\mu/\varepsilon}/\sqrt{\sigma/\omega\varepsilon}$	45°

These criteria lead to a skin depth, δ, magnitude of E_x to H_y, η_{eff}, relative phase angle of $\vec{E}$ to $\vec{H}$, φ, and phase velocity, u_p, of each material based on its material conductivity, as shown in Table 1.4.

Table 1.4 Electromagnetic wave properties in various propagating materials

Material	δ	η_{eff} (V/A)	φ	u_p (m/s)
Copper at 60 Hz	0.85 cm	2.9×10^{-6}	45°	3.2
Copper at 100 MHz	6.6×10^{-3} cm	3.7×10^{-3}	45°	4153
Seawater at 60 Hz	32.5 m	10.1×10^{-3}	45°	1.7×10^8
Seawater at 100 MHz	2.5 cm	14.1	39.3°	7.8×10^7
Distilled water at 60 Hz	4590 m	0.49	44.9°	5.5×10^5
Distilled water at 100 MHz	240 m	19.9	0.013°	5.7×10^7

1.5 DISPERSION AND GROUP VELOCITY

Thus far, we have dealt primarily with monochromatic waves with a definite frequency, ω_0, and the real part of a wave number, $\beta_0 = 2\pi/\lambda_0$, with the possibility that there will be an imaginary part, α_0 that causes waves to decay with propagation distance. We found that these waves have a phase velocity, $u_p = \left(c/\sqrt{\varepsilon_r}\right)$ for a nonconducting medium and $u_p = \left(c/\sqrt{\varepsilon_r}\right)\sqrt{2\omega_0\varepsilon/\sigma}$ for a strongly conducting medium. Thus, waves of definite frequency, ω_0, decay in amplitude with propagation distance, as shown in Figure 1.16, and move with phase velocity less than the speed of light, c, in a vacuum but they do not disperse.

Spread of Frequency Components for a Shaped Pulse

In most applications, electromagnetic waves are produced with a finite spread of frequencies or wavelengths because of the finite duration of a pulse, for example, in the packet shaping of a pulse of light for a fiber-optic transmission line. We can examine the Fourier transform for the packet in time to see the spread of frequencies. If the medium is dispersive (i.e., it is strongly conducting and has a complex permittivity or a complex permeability), the phase velocity of an individual frequency component of the wave will not be the same as those for other frequency components of the wave. In this case, high-frequency components will travel at different phase velocities than low-frequency components, and the pulse will disperse in space and time.

We will begin this characterization by examining the frequency spread of a shaped pulse. We will then examine the group velocity of the pulse through the different phase velocities of its frequency components. Finally, we will characterize the shape of the pulse as it propagates in a dispersive medium.

Frequency Spread of a Shaped Pulse

For simplicity, let us consider the propagation of an amplitude-modulated signal consisting of a carrier frequency signal (of frequency, ω_0) that is modulated by a pulse having a Gaussian distribution shape with time. At $z = 0$, the relative electric field would behave as

$$\vec{E}(0,t) = \vec{E}_0 \cos\omega_0 t\, e^{-t^2/2T_0^2}, \tag{1.47}$$

where T_0 is a measure of the width of the pulse.

The relative electric field intensity given by Equation 1.47 is shown in Figure 1.17. Also shown in this figure is the Gaussian envelope (in green).

The energy of the pulse is proportional to the square of the electric field intensity as

$$\left|\vec{E}(0,t)\right|^2 = \left|\vec{E}_0\right|^2 e^{-t^2/T_0^2} \cos^2\omega_0 t \tag{1.48}$$

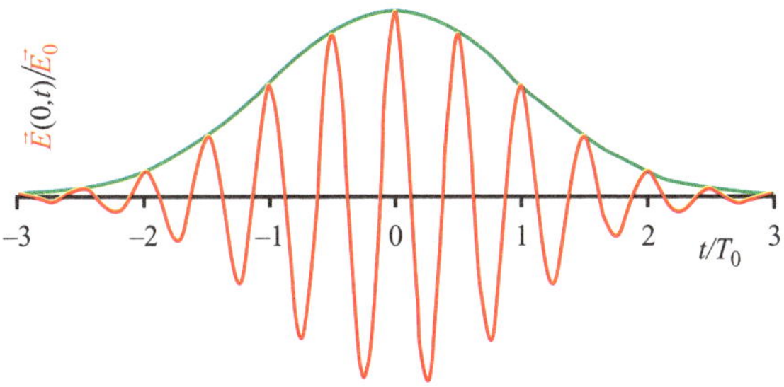

Figure 1.17 Electric field intensity $|\vec{E}(0, t)|/|\vec{E}_0|$ as a function of reduced time, t/T_0, for a Gaussian modulated pulse.

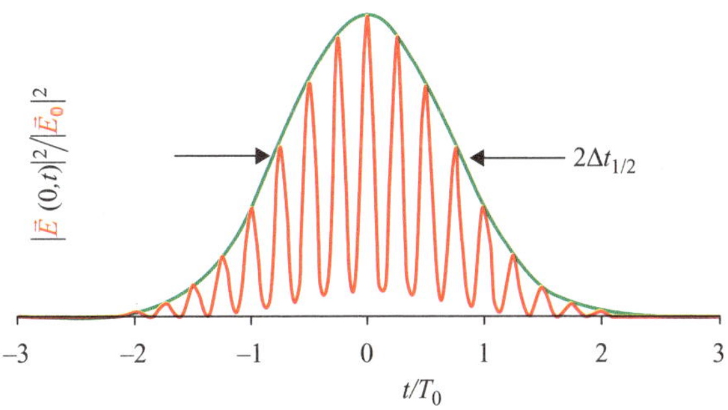

Figure 1.18 Square of the relative transverse electric field intensity, $|\vec{E}(0, t)|^2/|\vec{E}_0|^2$, as a function of reduced time, t/T_0, for a Gaussian modulated pulse.

The relative electric field intensity squared is shown (in red) in Figure 1.18. The envelope is another Gaussian distribution (in green).

Note that the half width of the Gaussian envelope shown in Figure 1.18 occurs at time $t^2 = T_0^2 \ln 0.5$ so the half width at half maximum (HWHM) is

$$\Delta t_{1/2} = \sqrt{0.693}\,T_0 \tag{1.49}$$

Example

The envelope in these figures relative to the carrier frequency is greatly exaggerated for a typical pulse application. For example, a typical optical pulse might have $2T_0 \approx 10^{-10}$ s and $\lambda_0 = 300$ nm ($f_0 = 10^{15}$ Hz) for ultraviolet (UV) light in a fiber-optic cable; thus, $n = c(2T_0)/\lambda_0 \approx 10^5$ cycles under the full width at half maximum (FWHM) of curve 1-18 (vs. ≈6 shown).

The Phasor electric field intensity is thus the real part of $E(0, t) = E_0 e^{-t^2/2T_0^2}\, e^{j\omega_0 t}$ whose Fourier transform is

$$\tilde{E}(0,\omega)=\int_{-\infty}^{\infty}E(0,t)e^{-j\omega t}\,dt \tag{1.50}$$

$$\text{or}\quad \tilde{E}(0,\omega)=\int_{-\infty}^{\infty}E_0e^{-t^2/2T_0^2}e^{-j(\omega-\omega_0)t}\,dt \tag{1.51}$$

which can be written

$$\tilde{E}(0,\omega)=2E_0\int_0^{\infty}e^{-t^2/2T_0^2}\cos(\omega-\omega_0)t\,dt=\sqrt{\pi}E_0T_0e^{-(\omega-\omega_0)^2T_0^2/2} \tag{1.52}$$

Conclusion

The Fourier transform of a Gaussian modulated electric field intensity oscillating at carrier frequency ω_0 in time is another Gaussian in frequency space centered about $\omega = \omega_0$; that is, modulating the pure electromagnetic wave with a single carrier frequency with a Gaussian envelope yields an electromagnetic wave with a distribution of frequencies (also in a Gaussian fashion) about the carrier frequency.

If we express the relative energy of this electric field intensity in frequency space as proportional to

$$\tilde{E}^2(0,\omega)=\pi E_0^2T_0^2e^{-(\omega-\omega_0)^2T_0^2}, \tag{1.53}$$

and we can plot the energy spectrum, as shown in Figure 1.19.

We can see that the energy spectrum falls to 0.5 for $\omega=\omega_0\pm\sqrt{0.693}/T_0$; that is,

$$\Delta\omega_{1/2}=\sqrt{0.693}/T_0 \tag{1.54}$$

It is interesting to note that the product of Equations 1.49 and 1.54 yields

$$\Delta t_{1/2}\Delta\omega_{1/2}=0.693 \tag{1.55}$$

We can multiply Equation 1.55 through by $\hbar$ and compare this result with that of the Heisenberg Uncertainty principle:

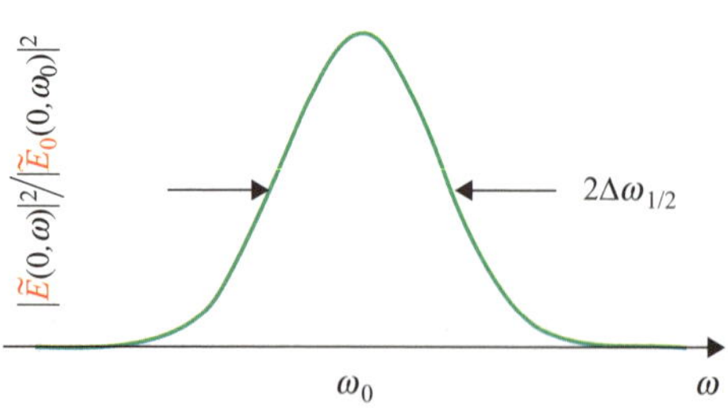

Figure 1.19 Relative energy spectrum, $|\tilde{E}(0,\omega)|^2/|\tilde{E}(0,\omega_0)|^2$, of a Gaussian modulated pulse of a pure electromagnetic wave, with carrier frequency ω_0 as a function of frequency ω.

$$\Delta t_{1/2}\hbar\Delta\omega_{1/2} \geq 0.500\,\hbar \tag{1.56}$$

to see that the uncertainty principle is a mathematical statement about the shape of pulses and their Fourier transforms; that is, for Gaussian modulated pulses, the uncertainty in time (HWHM of the energy spectrum) times the uncertainty in frequency (HWHM in energy divided by $\hbar$) is a constant (0.693).

Example

For the example above, with $\lambda_0 = 300$ nm, $\omega_0 = 2\pi \times 10^{15}$ Hz, and $\Delta t_{1/2} = \sqrt{0.693}\,T_0 = \sqrt{0.693}\,10^{-10}\,\mathrm{s}/2$, $\Delta\omega_{1/2} = 1.66 \times 10^{10}\ \mathrm{s}^{-1}$ and $\Delta\omega_{1/2}/\omega_0 = 2.65 \times 10^{-6}$.

We can conclude that a Gaussian-modulated electromagnetic wave will have components of frequency about the carrier frequency in a Gaussian distribution and that the width of the pulse in time will govern the width of the pulse in frequency space, according to Equation 1.55. For long pulses in time the width of the frequency, Gaussian will be very narrow; we can even take the $\lim \Delta t_{1/2} \rightarrow \infty$ to see that the Gaussian will become a delta function (i.e., only one frequency at ω_0). However, for very short pulses of an electromagnetic wave in time, we can see from Equation 1.55 that the frequency distribution will be very broad; that is, there will be a broad distribution of frequency components in the resulting frequency Gaussian.

If the medium in which the pulse is propagating has a dispersive character, that is, the phase velocity depends positively on the frequency of the wave, then the Gaussian distribution will spread out as the wave propagates because the higher frequency components will move with a higher velocity from the lower frequency components. Almost all media exhibit this property so it is called ***normal dispersion***. In rare cases, the Gaussian distribution will narrow as the wave propagates because the higher frequency components move with a lower velocity for the lower frequency components. Pulses that exhibit this property in a medium are said to have ***anomalous dispersion***. In either case, we need to define a new velocity of propagation of the group of wave components that constitute a pulse, which we will call u_g, the group velocity.

Group Velocity

As mentioned above, we found that electromagnetic waves traveling in a medium have a phase velocity, $u_p = c/\sqrt{\varepsilon_r}$ if the medium is nonconducting and $u_p = \left(c/\sqrt{\varepsilon_r}\right)\sqrt{2\omega_0\varepsilon/\sigma}$ if the medium is strongly conducting. We have also shown that information pulses shaped by a Gaussian envelope in time are described as a Gaussian distribution of frequencies. As shown in Figure 1.15, dielectrics generally exhibit decreasing permittivity with frequency (ignoring the resonances, which are discussed in Chapter 5) so the higher frequency components of those pulses will travel with higher phase velocity and the amplitude of each Fourier component will decay as $e^{-\alpha z} = e^{-\sqrt{\mu/\varepsilon}\,\sigma z/2}$, where σ is very small so they will separate in time (i.e., they will disperse) but will not decay very much with distance. In *good* conductors, we observed in Table 1.4 that phase velocity is frequency dependent, and, in addition, the Fourier components will decay in amplitude, with z as $e^{-\alpha z} = e^{-\sqrt{\mu/\varepsilon}\sqrt{2\omega\sigma}\,z/2}$, where σ is relatively large so the frequency components will move with different phase velocities (i.e., they will disperse) and they will decay rapidly with distance.

We thus need a new velocity to define the "average" velocity of a group of plane wave components with different frequencies. To simplify our understanding of the new definition, we will begin with only ***two*** plane waves with different frequencies $\omega + \Delta\omega/2$ and $\omega - \Delta\omega/2$ that differ in frequency by a small amount, $\Delta\omega$, and we will take the limit as $\Delta\omega \to 0$. We will then extend the argument to include additional different frequencies (always taking the limit as $\Delta\omega \to 0$) and thereby build up a continuum of frequencies in a pulse.

Figure 1.20 shows the linear combination of two plane waves of equal amplitude but slightly different frequencies, $\omega + \Delta\omega/2$ and $\omega - \Delta\omega/2$, in time. We have seen that the real part of the propagation constant, β, also depends on frequency in any conductor (good or bad) so there will be a corresponding $\beta + \Delta\beta/2$ and $\beta - \Delta\beta/2$ for each of the two waves.

The mathematical sum of the two electric field waves is given by the linear combination

$$\vec{E}_{total}(\vec{x}, t) = \vec{E}_{01}(z) e^{-\alpha z} e^{j\left[\left(\beta+\frac{\Delta\beta}{2}\right)z-\left(\omega+\frac{\Delta\omega}{2}\right)t\right]} + \vec{E}_{02}(z) e^{-\alpha z} e^{j\left[\left(\beta-\frac{\Delta\beta}{2}\right)z-\left(\omega-\frac{\Delta\omega}{2}\right)t\right]} \tag{1.57}$$

NOTE In Equation 1.57, we have assumed that the two waves are decaying with z by the same decay constant, α. In a rigorous treatment, we would have to take this into account for strong conductors by considering two different values of α but would eventually take the limit as the two values of α approached one another and the outcome would be the same.

If the two waves have equal amplitude, $\vec{E}_0(z)$, then the real part of Equation 1.57 may be written as

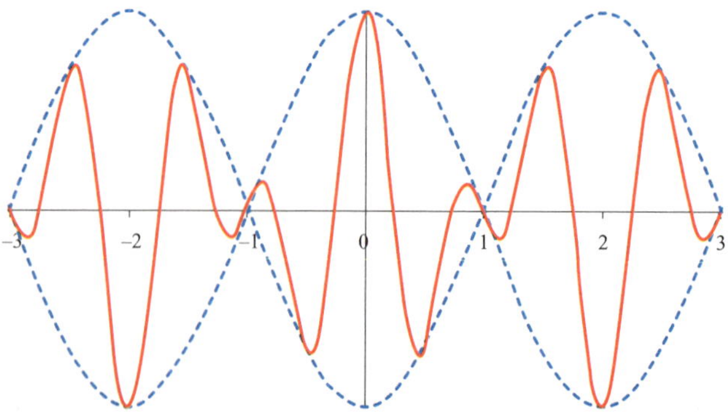

Figure 1.20 Sum of two plane waves with equal amplitude but different base frequency ω by $\Delta\omega$.

$$\operatorname{Re}\vec{E}_{total}(\vec{x},t)$$
$$=\vec{E}_0(z)e^{-\alpha z}\left\{\cos\left[\left(\beta+\frac{\Delta\beta}{2}\right)z-\left(\omega+\frac{\Delta\omega}{2}\right)t\right]+\cos\left[\left(\beta-\frac{\Delta\beta}{2}\right)z-\left(\omega-\frac{\Delta\omega}{2}\right)t\right]\right\},$$

and, by using the mathematical identity for the product of cosines,

$$\operatorname{Re}\vec{E}_{total}(\vec{x},t)=2\vec{E}_0(z)e^{-\alpha z}\cos(\Delta\beta z-\Delta\omega t)\cos(\beta z-\omega t) \tag{1.58}$$

Assuming $\Delta\omega << \omega$, the term $\cos(\Delta\beta z - \Delta\omega t)$ in Equation 1.58 describes the envelope of the base plane wave shown in Figure 1.19. The velocity of the envelope is called the "group velocity," u_g. We can find the velocity of the envelope by considering a point of constant envelope amplitude, $(\Delta\beta z - \Delta\omega t)$ = constant, and then take the derivative with respect to time to obtain $u_g = dz/dt = \Delta\omega/\Delta\beta$, and, taking the limit as $\Delta\omega \to 0$,

$$u_g=\frac{dz}{dt}=\lim_{\Delta\omega\to 0}\frac{\Delta\omega}{\Delta\beta}=\frac{1}{d\beta/d\omega}=\left.\frac{d\omega}{d\beta}\right|_{avg} \tag{1.59}$$

The group velocity, u_g, of the waves is thus the derivative of the ω versus β curve evaluated at the point ω. The group velocity is in contrast to the phase velocity of the waves, u_p, which is just the ratio of ω to β.

If the group velocity of two waves that are infinitesimally close in frequency is less than the phase velocity, we call the propagation and the dispersion normal. If the group velocity of the two waves equals the phase velocity, there will be no difference in their speed and there will be no dispersion. If the group velocity is greater than the phase velocity, the propagation and dispersion are said to be anomalous.

$$\begin{array}{ll} u_g < u_p & \text{Normal dispersion} \\ u_g = u_p & \text{No dispersion} \\ u_g > u_p & \text{Anomalous dispersion} \end{array} \tag{1.60}$$

Shape of the Pulse

For the pulse modulated by a Gaussian distribution above, a relatively long pulse will produce a relatively narrow frequency band. By comparison, β will be a slowly varying function of ω so we can expand β in a Taylor's series about the carrier frequency ω_0 and keep only the first three terms:

$$\beta(\omega)\approx\beta(\omega_0)+(\omega-\omega_0)\left.\frac{d\beta}{d\omega}\right|_{\omega_0}+\frac{1}{2}(\omega-\omega_0)^2\left.\frac{d^2\beta}{d\omega^2}\right|_{\omega_0} \tag{1.61}$$

Substituting this approximation into the inverse Fourier transform,

$$\vec{E}(z,t) = \frac{1}{2\pi}\int_{-\infty}^{\infty} \tilde{E}(z,\omega)e^{j\omega t}\,d\omega$$
$$= \frac{1}{2\pi}\int_{-\infty}^{\infty} \tilde{E}(0,\omega)e^{-(\alpha+j\beta)z}\,e^{j\omega t}\,d\omega \tag{1.62}$$

$$\vec{E}(z,t) \approx \frac{1}{2\pi}\int_{-\infty}^{\infty} \tilde{E}(0,\omega')e^{-\alpha z}\,e^{-j\beta(\omega_0)z}\,e^{-j\omega'\beta' z}\,e^{-j\omega'^2\beta'' z}\,e^{j\omega_0 t}\,d\omega'$$
$$\text{where } \omega' = (\omega-\omega_0),\ \beta' = (d\beta/d\omega)|_{\omega_0},\ and\ \beta'' = \left(d^2\beta/d\omega^2\right)_{\omega_0} \tag{1.63}$$

If we now employ Equation 1.53 $\tilde{E}^2(0,\omega) = \pi\tilde{E}_0^2T_0^2e^{-(\omega-\omega_0)^2T_0^2}$ in Equation 1.63, we obtain

$$\vec{E}(z,t) \approx \frac{\sqrt{\pi}\vec{E}_0T_0}{2\pi}\int_{-\infty}^{\infty} e^{-\frac{\omega'^2T_0^2}{2}}\,e^{-\alpha z}\,e^{-j\beta(\omega_0)z}\,e^{-j\omega'\beta' z}\,e^{-j\omega'^2\beta'' z/2}\,e^{j\omega_0 t}\,d\omega' \tag{1.64}$$

Taking the terms that are independent of ω' out of the integral, we get

$$\vec{E}(z,t) \approx \frac{\vec{E}_0T_0}{2\sqrt{\pi}}e^{-\alpha z}e^{-j[\beta(\omega_0)z-\omega_0 t]}\int_{-\infty}^{\infty} e^{-\omega'^2\frac{T_0^2}{2}}\,e^{-j\omega'^2\frac{\beta'' z}{2}}\,e^{j\omega'(t-\beta' z)}d\omega' \tag{1.65}$$

$$\text{or}\quad \vec{E}(z,t) \approx \frac{\vec{E}_0}{\sqrt{1+j\beta'' z/T_0^2}}e^{-\alpha z-j[\beta(\omega_0)t-\omega_0 t]}\,e^{-\left\{(t-\beta' z)^2\left[1-j\beta'' z/T_0^2\right]\right\}/\left(T_0^2+(\beta'' z/T_0)^2\right)} \tag{1.66}$$

Now, if we let

$$t_0 = \beta' z = \frac{z}{u_g|_{\omega_0}} = \text{group delay at } \omega_0 \tag{1.67}$$

$$\text{and}\quad S = \beta'' z/T_0 \tag{1.68}$$

$$\vec{E}(z,t) \approx \frac{\vec{E}_0}{\sqrt{1+jS/T_0}}e^{-\alpha z-j[\beta(\omega_0)t-\omega_0 t]}\,e^{-\left\{(t-t_0)^2[1-jS/T_0]\right\}/\left(T_0^2+S^2\right)} \tag{1.69}$$

Equation 1.66 or 1.69 may be interpreted as a plane propagating in the z-direction at the carrier frequency modified by a pulse envelope:

$$\text{Envelope} = e^{-(t-t_0)^2/T_0^2\sigma^2}\Big/\sigma \tag{1.70a}$$

where

$$\sigma = \sqrt{1+jS/T_0} \tag{1.70b}$$

at $z = 0$, $t_0 = 0$ and $S = 0$ so we can see that Equation 1.70 reduces to Equation 1.48, our initial Gaussian-shaped pulse. As the pulse propagates to some positive value

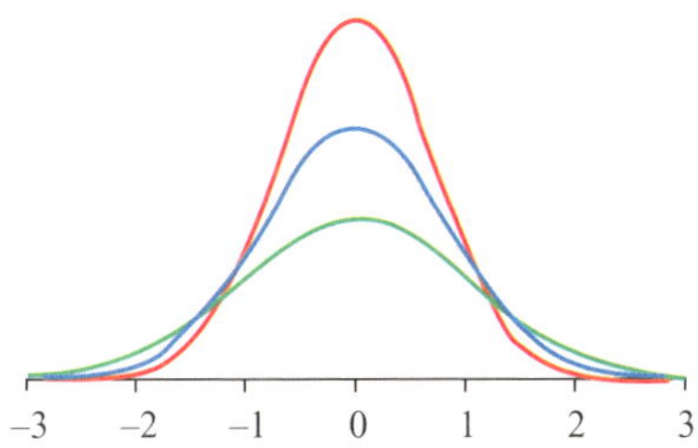

Figure 1.21 Amplitude of the Gaussian-shaped pulse envelope for a carrier wave propagating along the z-axis and centered at $z_0 = 0$, $z_1 = T_0/\beta''$, and $z_2 = 2T_0/\beta''$ for corresponding times $t_0 = 0$, $t_1 = (\beta'/\beta'')T_0$, and $t_2 = 2(\beta'/\beta'')T_0$.

of z, Equation 1.70 shows that the initial pulse is decreasing in amplitude, changing in phase, and becoming broader. The envelope of the pulse is shown in Figure 1.21.

In Figure 1.21, we can interpret the horizontal axis as corresponding either to the width of the pulse in physical space along the z-axis or to the width of the pulse in time. The height is the amplitude of the envelope that constrains the electric field oscillating at the carrier frequency. We can see from Figure 1.21 that the amplitude is quickly dying with propagation distance (or time) and that the pulse width is slowly increasing with propagation distance (or time).

For a plane wave propagating in a uniform material medium, we can define a phase refractive index, n, for the phase velocity relative to the speed of light in a vacuum by

$$u_p\big|_{\omega_0} = \omega_0/\beta(\omega_0) = c/n \tag{1.71a}$$

and a group refractive index, N, by

$$u_g\big|_{\omega_0} = (d\omega/d\beta)_{\omega_0} = 1/\beta' = c/N \tag{1.71b}$$

If we expand n as a Taylor series in frequency about ω_0, we can write $n \approx n(\omega_0) + (\omega - \omega_0)dn/d\omega$ and, using $(\omega/2\pi)\lambda_0 = c$, we can see

$$\frac{dn}{d\omega} = \frac{dn}{d\lambda_0}\frac{d\lambda_0}{d\omega} = \frac{dn}{d\lambda_0}\left(-\frac{2\pi c}{\omega^2}\right) \tag{1.72a}$$

so that

$$\frac{d\beta}{d\omega} = \frac{1}{c}\left(n - \lambda_0\frac{dn}{d\lambda_0}\right) \tag{1.72b}$$

and thus

$$N = n - \lambda_0\, dn/d\lambda_0 \tag{1.72c}$$

In Cartesian coordinates, we can show by brute force that

$$\vec{\nabla}\cdot(\vec{E}\times\vec{H})=\vec{H}\cdot(\vec{\nabla}\times\vec{E})-\vec{E}\cdot(\vec{\nabla}\times\vec{H}) \tag{1.75}$$

One of our fundamental theorems is that the choice of coordinate system cannot influence the outcome of a physical principle, so we can assert that Equation 1.75 is valid in any coordinate system.

Thus,

$$\vec{\nabla}\cdot(\vec{E}\times\vec{H})=\vec{H}\cdot(-\partial\vec{B}/\partial t)-\vec{E}\cdot(\vec{J}+\partial\vec{D}/\partial t), \tag{1.76}$$

and, by using the constitutive relations $\vec{B}=\mu\vec{H}$, $\vec{D}=\varepsilon\vec{E}$, and $\vec{J}=\sigma\vec{E}$ for a linear, isotropic medium,

$$\vec{\nabla}\cdot(\vec{E}\times\vec{H})=-\mu\frac{\partial}{\partial t}\left(\frac{1}{2}\vec{H}\cdot\vec{H}\right)-\sigma(\vec{E}\cdot\vec{E})-\varepsilon\frac{\partial}{\partial t}\left(\frac{1}{2}\vec{E}\cdot\vec{E}\right) \tag{1.77}$$

If we now use the Divergence Theorem to integrate Equation 1.77 over a surface S that surrounds volume V,

$$\oint_S(\vec{E}\times\vec{H})\cdot d\vec{\sigma}=-\frac{\partial}{\partial t}\left(\iiint_V\frac{1}{2}\mu H^2d^3x\right)-\iiint_V\sigma E^2d^3x-\frac{\partial}{\partial t}\left(\iiint_V\frac{1}{2}\varepsilon E^2d^3x\right) \tag{1.78}$$

In Equation 1.78, we can recognize the terms $\mu H^2/2$ and $\varepsilon E^2/2$ as the energy density of an electromagnetic field. In addition, we can recognize the term $\sigma E^2 = J^2/\sigma = I^2/\sigma A^2$, where I/A is the current per unit area (the current density) so we can express $-\iiint_V(1/\sigma A)(I^2/A)d^3x=-\int_L I^2(\rho/A)dx=-I^2R$ as the energy loss due to current, I, in a resistor, R, formed by the conducting material in volume V.

We can see that the left-hand side of Equation 1.78 is the integral of the Poynting vector, $\vec{\mathcal{P}}$, over the surface, S, that encloses the volume V. The right-hand side of Equation 1.78 is the negative time rate of change of energy (in the electric field, the magnetic field, and the ohmic conductor) in the volume V. Time rate of change is power so Equation 1.78 states in words

> *"The energy flowing out of volume V through its surface S, is the integral over the surface S of the Poynting vector."*

We can thus interpret the Poynting vector, $\vec{\mathcal{P}}$, as the instantaneous power per unit area flowing across a surface. The units of $\vec{\mathcal{P}}$ should thus be *Watts* per unit area, as we can confirm by $\vec{E}\times\vec{H}$, which would be in $(V/m)(A/m)$.

Time Average Power Density

Figure 1.23 shows that $\vec{\mathcal{P}}$ is a time-dependent quantity for the special case of TEM waves propagating in a lossless medium. In the case of a lossy medium, we have seen from Figure 1.16 that there is a phase shift between the magnetic field intensity and electric field intensity, and Equation 1.43 showed that, in the general case of TEM waves with no finite boundary conditions (in otherwise free space), $\vec{E}$ and $\vec{H}$ are related by the complex propagation constant $\tilde{k} = \beta - j\alpha$ as $(\beta - j\alpha)\hat{a}_k \times \vec{E}_0 = \omega\mu\vec{H}_0$ or

$$\vec{H}_0 = [(\beta - j\alpha)/\omega\mu]\hat{a}_k \times \vec{E}_0 = (1/\tilde{\eta})\,\hat{a}_k \times \vec{E}_0 \tag{1.79}$$

where

$$\tilde{\eta} \equiv \omega\mu/(\beta - j\alpha) = |\tilde{\eta}|\,e^{j\theta_\eta} \tag{1.80}$$

is a complex quantity (signified by the tilde) called the ***complex intrinsic impedance***.

To evaluate $\vec{\mathcal{P}}$, we can express the electric and magnetic field intensity as

$$\vec{E}(z,t) = \mathrm{Re}\left[\vec{E}_S(z)e^{j\omega t}\right] = \hat{a}_x E_0 e^{-\alpha z}\,\mathrm{Re}\left[e^{j(\beta z - \omega t)}\right] = \hat{a}_x E_0 e^{-\alpha z}\cos(\beta z - \omega t)$$

and

$$\vec{H}(z,t) = \mathrm{Re}\left[(E_0/|\tilde{\eta}|)e^{-\alpha z}\,e^{j(\beta z - \omega t)}\,e^{-j\theta_\eta}\right]\hat{a}_y = \hat{a}_y(E_0/|\tilde{\eta}|)e^{-\alpha z}\cos(\beta z - \omega t - \theta_\eta)$$

to write the Poynting vector as

$$\vec{\mathcal{P}}(z,t) = \vec{E} \times \vec{H} = \left(E_0^2/|\tilde{\eta}|\right)e^{-2\alpha z}\cos(\beta z - \omega t)\cos(\beta z - \omega t - \theta_\eta)\vec{a}_z \tag{1.81}$$

or with a mathematical identity for the cosine

$$\vec{\mathcal{P}}(z,t) = \left(E_0^2/|\tilde{\eta}|\right)e^{-2\alpha z}(1/2)[\cos\theta_\eta + \cos(2\beta z - 2\omega t - \theta_\eta)]\hat{a}_z \tag{1.82}$$

We can see that this quantity varies with time with a frequency $2\omega = 4\pi/T$ so, if we take the time average, evaluated at a particular point z, by averaging over an integer number of n cycles,

$$\vec{\mathcal{P}}_{Avg} = (1/nT)\int_0^{nT}\left(E_0^2/|\tilde{\eta}|\right)e^{-2\alpha z}(1/2)[\cos\theta_\eta + \cos(2\beta z - 2\omega t - \theta_\eta)]\hat{a}_z\,dt \tag{1.83}$$

or

$$\vec{\mathcal{P}}_{Avg} = \left(E_0^2/2|\tilde{\eta}|\right)e^{-2\alpha z}\cos\theta_\eta\,\hat{a}_z\left(W/m^2\right) \tag{1.84}$$

NOTE Equation 1.84 is equivalent to

$$\vec{\mathcal{P}}_{Avg} = (1/2)\,\mathrm{Re}\left[\vec{E} \times \vec{H}^*\right]. \tag{1.85}$$

PROBLEM

1.6 Compute $\vec{E} \times \vec{H}^*$ for E&M fields in a lossy medium to show that Equations 1.84 and 1.85 yield the same values.

1.7 MOMENTUM PROPAGATION

We can see that an electromagnetic field in the form of light has momentum by observing the effect of radiation pressure on a balanced vane that has a silver reflecting surface on one side and a black absorbing surface on the other. Such a device is called a Crooke's radiometer and is shown in Figure 1.24.

Normally, the rotation of the vane is explained by the photon concept of light in which photons are merely absorbed by the black surface, while photons are reflected by the silver surface. It is argued that the momentum change for the absorbed photon is thus half as much as the momentum change for the reflected photon and thus greater pressure is applied to the silver side of the vane than to the black side. Unless the glass bulb is highly evacuated, the molecules on the dark surface cause heating of the residual gas in the bulb, which in turn exerts a greater pressure on the black surface, causing the vane to turn in the wrong direction.

Electromagnetic pressure was predicted by James Maxwell in 1899 and demonstrated experimentally by Peter Lebedev. NASA has even proposed to use the pressure of sunlight to accelerate a solar sail-ship fitted with a huge silver sail. A variation of the space sail has been proposed in the form of a magnetic sail (magsail) which would cause a superconducting current loop to be oriented normal to charged particles in the solar wind to impart momentum and thus accelerate the spacecraft in the direction of the wind.

Some scientists, including Lord Kelvin and Helmholtz in 1871, have proposed[4] a "radio-panspermia theory" that radiation pressure could be the propulsion system

Figure 1.24 Crooke's radiometer.

that transported spores and bacteria through interstellar space to populate the first life forms on earth.[5]

ENDNOTES

1. E. G. Harris, Paul G. Huray, F. E. Obenshain, J. O. Thompson, and R. A. Villecco, "Experimental Test of Weyl's Gauge-Invariant Geometry," *Physical Review* D 7, no. 8 (April 1973): 15.
2. R. V. Pound, and G. A. Rebka Jr., "Gravitational Red-Shift in Nuclear Resonance," *Physical Review Letters* 3, no. 9 (1959): 439–41.
3. Hans C. Ohanian, *Gravitation and Spacetime* (New York: W. W. Norton, 1976).
4. S. Arrhenius, "The Propagation of Life in Space," *Die Umschau* 7 (1903): 481.
5. A website on the Panspermia theory can be found at http://www.panspermia.org/.

Chapter 2

Plane Waves in Compound Media

LEARNING OBJECTIVES

- Mathematically describe the character of plane waves as they propagate from one medium into another
- Relate the reflection and refraction of plane waves at the medium boundary to material properties
- Describe the changing character of waves as they propagate through a series of different media, including media that continuously change with distance
- Mathematically describe the change of polarization of waves as they interact with compound media

INTRODUCTION

Electromagnetic waves typically propagate from one medium to another in practical applications. For example, a pulsed carrier frequency from a modulated laser diode enters and departs from a glass fiber-optic transmission line; light is reflected and refracted as it passes from air to water; electromagnetic waves are incident on a conducting boundary; waves scatter from small dielectric impurities in a uniform medium. In some cases, the effects of a series of material boundaries must be considered. For example, information pulses are contained within a core glass fiber with a surrounding sheath with a different index of refraction; oil is on the surface of water from which light is reflecting/refracting; waves approach a conductor covered by an insulating material. In some cases, a continuum of changing media is considered. For example, electromagnetic waves are incident on the earth's atmosphere from outer space, and communication signals are reflected from the ionosphere to overcome the curvature of the earth.

These cases all assume that an electromagnetic wave with given amplitude and frequency, caused by an external source, is propagating in a material with properties that are defined by the character of the medium (permittivity, permeability,

The Foundations of Signal Integrity, by Paul G. Huray

conductivity), as described in Chapter 1. To understand how these E&M waves change when they encounter another medium, we consider the following sections under progressively more complicated scenarios:

1. The second medium has a different permittivity and permeability and is oriented normal to the direction of propagation.
2. The second medium has a different permittivity, permeability, and conductivity and is oriented normal to the direction of propagation.
3. The second medium has a different permittivity, permeability, and conductivity and is oriented obliquely to the propagation direction.
4. The incident waves have a linear or circular polarization.
5. The second material has a permittivity, permeability, or conduction that varies continuously with distance.

2.1 PLANE WAVE PROPAGATING IN A MATERIAL AS IT ORTHOGONALLY INTERACTS WITH A SECOND MATERIAL

As shorthand, we will consider the propagation of an electromagnetic wave in a medium to be represented by three coordinate axes as shown in Figure 2.1.

In Figure 2.1, the subscript i indicates that the wave is *incident* on some surface. In Figure 2.2, the wave is incident on a boundary between two media with differing values of permittivity, ε, and permeability, μ. In this drawing, the subscript $\boldsymbol{t}$ is used to indicate the part of the wave that has been *transmitted* into the second medium, and the subscript $\boldsymbol{r}$ is to indicate the part of the wave that has been *reflected* from the boundary between the first and the second media.

In Figure 2.2, there is no significance to the vertical location of the axes because the incident wave is assumed to be uniformly distributed in a vertical (and perpendicular to the page) direction. The parts of the wave that are transmitted and reflected at the boundary between medium 1 and medium 2 are also uniformly distributed in a vertical (and perpendicular to the page) direction. However, note that the direction of the magnetic field intensity for each of these parts is chosen to satisfy the right hand rule.

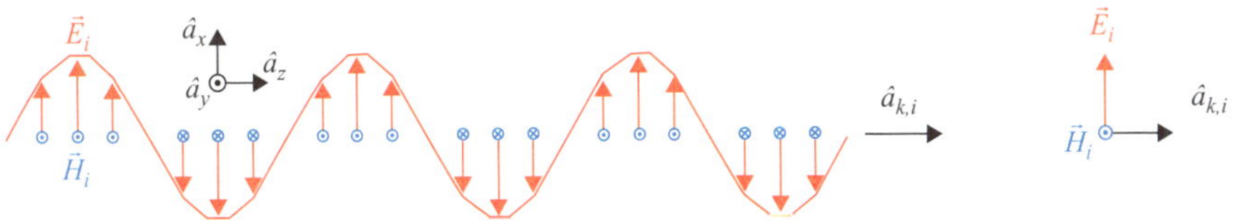

Figure 2.1 An incident electromagnetic wave is shown on the left with its electric field intensity, $\vec{E}_i$, in the $\pm x$-direction, its magnetic field intensity, $\vec{H}_i$, in the $\pm y$-direction, and its direction of propagation, $\hat{a}_{k,i}$, in the z-direction. The coordinate axes to the right will be used to represent this wave in the following figures.

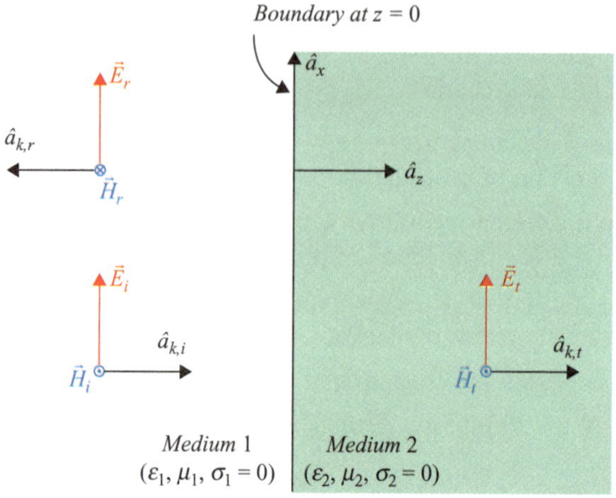

Figure 2.2 Incident electromagnetic wave propagating toward an orthogonal boundary between two media with differing permittivity and permeability (lower left), the transmitted part of that wave into medium 2 (right), and the reflected part of the incident wave back into medium 1 (upper left). The origin of Cartesian coordinates is on the boundary surface.

In constructing Figure 2.2, three assumptions have been made for the electric field intensity:

- Linear polarization is in the vertical (x)-direction.
- The transmitted part of the incident wave has the same polarization
- The reflected part of the incident wave has the same polarization

In the following mathematical analysis, we will use these assumptions to find the relative magnitude of the components of the transmitted and reflected electric and magnetic field intensities, as prescribed by the boundary conditions at the interface ($z = 0$) between the two media. If the equations yield a negative number for any one of the components, we will know we have assumed the direction of that component incorrectly.

2.2 ELECTROMAGNETIC BOUNDARY CONDITIONS

We have shown[1] that the boundary conditions for time-varying fields at a boundary between two media can be expressed as

$$E_{1\,\text{tangent}} = E_{2\,\text{tangent}}$$

$$\hat{a}_{2\,\text{normal}} \times \left(\vec{H}_1 - \vec{H}_2\right) = \vec{J}_s$$

$$B_{1\,\text{normal}} = B_{2\,\text{normal}}$$

$$\hat{a}_{2\,\text{normal}} \cdot \left(\vec{D}_1 - \vec{D}_2\right) = \Sigma_{e,s}$$

Interface between Two Linear* Dielectric Media

For a homogeneous, linear dielectric medium specified by a complex permittivity, $\tilde{\varepsilon}$, and a complex permeability, $\tilde{\mu}$, with $\sigma = 0$, there are no free charges at the boundary ($\Sigma_{e,s} = 0$) and no surface currents at the boundary ($\vec{J}_s = 0$) between two media, so, if we evaluate the fields at $z = 0$,

$$E_{1i}(0) + E_{1r}(0) = E_{2t}(0) \tag{2.1a}$$

$$H_{1i}(0) - H_{1r}(0) = H_{2t}(0) \tag{2.1b}$$

The minus sign in Equation 2.1b is consistent with our polarization assumptions above. Furthermore, we can relate the magnitude of the magnetic field intensity to the electric field intensity in the two media by the complex impedance, $\tilde{\eta}$, in each material:

$$\vec{H}_i(z)\big|_{z=0} = \hat{a}_y(E_{i0}/\tilde{\eta}_1)e^{j\beta z}\big|_{z=0} \text{ so} \tag{2.2a}$$

$$\left|\vec{H}_i(0)\right| = H_i(0) = E_{i0}/\tilde{\eta}_1 \tag{2.2b}$$

Using the corresponding relations for the transmitted and reflected parts in Equation 2.1, we get

$$E_{i0} + E_{r0} = E_{t0} \tag{2.3a}$$

$$E_{i0}/\tilde{\eta}_1 - E_{r0}/\tilde{\eta}_1 = E_{t0}/\tilde{\eta}_2 \tag{2.3b}$$

Equations 2.3a and 2.3b are two equations in three unknowns (assuming we know the values of $\tilde{\eta}_1$ and $\tilde{\eta}_2$) so we can find the two ratios:

$$\tilde{\Gamma}_{12} \equiv E_{r0}/E_{i0} = (\tilde{\eta}_2 - \tilde{\eta}_1)/(\tilde{\eta}_2 + \tilde{\eta}_1) \tag{2.4a}$$

$$\tilde{\tau}_{12} \equiv E_{t0}/E_{i0} = 2\tilde{\eta}_2/(\tilde{\eta}_2 + \tilde{\eta}_1) \tag{2.4b}$$

The quantity $\tilde{\Gamma}_{12}$ is called the complex ***reflection coefficient*** because it gives the magnitude of the electric field intensity of the scattered wave relative to the magnitude of the electric field intensity of the incident wave. The quantity $\tilde{\tau}_{12}$ is called the complex ***transmission coefficient*** because it gives the magnitude of the electric field intensity of the transmitted wave relative to the magnitude of the electric field intensity of the incident wave.

We can see from Equation 2.4 that $-1 < |\tilde{\Gamma}_{12}| < 1$ and $0 < |\tilde{\tau}_{12}| < 2$. Furthermore, we can see that, for normal incidence,

*__Note:__ If medium 2 were a perfect conductor (having infinite conductivity, $\sigma_2 \to \infty$), the values of the electric field intensity and magnetic field intensity in medium 2 would be zero.

$$1+\tilde{\Gamma}_{12}=\left(\frac{\tilde{\eta}_2+\tilde{\eta}_1}{\tilde{\eta}_2+\tilde{\eta}_1}\right)+\left(\frac{\tilde{\eta}_2-\tilde{\eta}_1}{\tilde{\eta}_2+\tilde{\eta}_1}\right)=\left(\frac{2\tilde{\eta}_2}{\tilde{\eta}_2+\tilde{\eta}_1}\right)=\tilde{\tau}_{12} \tag{2.5}$$

Because the medium wave impedances may be complex numbers, Equation 2.4a shows us that the reflected electric field intensity can be phase shifted by comparison to the incident electric field intensity, and Equation 2.4b shows us that the transmitted electric field intensity can be phase shifted by comparison to the incident electric field intensity.

Other Conclusions

1. $|\tilde{\tau}_{12}|$ is positive so some of the wave is always transmitted into medium 2, and the polarization of the transmitted wave is always the same (although perhaps phase shifted) as that of the incident wave. We can say polarization of the electric field intensity in normal incidence is preserved by transmission.
2. $|\tilde{\Gamma}_{12}|$ is positive as long as $|\tilde{\eta}_2| > |\tilde{\eta}_1|$, in which case we have chosen the direction of the scattered electric field intensity relative to the direction of the incident electric field intensity (both in the x-direction) correctly; polarization has been preserved in normal incidence by reflection.
3. $|\tilde{\Gamma}_{12}|$ is negative when $|\tilde{\eta}_2| < |\tilde{\eta}_1|$, in which case we have assumed the direction of the scattered electric field intensity relative to the direction of the incident electric field intensity incorrectly; that is, if $\vec{E}_i$ is in the x-direction, then the direction of $\vec{E}_r$ will be in the negative x-direction. In this case, the electric field intensity of the reflected wave shown in Figure 2.2 is incorrect, and, because the magnetic field intensity must preserve the right hand rule, it is also incorrect; polarization has not been preserved in normal incidence by reflection.

Mathematical Interpretation

We may use Equations 2.4a and 2.4b to express the incident, reflected, and transmitted components as

$$\vec{E}_i(z,t)=E_{i0}e^{-j(\tilde{k}_1 z-\omega t)}\hat{a}_x \tag{2.6a}$$

$$\vec{H}_i(z,t)=(E_{i0}/\tilde{\eta}_1)e^{-j(\tilde{k}_1 z-\omega t)}\hat{a}_y \tag{2.6b}$$

$$\vec{E}_r(z,t)=\tilde{\Gamma}_{12}E_{i0}e^{j(\tilde{k}_1 z+\omega t)}\hat{a}_x \tag{2.6c}$$

$$\vec{H}_r(z,t)=-\tilde{\Gamma}_{12}(E_{i0}/\tilde{\eta}_1)e^{j(\tilde{k}_1 z+\omega t)}\hat{a}_y \tag{2.6d}$$

$$\vec{E}_t(z,t)=\tilde{\tau}_{12}E_{i0}e^{-j(\tilde{k}_2 z-\omega t)}\hat{a}_x \tag{2.6e}$$

$$\vec{H}_t(z,t)=\tilde{\tau}_{12}(E_{i0}/\tilde{\eta}_2)e^{-j(\tilde{k}_2 z-\omega t)}\hat{a}_y \tag{2.6f}$$

In medium 1, the incident and reflected fields at various points z must be added so that

$$\vec{E}_1(z,t) = \vec{E}_i(z,t) + \vec{E}_r(z,t) = E_{i0}e^{-j(\tilde{k}_1 z-\omega t)}\hat{a}_x + \tilde{\Gamma}_{12}E_{i0}e^{j(\tilde{k}_1 z+\omega t)}\hat{a}_x \tag{2.7a}$$

$$\vec{H}_1(z,t) = \vec{H}_i(z,t) + \vec{H}_r(z,t) = (E_{i0}/\tilde{\eta}_1)e^{-j(\tilde{k}_1 z-\omega t)}\hat{a}_y - \tilde{\Gamma}_{12}(E_{i0}/\tilde{\eta}_1)e^{j(\tilde{k}_1 z+\omega t)}\hat{a}_y \tag{2.7b}$$

or

$$\vec{E}_1(z,t) = E_{i0}e^{-j(\tilde{k}_1 z-\omega t)}\hat{a}_x\left[1+\tilde{\Gamma}_{12}e^{2j\tilde{k}_1 z}\right] \tag{2.8a}$$

$$\vec{H}_1(z,t) = (E_{i0}/\tilde{\eta}_1)e^{-j(\tilde{k}_1 z-\omega t)}\hat{a}_y\left[1-\tilde{\Gamma}_{12}e^{2j\tilde{k}_1 z}\right] \tag{2.8b}$$

We can interpret Equation 2.8a as a transverse electric (TE^z) wave traveling in the z-direction whose amplitude is modulated by the factor $[1 + \tilde{\Gamma}_{12}e^{2j\tilde{k}_1 z}]$. Likewise, we can interpret Equation 2.8b as a transverse magnetic (TM^z) wave traveling in the z-direction whose amplitude is modulated by the factor $[1 - \tilde{\Gamma}_{12}e^{2j\tilde{k}_1 z}]$. The modulation factors are not a function of time but they are a function of the spatial variable z; that is, they are stationary in space. We can thus interpret the results Equations 2.8a and 2.8b as a standing modulation envelope under which the (TE^z) and (TM^z) waves propagate.

In all of the above equations, the tilde reminds us that the quantities may be a complex number. As stated previously, our conventions are $\tilde{k} \equiv \beta - j\alpha$, $\tilde{\eta} = |\tilde{\eta}|e^{j\phi_\eta} = \eta e^{j\phi_\eta}$ so now we permit a complex reflection coefficient $\tilde{\Gamma}_{12} = |\tilde{\Gamma}_{12}|e^{j\phi_\Gamma} = \Gamma_{12}e^{j\phi_\Gamma}$.

Special Case #1: A Lossless Medium 1 and Medium 2

For the special case of a dielectric medium 1 in which the electrical conductivity, $\sigma_l = 0$, and ε and μ are both real quantities, $\tilde{k}_1 = \beta_1$, and $\tilde{\eta}_1 = \eta_1$. If medium 2 is also lossless, $\tilde{\Gamma}_{12} = \Gamma_{12}$ is also a real quantity. In this case, the values of the electric and magnetic field intensities are given by

$$\vec{E}_1(z,t) = E_{i0}e^{-j(\beta_1 z-\omega t)}\hat{a}_x\left[1+\Gamma_{12}e^{j2\beta_1 z}\right] \tag{2.9a}$$

$$\vec{H}_1(z,t) = (E_{i0}/\eta_1)e^{-j(\beta_1 z-\omega t)}\hat{a}_y\left[1-\Gamma_{12}e^{j2\beta_1 z}\right] \tag{2.9b}$$

Taking the real part of these quantities,

$$\mathrm{Re}\left[\vec{E}_1(z,t)\right] = \hat{a}_x E_{i0}\cos(\beta_1 z-\omega t) + \Gamma_{12}\hat{a}_x E_{i0}\cos(\beta_1 z+\omega t) \tag{2.10a}$$

$$\mathrm{Re}\left[\vec{H}_1(z,t)\right] = \hat{a}_y(E_{i0}/\eta_1)\cos(\beta_1 z-\omega t) - \hat{a}_y\Gamma_{12}(E_{i0}/\eta_1)\cos(\beta_1 z+\omega t) \tag{2.10b}$$

We see the fields in medium 1 are produced by the interference between two waves; one propagating in the positive z-direction and another of smaller magnitude propagating in the negative z-direction.

If we subtract the term $\Gamma_{12}\hat{a}_x E_{i0}\cos(\beta_1 z - \omega t)$ from the first term in Equation 2.10a and add it to the second, we get

$$\mathrm{Re}\left[\vec{E}_1(z,t)\right] = \hat{a}_x E_{i0}\cos(\beta_1 z - \omega t)[1-\Gamma_{12}] + \Gamma_{12}\hat{a}_x E_{i0}[\cos(\beta_1 z + \omega t) + \cos(\beta_1 z - \omega t)]$$
$$\text{or}\quad \mathrm{Re}\left[\vec{E}_1(z,t)\right] = \hat{a}_x E_{i0}\cos(\beta_1 z - \omega t)[1-\Gamma_{12}] + 2\Gamma_{12}\hat{a}_x E_{i0}\cos(\beta_1 z)\cos(\omega t) \tag{2.11a}$$

Using $\beta_1 = 2\pi/\lambda_1$, we can see that the left-hand term of Equation 2.11a represents a TE^z wave propagating in the positive z-direction with amplitude $E_{i0}[1 - \Gamma_{12}]$, and the second term is a standing wave that (for a given value of z) oscillates in time between $\pm 2\Gamma_{12}E_{i0}\cos(2\pi/\lambda_1)z$. Thus, we see that the portion of the incident wave that reflects from the boundary interferes with that same amount of the incoming wave to form a standing wave, while the remainder of the wave propagates in the positive z-direction.

From Equation 2.6e, the magnitude of the transmitted wave with a real value of τ_{12} and k_2, is

$$\mathrm{Re}\left[\vec{E}_t(z,t)\right] = \mathrm{Re}\left[\tau_{12}E_{i0}e^{-j(\beta_2 z - \omega t)}\hat{a}_x\right] = [1+\Gamma_{12}]E_{i0}\cos(\beta_2 z - \omega t) \tag{2.12a}$$

It is comforting to note from Equations 2.11a and 2.12a that the boundary condition at $z = 0$ provides the same value of the electric field intensity in medium 1 and in medium 2; that is, the electric field intensity is continuous across the boundary between the two media as required. Note that the phase velocity of the TE^z wave in medium 1 is $u_{p1} = \omega/\beta_1 = 1/\sqrt{\varepsilon_1\mu_1}$, while that for medium 2 is $u_{p2} = \omega/\beta_2 = 1/\sqrt{\varepsilon_2\mu_2}$. We can also see from Equation 2.7a that, *for this special case of lossless media*, $\vec{E}_1(z, t) = E_{i0}\hat{a}_x[e^{-j(\beta_1 z)} + \Gamma_{12}e^{j(\beta_1 z)}]e^{j\omega t}$ and the bracket has a maximum of $|1 + \Gamma|$ where $\beta_1 z_{Max} = \pi n$ (n = integer) and a minimum of $|1 - \Gamma_{12}|$, where $\beta_1 z_{Min} = (2n + 1)\pi/2$. Thus, the amplitude of the electric field intensity in medium 1 changes between a maximum and a minimum between the two locations $z_{Max} = \lambda_1 n/2$ and $z_{Min} = \lambda_1 n/2 + \lambda_1/4$. For these two locations, separated by $\lambda_1/4$, the ***standing wave ratio***, S, is defined as

$$S \equiv \frac{\left|\vec{E}_1(z,t)\right|_{Max}}{\left|\vec{E}_1(z,t)\right|_{Min}} = \frac{1+\Gamma_{12}}{1-\Gamma_{12}} \tag{2.13}$$

We can see that Equations 2.10a and 2.12a can be represented by the instantaneous electric field envelope shown plotted in Figure 2.3.

Special Case #2: Lossless Medium 1 and *Conducting* Medium 2

For the special case of a dielectric medium 1, in which the electrical conductivity, $\sigma_1 = 0$ and in which ε and μ are both real quantities, $\tilde{k}_1 = \beta_1$ and $\tilde{\eta}_1 = \eta_1$. However, if $\sigma_2 \neq 0$, then $\tilde{k}_2 \equiv \beta_2 - j\alpha_2$, and both $\tilde{\eta}_2 = \eta_2 e^{j\phi_\eta}$ and $\tilde{\Gamma}_{12} = (\tilde{\eta}_2 - \eta_1)/(\tilde{\eta}_2 + \eta_1) = \Gamma_{12}e^{j\varphi_\Gamma}$

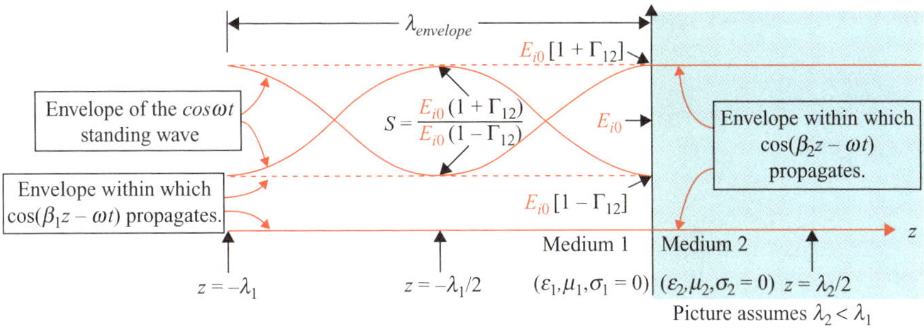

Figure 2.3 Plot of the magnitude of the electric field intensity envelopes of a transverse electric wave propagating in *lossless* Medium 1 as it interacts with *lossless* Medium 2 at an orthogonal boundary (at $z = 0$).

are complex quantities. In this case, medium 2 is lossy, and the value of the electric field intensity is given by Equation 2.9a with a complex reflection cefficient

$$\vec{E}_1(z,t)=E_{i0}e^{-j(\beta_1 z-\omega t)}\hat{a}_x\left[1+\tilde{\Gamma}_{12}e^{j2\beta_1 z}\right] \tag{2.14}$$

In this case,

$$\vec{E}_1(z,t)=\hat{a}_x E_{i0}e^{j\omega t}e^{j\frac{\phi_\Gamma}{2}}\left[e^{-j\beta_1 z}e^{-j\frac{\phi_\Gamma}{2}}+\Gamma_{12}e^{j\beta_1 z}e^{j\frac{\phi_\Gamma}{2}}\right] \tag{2.15}$$

If we add and subtract the term $[1-\Gamma_{12}]\hat{a}_x E_{i0}e^{j\omega t}e^{-j\beta_1 z}$ to this equation, we get

$$\vec{E}_1(z,t)=\hat{a}_x E_{i0}[1-\Gamma_{12}]e^{-j(\beta_1 z-\omega t)}+\Gamma_{12}\hat{a}_x E_{i0}e^{j\omega t}e^{j\frac{\phi_\Gamma}{2}}\left[e^{-j\beta_1 z}e^{-j\frac{\phi_\Gamma}{2}}+e^{j\beta_1 z}e^{j\frac{\phi_\Gamma}{2}}\right]$$

$$\mathrm{Re}\left[\vec{E}_1(z,t)\right]=\hat{a}_x E_{i0}[1-\Gamma_{12}]\cos(\beta_1 z-\omega t)+2\Gamma_{12}\hat{a}_x E_{i0}\cos\left(\beta_1 z-\frac{\phi_\Gamma}{2}\right)\cos\left(\omega t-\frac{\phi_\Gamma}{2}\right) \tag{2.16}$$

Using $\beta_1 = 2\pi/\lambda_1$, we can again see that the left-hand term represents a TE^z wave propagating in the positive z-direction with amplitude $E_{i0}[1-\Gamma_{12}]$, and the second term is a standing wave that (for a given value of z) oscillates in time. Again, we see that the portion of the incident wave that reflects from the boundary interferes with that same amount of the incoming wave to form a standing wave, while the remainder of the wave propagates in the positive z-direction. Equation 2.6e thus becomes

$$\vec{E}_t(z,t)=\tilde{\tau}_{12}E_{i0}e^{-j(\tilde{k}_2 z-\omega t)}\hat{a}_x=\left[1+\Gamma_{12}e^{j\phi_\Gamma}\right]E_{i0}e^{-\alpha_2 z}e^{-j(\beta_2 z-\omega t)}\hat{a}_x, \tag{2.17}$$

and we see that the wave dies exponentially in medium 2 as $e^{-\alpha_2 z}$. At the point $z = 0$, Equation 2.17 becomes

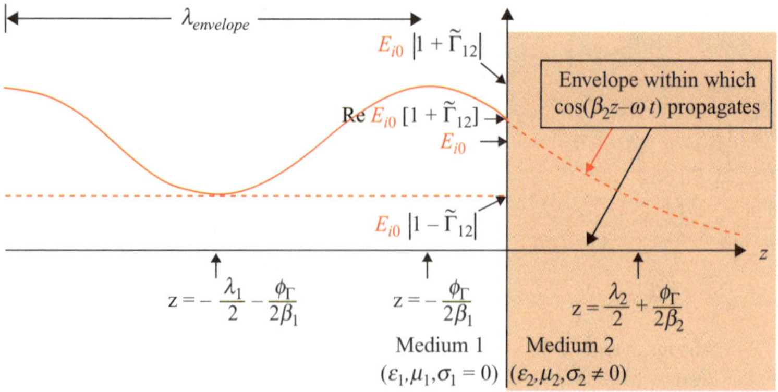

Figure 2.4 Plot of the magnitude of the electric field intensity envelope of a transverse electric wave propagating in a *lossless* medium 1 as it interacts with an orthogonal boundary (at z = 0) at lossy medium 2.

$$\bar{E}_t(0,t)=\left[1+\tilde{\Gamma}_{12}\right]E_{i0}e^{-j\omega t}\hat{a}_x, \tag{2.18}$$

which is the same as Equation 2.14. Thus, the electric field intensity at z = 0 is continuous.

We can see that Equations 2.16 and 2.17 can be represented by the instantaneous electric field intensity envelope shown plotted in Figure 2.4.

Very Special Case #2: A Lossless Medium 1 and a Perfect Conducting Medium 2

In the very special case of a propagating wave in lossless medium 1 interacting orthogonally with a perfect electric conductor ($\sigma_2=\infty$), we note that $\alpha=\sqrt{\mu\sigma\omega/2}=\infty$ so there is ***no*** penetration of the wave into medium 2. In this very special case, the boundary condition on the electric field intensity in medium 1 is that its amplitude must also go to zero. For this case, the incident traveling wave is totally reflected so that Γ_{12} = 1, which yields a standing wave ratio of ∞ because there is total constructive or destructive interference between the incident and the reflected waves. For this case, the electric and magnetic fields form a standing wave pattern, as shown in Figure 2.5.

2.3 PLANE WAVE PROPAGATING IN A MATERIAL AS IT ORTHOGONALLY INTERACTS WITH TWO BOUNDARIES

In many physical problems, plane waves interact with two boundaries between materials with differing physical properties, as indicated by Figure 2.6.

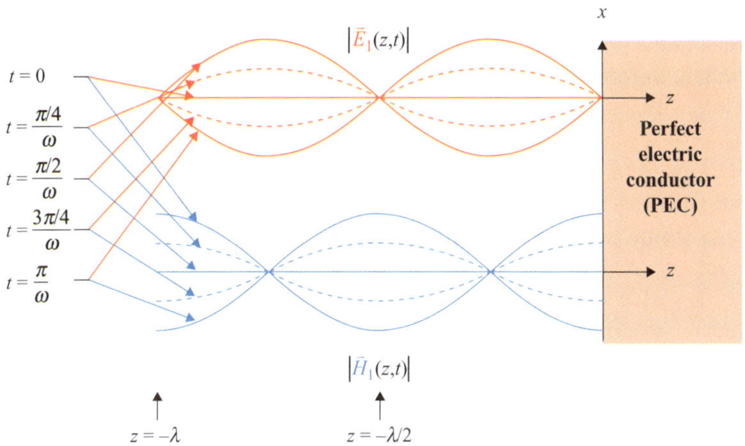

Figure 2.5 Standing wave patterns for the electric and magnetic field intensities for incident waves in a *lossless* medium incident on a perfect conductor.

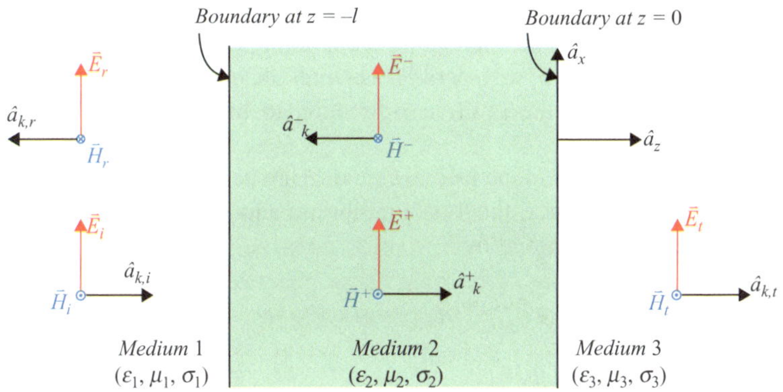

Figure 2.6 Electric and magnetic field intensities (a) that are incident and reflected in medium 1 ($z < -l$); (b) in medium 2 ($-l < z < 0$) that propagate in the positive and negative z-direction; and (c) that are transmitted into medium 3 ($z > 0$).

In effect, the field intensity shown in Figure 2.6 in medium 1 is incident to a slab of material between $-l < z < 0$. At the boundary between medium 1 and medium 2 (at $z = -l$), some of the incident field is transmitted into medium 2. The transmitted part of that incident field is consequently incident on the boundary between medium 2 and medium 3 (at $z = 0$), where some of it is transmitted and some of it is reflected back in the negative z-direction. The reflected part of the wave from the boundary at $z = 0$ in medium 2 is subsequently incident again on the boundary at $z = -l$, where part reflects back in the positive z-direction and part is transmitted back into medium 1. The total amount of the field intensity in medium 1 is thus the sum of the original incident field intensity and that reflected from the boundary at $z = -l$ plus that transmitted into medium 2, reflected from the boundary at $z = 0$, and subsequently transmitted back into medium 1.

We can see that the field intensities in medium 2 are made up of many such reflections and transmissions between the two boundaries at $z = -l$ and $z = 0$ (in principle, an infinite number). Furthermore, the total amount of field intensity reflected from the surface at $z = -l$ is the sum of the original reflected part plus the subsequent transmission back into medium 1 from fields in medium 2 traveling in the negative z-direction (in principle, an infinite number). Finally, we can see that the total amount of field intensity in medium 3 is a combination of all of the fields in medium 2 that are transmitted through the boundary at $z = 0$ (in principle, an infinite number).

Because the reflection and transmission coefficients at any boundary are restricted to be in the range from $-1 < |\tilde{\Gamma}_{ij}| < 1$ to $0 < |\tilde{\tau}_{ij}| < 2$, we can see that the series of additions will constitute a convergent sum (as we would expect from physical principles). We will thus permit the *total* field intensity additions in medium 1 that propagate in the negative z-direction to be represented by $\vec{E}_r$ and $\vec{H}_r$. We will also permit the *total* field intensity additions in medium 2 that propagate in the positive (and negative) z-direction to be represented by $\vec{E}^+$ and $\vec{H}^+$ (and $\vec{E}^-$ and $\vec{H}^-$). Finally, we will permit the *total* field intensity additions in medium 3 that propagate in the positive z-direction to be represented by $\vec{E}_t$ and $\vec{H}_t$.

Assuming that the incident field intensities $\vec{E}_i$ and $\vec{H}_i$ are harmonic (not a pulse) variations with a constant frequency, the total equilibrium equations for the quantities shown in Figure 2.6 are expressed by

$$\vec{E}_i(z,t) = E_{i0} e^{-j(\tilde{k}_1 z - \omega t)} \hat{a}_x \quad (-\infty < z < -l) \tag{2.19a}$$

$$\vec{H}_i(z,t) = (E_{i0}/\tilde{\eta}_1) e^{-j(\tilde{k}_1 z - \omega t)} \hat{a}_y \quad (-\infty < z < -l) \tag{2.19b}$$

$$\vec{E}_r(z,t) = \tilde{\Gamma}_{eff} E_{i0} e^{j(\tilde{k}_1 z + \omega t)} \hat{a}_x \quad (-\infty < z < -l) \tag{2.19c}$$

$$\vec{H}_r(z,t) = -\tilde{\Gamma}_{eff} (E_{i0}/\tilde{\eta}_1) e^{j(\tilde{k}_1 z + \omega t)} \hat{a}_y \quad (-\infty < z < -l) \tag{2.19d}$$

$$\vec{E}^+(z,t) = E_0^+ e^{-j(\tilde{k}_2 z - \omega t)} \hat{a}_x \quad (-l < z < 0) \tag{2.19e}$$

$$\vec{H}^+(z,t) = (E_0^+/\tilde{\eta}_2) e^{-j(\tilde{k}_2 z - \omega t)} \hat{a}_y \quad (-l < z < 0) \tag{2.19f}$$

$$\vec{E}^-(z,t) = E_0^- e^{j(\tilde{k}_2 z + \omega t)} \hat{a}_x \quad (-l < z < 0) \tag{2.19g}$$

$$\vec{H}^-(z,t) = -(E_0^-/\tilde{\eta}_2) e^{j(\tilde{k}_2 z + \omega t)} \hat{a}_y \quad (-l < z < 0) \tag{2.19h}$$

$$\vec{E}_t(z,t) = \tilde{\tau}_{23}|_{z=0} E_0^+ e^{-j(\tilde{k}_3 z - \omega t)} \hat{a}_x \quad (0 < z < \infty) \tag{2.19i}$$

$$\vec{H}_t(z,t) = \tilde{\tau}_{23}|_{z=0} (E_0^+/\tilde{\eta}_3) e^{-j(\tilde{k}_3 z - \omega t)} \hat{a}_y \quad (0 < z < \infty) \tag{2.19j}$$

where

$$\tilde{\Gamma}_{12}\Big|_{z=-l} = \frac{\vec{E}_r(z)}{\vec{E}_i(z)}\Bigg|_{z=-l} = \frac{\tilde{\Gamma}_{eff} E_{i0} e^{j\tilde{k}_1 z}}{E_{i0} e^{-j\tilde{k}_1 z}}\Bigg|_{z=-l} = \tilde{\Gamma}_{eff} e^{-j2\tilde{k}_1 l} \quad (2.20)$$

takes into account the fact that the 1,2 boundary is located at $z = -l$.

Note that $\tilde{\Gamma}_{eff}$ in Equation 2.20 will be

$$\tilde{\Gamma}_{eff} = \tilde{\Gamma}_{12} + \tilde{\tau}_{12} e^{-j(\tilde{k}_2+\tilde{k}_1)l} \tilde{\Gamma}_{23} \tilde{\tau}_{12} e^{-j(\tilde{k}_2+\tilde{k}_1)l} + \cdots \quad (2.21)$$

In Equation 2.21, the factor $e^{-j(\tilde{k}_2+\tilde{k}_1)l}$ takes into account the fact that $\tilde{\tau}_{12}$ is to be evaluated at $z = -l$. The first term on the right is the amount of incident electric field that is reflected from the 1,2 boundary. The second term on the right is the amount of incident electric field that is transmitted through the 1,2 boundary, reflected from the 2,3 boundary, and then is subsequently transmitted through the 1,2 boundary from medium 2. The … in Equation 2.21 recognizes the fact that there will be more field intensity that is reflected from boundary 2,3 and then transmitted through the boundary 1,2 from additional multiple processes.

Although Equations 2.19, 2.20, and 2.21 can be, in principle, solved analytically, it is more practical to use a numerical calculation to evaluate the convergent series given in Equation 2.21 and then find the effective values of the coefficients of reflection and transmission at the two boundaries, especially when the three media have lossy characteristics that result in complex quantities for the propagation constants and all other dependent terms.

Special Case: Lossless Media 1, 2, and 3

In the case of lossless dielectric media, Equations 2.19, 2.20, and 2.21 involve real constants, $\tilde{k}_i = \beta_i$, $\tilde{\Gamma}_{ij} = (\eta_j - \eta_i)/\eta_j + \eta_i$, and $\tilde{\tau}_{ij} = 2\eta_j/(\eta_j + \eta_i)$ and they are more conducive to an analytic solution. The solution is found by matching the boundary conditions at $z = -l$ and at $z = 0$ through the continuity of the tangential components of the electric and magnetic field intensities. The space-dependent terms in medium 1 at $z = -l$ are

$$\vec{E}_1^{total}\Big|_{z=-l} = \left(\vec{E}_i + \vec{E}_r\right)\Big|_{z=-l} = E_{i0} e^{j\beta_1 l}\left[1 + \Gamma_{eff} e^{-j2\beta_1 l}\right] \quad (2.22a)$$

$$\vec{H}_1^{total}\Big|_{z=-l} = \left(\vec{H}_i - \vec{H}_r\right)\Big|_{z=-l} = \frac{E_{i0}}{\eta_1} e^{j\beta_1 l}\left[1 - \Gamma_{eff} e^{-j2\beta_1 l}\right] \quad (2.22b)$$

The space dependent terms in medium 2 at $z = -l$ are

$$\vec{E}_2^{total}\Big|_{z=-l} = E_0^+ e^{j\beta_2 l}\hat{a}_x + E_0^- e^{-j\beta_2 l}\hat{a}_x \quad (2.22c)$$

$$\vec{H}_2^{total}\Big|_{z=-l} = \left(E_0^+/\eta_2\right) e^{j\beta_2 l}\hat{a}_y - \left(E_0^-/\eta_2\right) e^{-j\beta_2 l}\hat{a}_y, \quad (2.22d)$$

where we have used

$$H_0^+ = (E_0^+/\eta_2) \; and \; H_0^- = -(E_0^-/\eta_2) \tag{2.23a}$$

Furthermore,

$$E_0^- = \Gamma_{23} E_0^+ = [(\eta_3 - \eta_2)/(\eta_3 + \eta_2)] E_0^+ \tag{2.23b}$$

We can thus equate 2.22a to 2.22c to obtain

$$E_{i0} e^{j\beta_1 l}\left[1 + \Gamma_{eff} e^{-j2\beta_1 l}\right] = E_0^+ \left[e^{j\beta_2 l} + \Gamma_{23} e^{-j\beta_2 l}\right] \tag{2.24a}$$

and we can equate 2.22b to 2.22d to obtain

$$(E_{i0}/\eta_1) e^{j\beta_1 l}\left[1 - \Gamma_{eff} e^{-j2\beta_1 l}\right] = (E_0^+/\eta_2)\left[e^{j\beta_2 l} - \Gamma_{23} e^{-j\beta_2 l}\right] \tag{2.24b}$$

Dividing these two equations, we get

$$\eta_1 \left[\frac{1 + \Gamma_{eff} e^{-2j\beta_1 l}}{1 - \Gamma_{eff} e^{-2j\beta_1 l}}\right] = \eta_2 \left[\frac{e^{j\beta_2 l} + \Gamma_{23} e^{-j\beta_2 l}}{e^{j\beta_2 l} - \Gamma_{23} e^{-j\beta_2 l}}\right] \tag{2.25}$$

We can define the *wave impedance* of the field intensities in medium 2 as

$$\eta_{2,w}(z) = \frac{E_2^{total}}{H_2^{total}} = \frac{E_0^+ e^{-j\beta_2 z} + E_0^- e^{j\beta_2 z}}{H_0^+ e^{-j\beta_2 z} + H_0^- e^{j\beta_2 z}} \tag{2.26}$$

Using Equations 2.23a and 2.23b in Equation 2.26, we get

$$\eta_{2,w}(z) = \eta_2 \left[\frac{e^{-j\beta_2 z} + \Gamma_{23} e^{j\beta_2 z}}{e^{-j\beta_2 z} - \Gamma_{23} e^{j\beta_2 z}}\right] \tag{2.27}$$

Evaluating Equation 2.27 at $z = -l$, we get

$$\eta_{2,w}(-l) = \eta_2 \left[\frac{e^{j\beta_2 l} + \Gamma_{23} e^{-j\beta_2 l}}{e^{j\beta_2 l} - \Gamma_{23} e^{-j\beta_2 l}}\right], \tag{2.28}$$

which we can recognize to be the same as the right-hand side of Equation 2.25. Thus,

$$\eta_{2,w}(-l) = \eta_1 \left[\frac{1 + \Gamma_{eff} e^{-2j\beta_1 l}}{1 - \Gamma_{eff} e^{-2j\beta_1 l}}\right], \tag{2.29}$$

which we can solve for Γ_{eff} to be

$$\Gamma_{eff} = \left[\frac{\eta_{2,W}(-l) - \eta_1}{\eta_{2,W}(-l) + \eta_1} \right] e^{2j\beta_1 l} \tag{2.30}$$

We see that Equation 2.30 (like Equation 2.20) has a term $e^{2j\beta_1 l}$ that takes into account the fact that the 1,2 boundary is located at $z = -l$. The square bracket in Equation 2.30 is of the form of a traditional reflection coefficient, Γ_{12}, at the boundary between medium 1 and medium 2, except that η_2 has been replaced by $\eta_{2,W}(-l)$; that is, the reflection coefficient, Γ_{12}, at the 1,2 boundary for a three medium transmission problem is effectively altered to Γ_{eff} by the multiple internal reflections inside medium 2, in which η_2 is replaced by $\eta_{2,W}(-l)$.

Evaluation of $\eta_{2,W}$

Using Equation 2.27 and Euler's identity, we have

$$\eta_{2,W}(z) = \eta_2 \left(\frac{(\eta_3 + \eta_2)(\cos\beta_2 z + j\sin\beta_2 z) + (\eta_3 - \eta_2)(\cos\beta_2 z - j\sin\beta_2 z)}{(\eta_3 + \eta_2)(\cos\beta_2 z + j\sin\phi_2 z) - (\eta_3 - \eta_2)(\cos\beta_2 z - j\sin\beta_2 z)} \right) \tag{2.31}$$

which yields

$$\eta_{2,W}(z) = \eta_2 \left(\frac{\eta_3 \cos\beta_2 z + j\eta_2 \sin\beta_2 z}{\eta_2 \cos\beta_2 z + j\eta_3 \sin\beta_2 z)} \right) \tag{2.32}$$

We can evaluate Equation 2.32 at the point $z = 0$ to obtain $\eta_{2,W}(0) = \eta_3$; that is, the total *wave impedance* inside the region 2 (evaluated at the boundary between medium 2 and medium 3 is the same as the wave impedance in region 3 (evaluated at the same boundary). Because both the magnitudes of the electric field intensity and magnetic field intensity components are constrained to match at the 2,3 boundary, it is comforting that their ratio also matches.

We can evaluate Equation 2.32 at the point $z = -l$, to obtain

$$\eta_{2,W}(-l) = \eta_2 \left(\frac{\eta_3 \cos\beta_2 l - j\eta_2 \sin\beta_2 l}{\eta_2 \cos\beta_2 l - j\eta_3 \sin\beta_2 l)} \right) \tag{2.33}$$

Equations 2.30 and 2.33 constitute the general results for the net reflected electric field intensity, $\Gamma_{eff} = \vec{E}_r(-l, t)/\vec{E}_i(-l, t)$, at the point $z = -l$, between two lossless media when another material boundary exists at $z = 0$. These equations will be later seen to be the same as the equations for reflection of potentials in transmission lines that exist between $z = -l$ and $z = 0$ and so they are very important in applications when a transmission medium changes twice. This change is actually more common than a simple, isolated change at the boundary between two materials.

Applications of Reflection and Transmission through a Dielectric Slab

The preceding equations have shown us how much Γ_{eff} of the electric field intensity incident on a dielectric slab of material will reflect from that slab and how much, τ_{23} is transmitted through the slab. The internal reflections in the slab cause the equations to be messy (Equations 2.30 and 2.33) but analytic in terms of the material properties of the three media. One of the most important applications of this result is the transmission of radar waves through a protective material. In the case of stationary radar installations in Alaska, for example, radar equipment is protected from snow and other atmospheric elements by constructing a dome (called a *radome*) over the equipment. In the case of moving radar equipment inside the nose of a jet fighter aircraft, the radar equipment is protected from birds and other objects moving at high relative velocities. The object in both of these cases is to maximize the amount of electric and magnetic field intensity transmitted through the protective dielectric *radome*. In fact, by the careful construction of the thickness of the *radome* slab, it is possible (for a given radar frequency) to produce no reflection back into the radar equipment and to produce a full transmission of power through the protective material. We can see how this is accomplished by trying to make $\Gamma_{eff} = 0$ in Equation 2.30; this can be accomplished by making $\eta_{2,W}(-l) = \eta_1$, which we can calculate from Equation 2.33.

Special Case #1

Suppose the material in medium 1 and in medium 3 are the same; that is, $\eta_1 = \eta_3$. This would be the case for air inside and outside of a *radome*. Then we can see that Equation 2.33 yields the result $\eta_{2,W}(-l) = \eta_1 = \eta_3$ if $\beta_2 l = m\pi$, where m is an integer. Because we can write $\beta_2 = 2\pi/\lambda_2$, we see this condition can be satisfied if

$$l = (m/2)\lambda_2 \tag{2.34}$$

Conclusion

If the protective material surrounding a source of electromagnetic radiation is chosen to have a thickness that is a half integer multiple of the wavelength of the radiation (in that material, where $\lambda_2 = c/\sqrt{\varepsilon_r} f$), there will be no reflection of the waves, and the radiation will be transmitted (except for the internal reflections in the slab) as if the slab did not exist. For a 10-GHz frequency, and a *radome* material with $\varepsilon_r = 9$, this leads to $\lambda_2 = 3 \times 10^8$ m/s$/3 \times 10^{10}$ s^{-1} = 1 cm and requires slab thicknesses of 0.5, 1.0, 1.5, 2.0, 2.5 cm, etc. For structural integrity, the value of m may be chosen large enough so that the protective material is several multiples of the half-wavelength.

Special Case #2

Suppose the material in medium 1 and in medium 3 are the same, that is, $\eta_1 = \eta_3$, and we wish to minimize the transmission of electric field intensity that propagates through a material slab. An example of this case might occur for air inside and outside of a microwave oven in which we want to enclose the electromagnetic radiation for health purposes. A second example would be for the sensitive testing of components inside a room for electromagnetic compatibility in which we wish to eliminate external sources of radiation from propagating into our test volume. For both of these cases, we may choose to use a dielectric slab with an effective transmission coefficient, τ, as a minimum. We can see that, for three dielectric materials,

$$\tau_{12} \equiv \frac{E_0^+}{E_{i0}} = \left(\frac{2\eta_{2,w}(-l)}{\eta_{2,w}(-l)+\eta_1}\right) \quad \text{and} \quad \tau_{23} \equiv \frac{E_t}{E_0^+} = \left(\frac{2\eta_3}{\eta_3+\eta_{2,w}(0)}\right) = \left(\frac{2\eta_3}{\eta_3+\eta_3}\right) = 1$$

so, if we want to minimize the radiation in region 3, we must minimize the transmission into region 2 from 1. Because $\tau_{12} = 1+\Gamma_{eff} = 1+\left[\frac{\eta_{2,w}(-l)-\eta_1}{\eta_{2,w}(-l)+\eta_1}\right]e^{2j\beta_1 l}$, we can that see Γ_{eff} must be negative and that its magnitude must be as large as possible. This can occur for values of $\eta_{2,w}(-l) < \eta_1$ only if $2\beta_1 l = (2n)\pi$, where n is a positive integer. Thus, in the case of air in medium 1 (where η_1 is already the highest value possible = 377Ω), we should choose $l = (n/2)\lambda_1$ and $\eta_{2,w}(-l)$ should be as small as possible to minimize the total transmission (and maximize the total reflection) from the slab of protective material. This means that the thickness of the slab should be an integer number of half-wavelengths of the wavelength in medium 1 and that

$$\eta_{2,w}(-l) = \eta_2\left(\frac{\eta_1\cos[(2\pi/\lambda_2)(n\lambda_1/2)] - j\eta_2\sin[(2\pi/\lambda_2)(n\lambda_1/2)]}{\eta_2\cos[(2\pi/\lambda_2)(n\lambda_1/2)] - j\eta_1\sin[(2\pi/\lambda_2)(n\lambda_1/2)]}\right)$$

should be as small as possible.

PROBLEMS

2.1 Given that $\eta_1 = 377\ \Omega$ and that $\eta_2 = 377\ \Omega/\sqrt{\varepsilon_2}$, (a) find the properties of the glass, and (b) find the possible values of glass thicknesses for the window of a microwave oven that will minimize its external radiation.

2.2 Microwave ovens normally have a glass door with an imbedded conducting mesh so that the interior may be seen through the door. Discuss the rationale for including a conducting mesh in the window and explain how large the mesh size can be to safely ensure that microwaves remain primarily interior to the oven.

Figure 2.7 (Above) Photograph that includes flash reflection (glare) from a child's glasses. (Below) The same photograph with the glare removed.

Special Case #3

Suppose the materials in medium 1 and in medium 3 are different, that is, $\eta_1 \neq \eta_3$, and we wish to minimize the reflection of electric field intensity from a material slab. An example of this case occurs for photographic "flash" light propagating in air and striking glass at an orthogonal direction, as is shown in Figure 2.7.

The manufacturers of eyeglasses typically offer a nonglare option that minimizes the reflection from their products. This involves a surface coat of material (medium 2) on the lens glass (medium 3 with $\eta_3 = 260\ \Omega$) that reduces reflection back into air (medium 1 with $\eta_1 = 377\ \Omega$). How do they choose the coating material and thickness to best accomplish this nonglare option?

ANSWER We wish to minimize reflection in medium 1 so, if we choose $\beta_2 l = (2n + 1)\pi/2$, $\eta_{2,w}(-l) = \eta_2\left(\dfrac{\eta_3 \cos\beta_2 l - j\eta_2 \sin\beta_2 l}{\eta_2 \cos\beta_2 l - j\eta_3 \sin\beta_2 l)}\right) = \eta_2\left(\dfrac{\eta_2}{\eta_3}\right)$ and at the interface between air (medium 1) and the coating (medium 2), $\eta_2 = \sqrt{\eta_1\eta_3}$. Thus, the coating material should have an odd-integer number of quarter-wavelengths (in medium 2) of thickness and a wave impedance of 313 Ω or an $\varepsilon_r = 1.45$ for nonmagnetic materials. The minimum thickness of the coating is then $l = \lambda_2/4 = \lambda_1/4\sqrt{\varepsilon_{r,2}} = 0.21\,\lambda_1$. For white light in the visible spectrum, $\lambda_{1,\text{avg}} \approx 5000$ Å so $l \approx 1038$ Å or 3113 Å or 5189 Å, … .

Multiple Coatings

In special case #3, we have chosen a lens coating to minimize reflection at an average wavelength of light in a white spectrum, but the wavelengths of various colors of light above and below this average will not be so well matched. Thus, a single coating will act as a filter of light for one color at the average wavelength. If we choose a subsequent coating (with its appropriate thickness) or, better, a series of subsequent coatings such that the impedances progressively increase from layer to layer, we can tune the glass surface to minimally reflect light of any color. We can see how this series of coatings work in the following section.

2.4 PLANE WAVE PROPAGATING IN A MATERIAL AS IT ORTHOGONALLY INTERACTS WITH MULTIPLE BOUNDARIES

Figure 2.8 shows a series of n parallel slabs of equal thickness but with unequal material properties into which an electromagnetic wave propagates orthogonally from the left. We expect the electric field intensity and magnetic field intensity in each medium, in general, to be different from one another and we must keep track of the fields propagating in the positive z-direction and in the negative z-direction

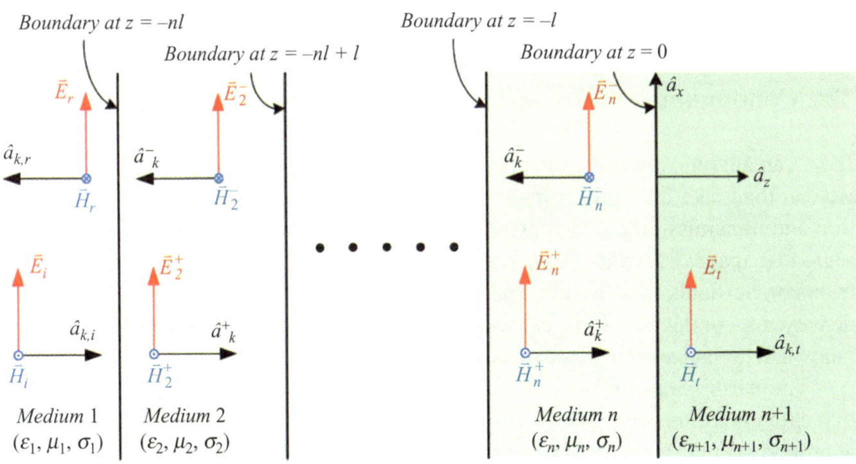

Figure 2.8 A multiple-interface problem geometry for the reflection and transmission of electromagnetic wave intensities in a sequence of materials with equal thickness but different permittivity, permeability, and conductivity.

in each medium. In principle, we could set up a set of equations for each medium and direction, match the boundary conditions (BC) at each boundary, and find an analytical solution for every unknown coefficient of reflection and transmission. Given the complexity of the previous section, we might expect this process to be burdensome but we could also teach a computer to make the BC matches and to solve the problem in each material. The answers, would give us the effective reflection coefficient in medium 1 and the effective transmission coefficient in medium $n + 1$, which we could then use in applications.

For the *special case* of no conducting materials, that is, $\sigma_i = 0$, we can use the results of the three-medium, dielectric-material problem in section 2.3 to deduce the solution. At the boundary #1 between medium 1 and medium 2, we will expect a reflection coefficient of the form

$$\Gamma_{1,eff} = \left[\frac{\eta_{2,w}(-nl) - \eta_1}{\eta_{2,w}(-nl) + \eta_1}\right] e^{-2j\beta_1 nl} \tag{2.35}$$

However, the value of $\eta_{2,w}(-nl)$ will depend upon the value of $\eta_{2,w}(-nl + l)$, and so forth, for each successive boundary. For the far-right slab, we should be able to write the value of $\eta_{n,w}(-nl + nl) = \eta_{n,w}(0)$ in terms of the value of $\eta_{n+1}(0)$ and thereby deduce

$$\eta_{n,w}(-l) = \eta_n \left(\frac{\eta_{n+1} \cos \beta_n l - j\eta_n \sin \beta_n l}{\eta_n \cos \beta_n l - j\eta_{n+1} \sin \beta_n l)} \right) \tag{2.36}$$

We can then work our way backward with each successive layer to eventually conclude an effective wave impedance on either side of each boundary. Putting these impedances into Equation 2.35 (and keeping track of the phase factors), we can, in principle, write the solution for n slabs.

The Continuum

If we can analytically solve the problem for n different materials of equal width, Δh, we can then take the limit as $n \to \infty$ for $\Delta h = L/n$ and compute the effective reflection and transmission coefficients for a continuously variable medium in a specified region of space. An example of such a problem would be the propagation of electromagnetic fields into the ionosphere, where the density of air molecules and the density of conducting ions cause a continuous variation of medium properties (shown schematically in Figure 2.9.

The earth's ionosphere is a plasma of dissociated positive ions and electrons that extends from roughly 50 to 1000 km above the surface of the earth (with a peak at around 300 km) and has a typical maximum density of free electrons of 10^{10}–10^{12} electrons/m^3 that are caused by solar radiation, catalytic compounds released in the atmosphere, and collisions with interstellar particles. The density of ions and electrons is typically measured by high-altitude balloons and is found to vary with the

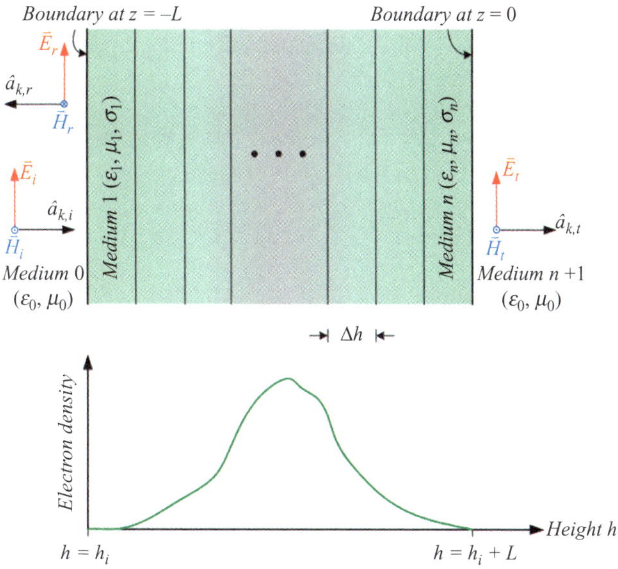

Figure 2.9 Schematic representation of the electron density in the ionosphere as a function of height above the surface of the earth.

time of day, the time of year, sunspot activity, variation with the earth's magnetic field, and location relative to the earth's poles. The plasma refracts light in the northern hemisphere as the aurora borealis, and it is so common that popular media refer to local reductions in the density as "holes in the ozone layer." Very-low-frequency (VLF) radio waves are reflected from the base of the ionosphere are useful for "bouncing" long wave radio waves over the horizon. As the frequency increases to about 30 MHz (depending on the angle of incidence), radio waves penetrate through the maximum density and most do not return to earth.

2.5 POLARIZED PLANE WAVES PROPAGATING IN A MATERIAL AS THEY INTERACT OBLIQUELY WITH A BOUNDARY

Parallel Polarization

Electromagnetic waves that approach a boundary between two media at an angle, θ_i, with respect to a normal to the boundary, are said to have an **angle-of-incidence** with the boundary. The **plane-of-incidence** is defined as the plane that contains both the propagation vector direction and the normal vector, as shown in Figure 2.10.

Figure 2.10 shows an incident TEM wave propagating in the $\hat{a}_{k,j} = \hat{a}_x \sin\theta_i + \hat{a}_z \cos\theta_i$ direction, with its electric field intensity vector parallel to the plane of

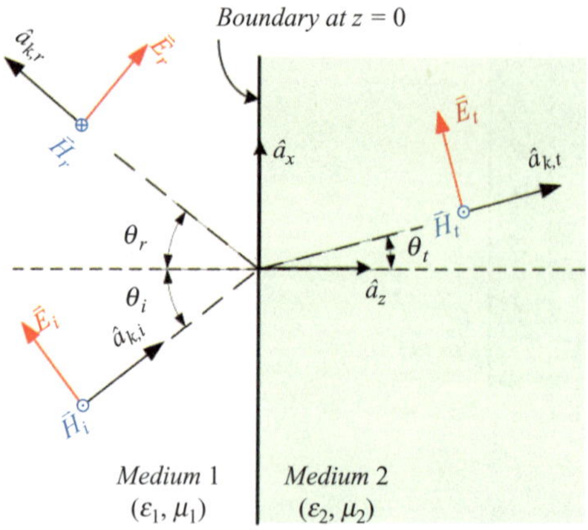

Figure 2.10 Polarized plane wave incident to the boundary between two media. In this case, the electric field intensity is parallel to the plane of incidence.

incidence. In this figure, $\vec{x} = x\hat{a}_x + z\hat{a}_z$ so $\vec{k} \cdot \vec{x} = k_1(x \sin\theta t + z \cos\theta i)$, and we may express the electric field intensity of the incident wave as

$$\vec{E}_i(\vec{x},t) = E_{i0}[\hat{a}_x \cos\theta_i - \hat{a}_z \sin\theta_i]e^{jk_1(x\sin\theta_i + z\cos\theta_i)}e^{-j\omega t} \tag{2.37a}$$

In this figure, the magnetic field intensity of the incident wave is in the $\hat{a}_y$ direction so

$$\vec{H}_i(\vec{x},t) = (E_{i0}/\eta_1)\hat{a}_y e^{jk_1(x\sin\theta_i + z\cos\theta_i)}e^{-j\omega t}, \tag{2.37b}$$

where $k_1 = \beta + j\alpha$ and $\alpha = 0$ for a *lossless* medium. The corresponding electric field intensity and magnetic field intensity of the reflected wave are

$$\vec{E}_r(\vec{x},t) = E_{r0}[\hat{a}_x \cos\theta_r + \hat{a}_z \sin\theta_r]e^{jk_1(x\sin\theta_r - z\cos\theta_r)}e^{-j\omega t} \tag{2.37c}$$

$$\vec{H}_r(\vec{x},t) = -(E_{r0}/\eta_1)\hat{a}_y e^{jk_1(x\sin\theta_r - z\cos\theta_r)}e^{-j\omega t} \tag{2.37d}$$

and the transmitted field intensities are

$$\vec{E}_t(\vec{x},t) = E_{t0}[\hat{a}_x \cos\theta_t - \hat{a}_z \sin\theta_t]e^{jk_2(x\sin\theta_t + z\cos\theta_t)}e^{-j\omega t} \tag{2.37e}$$

$$\vec{H}_t(\vec{x},t) = (E_{t0}/\eta_2)\hat{a}_y e^{jk_2(x\sin\theta_t + z\cos\theta_t)}e^{-j\omega t} \tag{2.37f}$$

Using Equations 2.37a, 2.37c, and 2.37e, we can equate the tangential components of the total electric field intensity at $z = 0$ in media 1 and 2:

$$E_{i0}\cos\theta_i e^{jk_1 x\sin\theta_i} + E_{r0}\cos\theta_r e^{jk_1 x\sin\theta_r} = E_{t0}\cos\theta_t e^{jk_2 x\sin\theta_t}, \tag{2.38}$$

which must be valid for all values of x. Thus, we can conclude that $k_1 \sin\theta_i = k_1 \sin\theta_r = k_2 \sin\theta_t$ or that $\theta_i = \theta_r$ and $k_1 \sin\theta_i = k_2 \sin\theta_t$. For a dielectric material in media 1 and 2, $k_1 = \beta_1 = \omega\sqrt{\mu\varepsilon} = \omega/u_{p1}$ and $k_2 = \beta_2 = \omega\sqrt{\mu\varepsilon} = \omega/u_{p2}$, so

$$\theta_i = \theta_r \quad \text{and} \quad \frac{\sin\theta_i}{\sin\theta_t} = \frac{u_{p1}}{u_{p2}} = \frac{\sqrt{\mu_2\varepsilon_2}}{\sqrt{\mu_1\varepsilon_1}} \tag{2.39}$$

Equation 2.39 is known as Snell's Law of Reflection, in honor of Willebrord van Roijen Snell (1580–1626). Of course, Snell was making measurements on reflection and refraction long before the speed of a wave in a medium had been determined. Snell, therefore, defined an "index-of-refraction," which he labeled n that we write today as

$$n_i = c/u_{pi} \tag{2.40}$$

Values of n were previously shown in Figure 1.20 as a function of wavelength for fused silica, where they ranged between 1.44 and 1.46 for wavelengths near the visible region of the electromagnetic spectrum. Several examples of the application of Snell's law are given below.

Using Snell's law in Equation 2.38, we conclude that $(E_{i0} + E_{r0})\cos\theta_i = E_{t0}\cos\theta_t$ or

$$E_{t0}/E_{i0} = (1 + E_{r0}/E_{i0})(\cos\theta_i/\cos\theta_t) \tag{2.41}$$

Using Equations 2.37b, 2.37d, and 2.37f, we can now equate the tangential components of the total magnetic field intensity at $z = 0$ in media 1 and 2:

$$(E_{i0}/\eta_1 - E_{r0}/\eta_1) = E_{t0}/\eta_2 \quad \text{or} \tag{2.42}$$

$$E_{t0}/E_{i0} = (\eta_2/\eta_1)(1 - E_{r0}/E_{i0}) \tag{2.43}$$

Now we can use Equations 2.41 and 2.43 to see

$$(1 + E_{r0}/E_{i0})(\cos\theta_i/\cos\theta_t) = (\eta_2/\eta_1)(1 - E_{r0}/E_{i0})$$

so

$$\Gamma_{\|} = \frac{E_{r0}}{E_{i0}} = \frac{\eta_2 \cos\theta_t - \eta_1 \cos\theta_i}{\eta_2 \cos\theta_t + \eta_1 \cos\theta_i} \tag{2.44a}$$

and

$$\tau_{\|} = \frac{E_{t0}}{E_{i0}} = \frac{2\eta_2 \cos\theta_i}{\eta_2 \cos\theta_t + \eta_1 \cos\theta_i} \tag{2.44b}$$

and

$$1 + \Gamma_{\|} = \tau_{\|}\left(\frac{\cos\theta_t}{\cos\theta_i}\right) \tag{2.44c}$$

We can use Snell's law to substitute the cosine ratio in all of these equations as

$$\frac{\cos\theta_t}{\cos\theta_i} = \frac{\sqrt{1-\sin^2\theta_t}}{\sqrt{1-\sin^2\theta_i}} = \frac{\sqrt{1-(\mu_1\varepsilon_1/\mu_2\varepsilon_2)\sin^2\theta_i}}{\sqrt{1-\sin^2\theta_i}} \tag{2.44d}$$

Perpendicular Polarization

TEM electromagnetic waves can also approach a boundary between two media at an angle, θ_i, with respect to a normal to the boundary with an electric field intensity polarization perpendicular to the "plane-of-incidence," as shown in Figure 2.11.

In this figure, we may express that the electric field intensity of the incident wave is only in the $\hat{a}_y$ direction, so

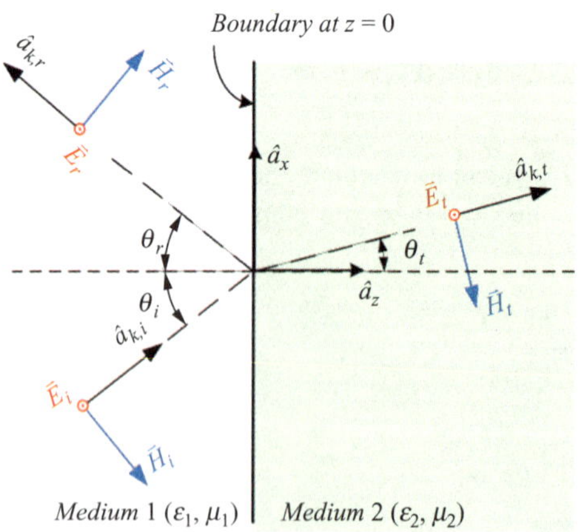

Figure 2.11 Polarized plane wave incident to the boundary between two media. In this case, the electric field intensity is perpendicular to the plane of incidence.

$$\vec{E}_i(\vec{x},t) = E_{i0}\,\hat{a}_y e^{jk_1(x\sin\theta_i + z\cos\theta_i)} e^{-j\omega t} \tag{2.45a}$$

and the magnetic field intensity of the incident wave is

$$\vec{H}_i(\vec{x},t) = (E_{i0}/\eta_1)[-\hat{a}_x\cos\theta_i + \hat{a}_z\sin\theta_i]e^{jk_1(x\sin\theta_i + z\cos\theta_i)} e^{-j\omega t} \tag{2.45b}$$

In a similar fashion, we can write the field intensities of the reflected waves as

$$\vec{E}_r(\vec{x},t) = E_{r0}\,\hat{a}_y e^{jk_1(x\sin\theta_i - z\cos\theta_i)} e^{-j\omega t} \quad \text{and} \tag{2.45c}$$

$$\vec{H}_r(\vec{x},t) = (E_{r0}/\eta_1)[\hat{a}_x\cos\theta_r + \hat{a}_z\sin\theta_r]e^{jk_1(x\sin\theta_i - z\cos\theta_i)} e^{-j\omega t} \tag{2.45d}$$

Finally, we can write the field intensities of the transmitted waves as

$$\vec{E}_t(\vec{x},t) = E_{t0}\,\hat{a}_y e^{jk_1(x\sin\theta_i + z\cos\theta_i)} e^{-j\omega t} \quad \text{and} \tag{2.45e}$$

$$\vec{H}_t(\vec{x},t) = (E_{t0}/\eta_1)[-\hat{a}_x\cos\theta_t + \hat{a}_z\sin\theta_t]e^{jk_1(x\sin\theta_i + z\cos\theta_i)} e^{-j\omega t} \tag{2.45f}$$

Using Equations 2.45a, 2.45c, and 2.45e, we can equate the tangential components of the total electric field intensity at $z = 0$ in media 1 and 2:

$$E_{i0}e^{jk_1 x\sin\theta_i} + E_{r0}e^{jk_1 x\sin\theta_r} = E_{t0}e^{jk_2 x\sin\theta_t}, \tag{2.46}$$

which must be valid for all values of x. Thus, we can conclude that Snell's law also holds for perpendicular polarization of the electric field intensity:

$$\theta_i = \theta_r \quad \text{and} \quad \frac{\sin\theta_i}{\sin\theta_t} = \frac{u_{p1}}{u_{p2}} = \frac{\sqrt{\mu_2\varepsilon_2}}{\sqrt{\mu_1\varepsilon_1}}$$

Using Snell's law in Equation 2.46, we conclude that

$$(E_{i0} + E_{r0}) = E_{t0} \tag{2.47a}$$

Using Equations 2.45b, 2.45d, and 2.45f, we can equate the tangential components of the total magnetic field intensity at $z = 0$ in media 1 and 2:

$$(E_{i0}/\eta_1)[-\hat{a}_x\cos\theta_i] + (E_{r0}/\eta_1)[\hat{a}_x\cos\theta_r] = (E_{t0}/\eta_2)[-\hat{a}_x\cos\theta_t] \tag{2.47b}$$

Solving Equations 2.47a and 2.47b for the electric field intensity ratios, we conclude that

$$\Gamma_{\perp} = \frac{E_{r0}}{E_{i0}} = \frac{\eta_2 \cos\theta_i - \eta_1 \cos\theta_t}{\eta_2 \cos\theta_i + \eta_1 \cos\theta_t} \tag{2.48a}$$

$$\tau_{\perp} = \frac{E_{t0}}{E_{i0}} = \frac{2\eta_2 \cos\theta_i}{\eta_2 \cos\theta_i + \eta_1 \cos\theta_t} \tag{2.48b}$$

$$1 + \Gamma_{\perp} = \tau_{\perp}, \tag{2.48c}$$

which we can write purely in terms of θ_i via Snell's law using

$$\frac{\cos\theta_t}{\cos\theta_i} = \frac{\sqrt{1-\sin^2\theta_t}}{\sqrt{1-\sin^2\theta_i}} = \frac{\sqrt{1-(\mu_1\varepsilon_1/\mu_2\varepsilon_2)\sin^2\theta_i}}{\sqrt{1-\sin^2\theta_i}} \tag{2.48d}$$

Conclusions

We can compare the two cases of parallel and perpendicular polarization reflection by comparing Equation 2.44a with 2.48a:

$$\Gamma_{\parallel} = \frac{\eta_2\sqrt{1-(\mu_1\varepsilon_1/\mu_2\varepsilon_2)\sin^2\theta_i} - \eta_1\cos\theta_i}{\eta_2\sqrt{1-(\mu_1\varepsilon_1/\mu_2\varepsilon_2)\sin^2\theta_i} + \eta_1\cos\theta_i} \tag{2.49a}$$

$$\Gamma_{\perp} = \frac{\eta_2\cos\theta_i - \eta_1\sqrt{1-(\mu_1\varepsilon_1/\mu_2\varepsilon_2)\sin^2\theta_i}}{\eta_2\cos\theta_i + \eta_1\sqrt{1-(\mu_1\varepsilon_1/\mu_2\varepsilon_2)\sin^2\theta_i}} \tag{2.49b}$$

PROBLEM

2.3 Show that, for nonmagnetic materials, Equations 2.49 can be plotted as shown in Figures 2.12 and 2.13.

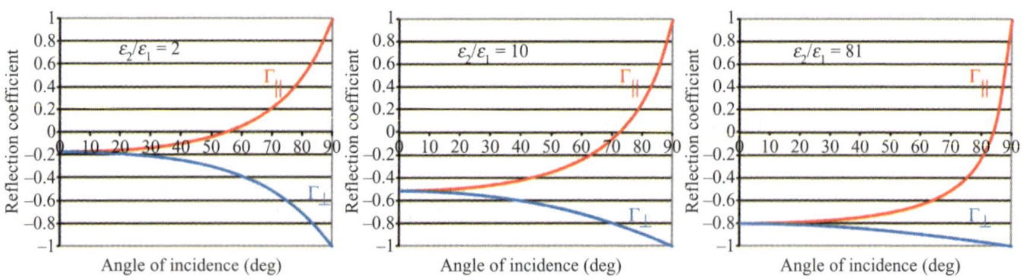

Figure 2.12 Parallel and perpendicular reflection coefficients $\Gamma_{\parallel}$ *and* $\Gamma_{\perp}$ for nonmagnetic materials as a function of the incident angle, θ_i for $\varepsilon_2/\varepsilon_1 = 2$, 10, and 81 as shown.

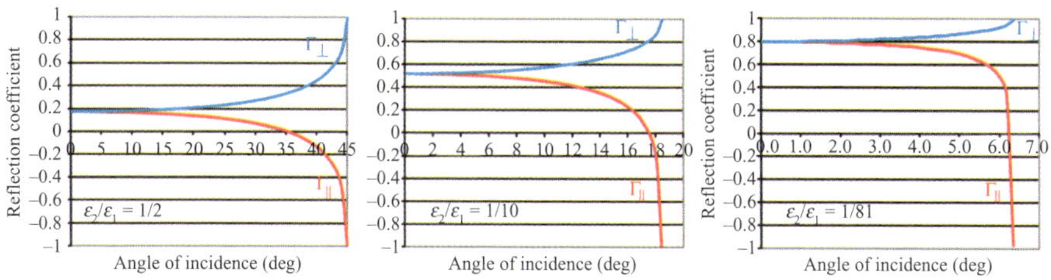

Figure 2.13 Parallel and perpendicular reflection coefficients $\Gamma_{\parallel}$ and $\Gamma_{\perp}$ for non-magnetic materials as a function of the incident angle, θ_i for $\varepsilon_2/\varepsilon_1 = 1/2$, 1/10, and 1/81 as shown.

2.6 BREWSTER'S LAW

Sir David Brewster† (1791–1868) first noticed that $\Gamma_{\parallel}$ can be zero but that $\Gamma_{\perp}$ is never zero; that is, there is an angle, θ_B (now called Brewster's angle), at which the electric field intensity component that is parallel to the plane-of-incidence is *not* reflected. At $\theta = \theta_B$ TEM$_{\parallel}$, waves are *only* transmitted. We can see that this happens when $\eta_2 \cos\theta_t = \eta_1 \cos\theta_i$ or, using Equation 2.48d,

$$\frac{\eta_1}{\eta_2} = \frac{\cos\theta_t}{\cos\theta_i} = \frac{\sqrt{1-(\mu_1\varepsilon_1/\mu_2\varepsilon_2)\sin^2\theta_i}}{\sqrt{1-\sin^2\theta_i}} \tag{2.50}$$

Squaring both sides of Equation 2.50 and solving for $\sin\theta_i$ yield

$$\sin^2\theta_B = \frac{1-\mu_2\varepsilon_1/\mu_1\varepsilon_2}{1-(\varepsilon_1/\varepsilon_2)^2} \tag{2.51}$$

PROBLEM

2.4 Show that, at the Brewster's angle, the reflected and transmitted waves propagate at right angles to one another, that is, $\theta_B + \theta_r = 90°$.

NOTE Snell and Brewster worked before Maxwell's equations were understood. They made measurements to show the principles named after them in terms of the "index of refraction" of materials. Brewster found the answer to Problem 2.4 experimentally and then used Snell's law to show

$$n_1 \sin\theta_B = n_1 \sin(90^\circ - \theta_B) = n_2 \cos\theta_B \quad \text{or}$$
$$\theta_B = \tan^{-1}\left(\frac{n_2}{n_1}\right) \quad \text{(Brewster's angle)} \tag{2.52}$$

† Brewster was a Scottish physicist who entered the University of Edinburgh at age 12. He wrote more than 400 scientific papers and books, mostly on optics. He invented the kaleidoscope and the stereoscope in 1816 and founded the British Association for the Advancement of Science in 1831. At the age of 75 he married for a second time and had a daughter at the age of 77.

Because Brewster always worked with polarized light in air (for which $n_l = 1$), he wrote the experimental law simply as $\theta_B = \tan^{-1} n_2$. This simple rule and the orthogonality of the reflected and transmitted were first known as Brewster's law.

2.7 APPLICATIONS OF SNELL'S LAW AND BREWSTER'S LAW

The relative permittivity, ε_r, of fused silica is about 2, flint glass is about 10, and water is about 81. Figure 2.12 shows us how polarized propagating electromagnetic waves in air will reflect from those nonmagnetic surfaces as a function of incident angle. We have also shown that Snell's law $\theta_i = \theta_r$ and $\dfrac{\sin\theta_i}{\sin\theta_t} = \dfrac{\sqrt{\varepsilon_2}}{\sqrt{\varepsilon_1}}$ holds for either polarization. These phenomena also pertain inversely to electromagnetic waves propagating in one of the three media as they encounter an air surface. These properties are used in many optical and communications applications. A few examples are described in the following sections.

E&M Waves Entering Water

A classic phenomenon of light entering water at an angle of incidence, θ_i, is illustrated in Figure 2.14.

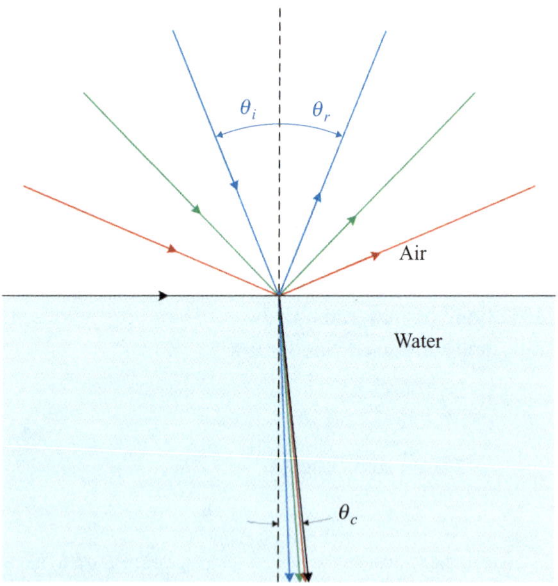

Figure 2.14 Light in air, incident upon the surface of water at an angle of incidence θ_i is refracted at angle θ_t, as shown for increasing angles of incidence $\theta_i = 22.5°$, $\theta_i = 45°$, $\theta_i = 67.5°$, $\theta_i = 90°$.

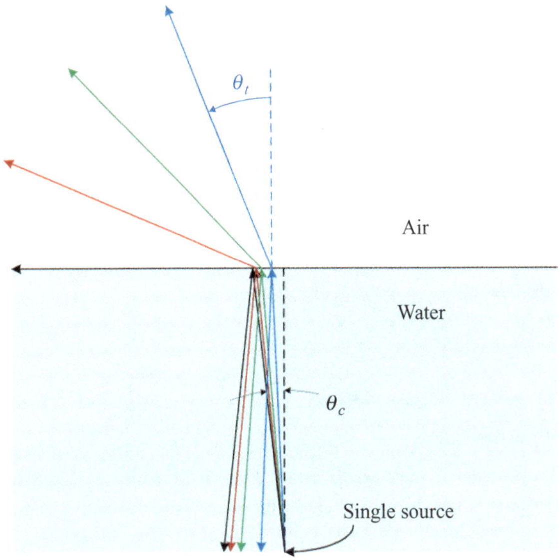

Figure 2.15 Light in water from a single point, incident upon the air–water interface at several angles of incidence θ_i is refracted at various angles θ_t as shown for increasing angles of refraction $\theta_t = 22.5°$, $\theta_t = 45°$, $\theta_t = 67.5°$, $\theta_i = 90°$. The maximum angle of refraction (the critical angle θ_C) occurs when the angle of water incidence is $\theta_C = 6.4°$.

The maximum angle of refraction (the critical angle $\theta_C = 6.4°$) occurs when the air incident angle is $\theta_i = 90°$ (solid black vector).

We can inversely use Snell's law to deduce that light being emitted from points below the surface of the water will be refracted in air, as shown in Figure 2.15, by reversing the arrows that indicate the propagation directions.

The maximum angle of refraction (the critical angle θ_C), occurs when the angle of water incidence is $\theta_C = 6.4°$.

E&M Waves Exiting Water

If we plot the light emitted from a single source below the surface of the water, we see a refraction scheme, as shown in Figure 2.15. The direction of propagation of the emitted E&M waves propagates in the directions shown for several angles chosen to yield refraction into air at angles $\theta_t = 22.5°$, $\theta_t = 45°$, $\theta_t = 67.5°$, $\theta_t = 90°$. We can see that angles of incidence greater than the critical angle, $\theta_C = 6.4°$, cannot be refracted into air and are thus totally reflected into the water at the air–water interface. The reflected intensity subsequently illuminates other portions of the structure containing the water. The waves also refract from surface points in a similar fashion to that shown in Figure 2.15. Architects employ this principle in water features, such as fountains or pools, by placing sources of visible light on the bottom

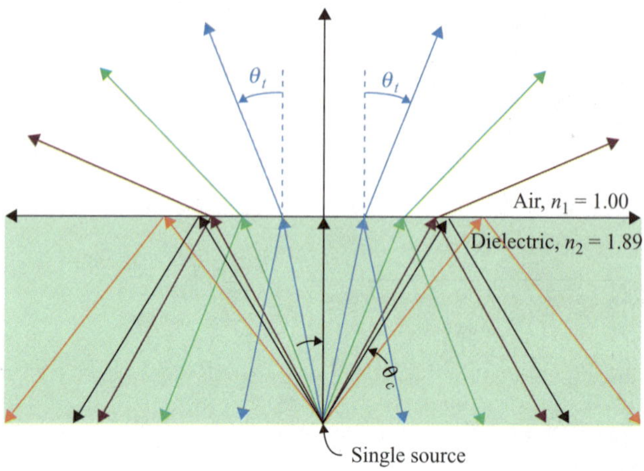

Figure 2.16 Refraction and reflection from an air–dielectric interface for which the dielectric index of refraction is 1.89. For incident angles greater than the critical angle, the E&M wave is totally reflected.

of the water containers. The subsequent multiple reflections from the bottom of the container appear to make the container luminescent and much larger than it actually is. The effect is especially dramatic when observed from an airplane while it is flying over a wealthy subdivision of a major city at night.

In Figure 2.16, the refracted propagations are clear, but the propagations that reflect from the interface are crowded because of the small critical angle (large permittivity difference) for an air–water interface. To see the reflected more clearly and to include incident angles on both sides of the normal to the interface, Figure 2.16 shows reflection and refraction for $n_1 = \sqrt{\varepsilon_1} = 1.0$ and $n_2 = \sqrt{\varepsilon_2} = 1.89$.

PROBLEMS

2.5 What fraction of the light intensity from a single source is refracted into air at the first interaction with the surface? If the bottom of the container were painted a totally absorbing black color, would there be any other light emitted from the pool?

2.6 Should architects employ a parabolic reflector below the single source of light? Carefully describe the effect as seen from an observer standing at the edge of a pool at night. If there were a concentric set of flat mirrors below the single source of light, arranged on the surface of a parabolic bowl, how would the effect change?

2.7 What properties of glass would you choose to retain most of the light inside? Could you coat glass with an intermediate medium of a certain thickness of a third material that would prevent light from being refracted into the surrounding air or water?

Figure 2.17 Near Field scanner with an H-probe and an E-probe that measure the magnetic field intensity and electric field intensity as a function of position (x, y, z).

2.8 If there were a small hole in the side of a closed, perfectly reflecting container that permitted one to arbitrarily add electromagnetic energy inside, would there be any rules like the Fermi exclusion principle for the photon gas inside? What would limit the total amount of energy inside the container?

Snell's Law at Lower Frequencies

Almost all of the applications of Snell's law are made with visible light (~10^{15} Hz). We were unable to find any proof that Snell's law is valid at other frequencies, so we[‡] made measurements of the critical angle at 300 MHz and 3 GHz to confirm the principle. The equipment used for these measurements is shown in Figure 2.17.

[‡] Jason Douglas Ramage, Proof of Snell's Law at Lower Frequencies, Thesis in partial fulfillment of the MS degree in Electrical Engineering at the University of South Carolina, Dec. 2007. The equipment used for these measurements was provided by Kevin Slattery and with the assistance of Xiaopeng Dong at the Jones Farm Intel facilities in Hillsboro, OR.

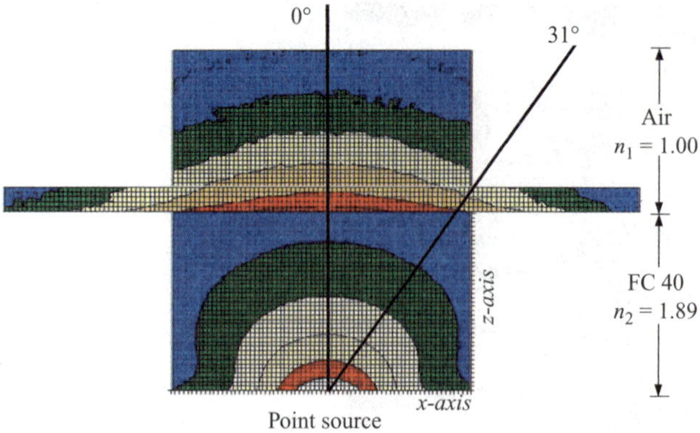

Figure 2.18 Magnitude of the electric field intensity measured in an FC 40 liquid dielectric liquid with an effective constant index of refraction of 1.89 at 300 MHz and 3 GHz.

In our measurements, the E-probe measured the electric field intensity produced by a unipole electric field radiator located at the bottom of a container filled with a liquid dielectric called FC 40 with an effective index of refraction of 1.89. By physically scanning the E-probe in increments along an x- and z-direction, we were able to plot the cross section of the electric field intensity below the liquid surface at the $y = 0$ location. The results of the measurements are shown in Figure 2.18.

Also shown in Figure 2.18 are the measurements taken in the air above the liquid–air interface of FC 40 at a frequency of 3 GHz. Note that the intensity scale of the two regions are different to maximize the most intense electric field intensity in each region, with dark red being the most intense. We can see that the dark red measurements in air form a dome-shaped bowl over the liquid and that their intensity becomes small for values of x that correspond to a critical angle of 31°. This value compares well with the theoretical value obtained from Snell's law that $\theta_C = \sin^{-1}[(1/1.89)\sin 90°] = 31.9°$. Similar values were measured for a frequency of 300 MHz.

Brewster's Law Applications

One of the old applications of Brewster's law involved projection spotlights used in theaters. Before the advent of the halogen lamp, very bright lights were often created by a carbon-arc lamp that projected a spot of light onto a stage from the back of an auditorium. The carbon-arc lamp involved the breakdown of the gases between two closely spaced, pointed carbon rods that are connected in series with a limiting resistor to a high-current AC transformer with a small spacing between them, much like today's welding rods. By striking an arc between the two rods, the resulting light was almost the same visible spectrum of the sun that was

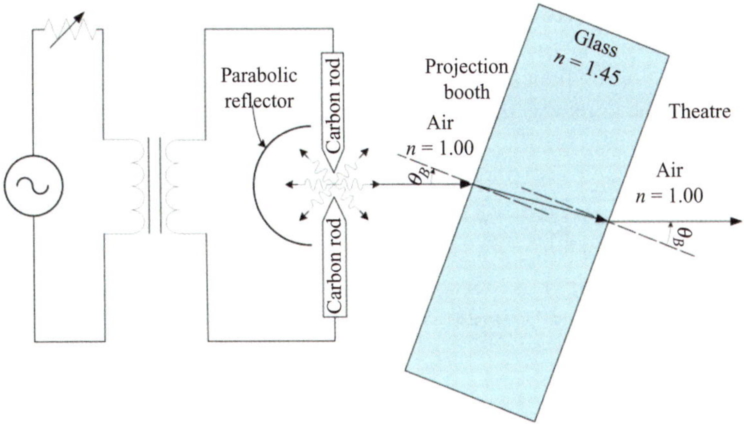

Figure 2.19 Spotlight in a "sound proof" theater projection booth passing through a glass window.

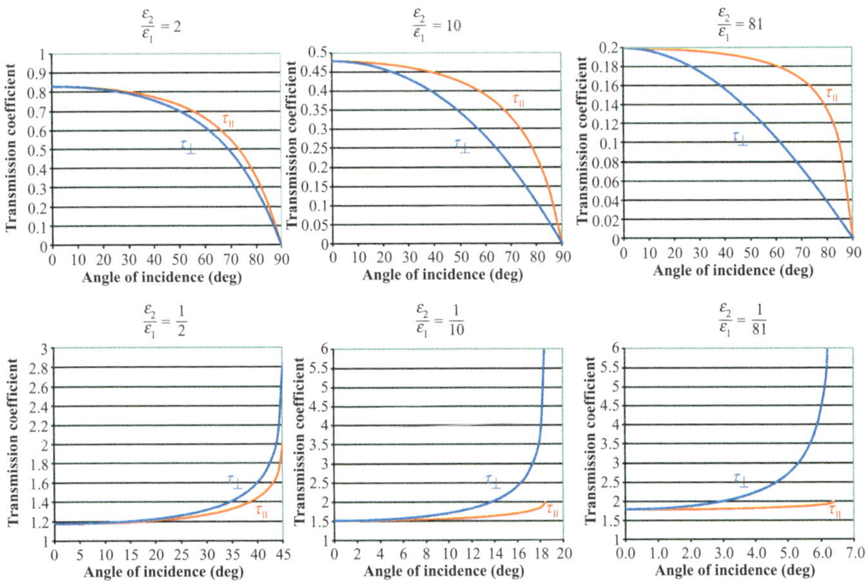

Figure 2.20 Transmission coefficients for various values of permittivity ratios at the interface between two media.

partially polarized, but the arc produced loud noises that could mask audio from the stage to the audience. The solution was to enclose the projector booth in a soundproof medium and to let the light project through a glass window, as shown in Figure 2.19.

Architects often oriented the glass transmission window so that its angle of incidence was at the Brewster angle, θ_B, both to contain the sound of the arc and to minimize the reflection of light into the projection booth. The partially polarized light incident upon the glass was transmitted into the theater, according to the transmission coefficient at each interface, as shown in Figure 2.20.

We can see that the closest graph for light propagating from air to glass, $n_2/n_1 = 1.45$, is the graph for $\sqrt{\varepsilon_2/\varepsilon_1} = \sqrt{2} = 1.41$. The light subsequently exits the glass propagating from $\sqrt{\varepsilon_2/\varepsilon_1} = \sqrt{1/2} = 0.707$. We can see that the transmission coefficients at Brewsters angle for glass (55.4°) from the above graphs are $\tau_{\parallel}^{1,2}\,\tau_{\parallel}^{2,1} \approx (0.71)(2.0) = 1.42$ and $\tau_{\perp}^{1,2} > \tau_{\perp}^{2,1} \approx (0.66)\,(2.0) = 1.32$.

PROBLEMS

2.9 Rather than use the approximate values from Figure 2.20, compute the exact transmission coefficients for $n_{glass} = 1.45$ and $n_{air} = 1.00$ and find the resultant the intensity of light that propagates through the projection room glass assuming a 50%/50% polarization from the carbon-arc lamp.

2.10 Find the value of glass thickness using the principles of section 9.2 for a radome transmitter that will pass a maximum intensity of light through the projection booth window at Brewster's angle.

ENDNOTE

1. Paul G. Huray, *Maxwell's Equations* (Hoboken, NJ: John Wiley & Sons, 2009), Chapter 7.

Chapter 3

Transmission Lines and Waveguides

LEARNING OBJECTIVES

- Relate voltages and currents associated with electromagnetic waves as they propagate in an infinitely long transmission line to material and geometric properties
- Relate voltages and currents as they propagate from one end of a finite transmission line to the other, their interference, and their relative times of propagation
- Relate reflection and transmission of electromagnetic waves at transmission line loads by analytic and computational means
- Relate the changing character of electromagnetic waves and their losses as they propagate through a series of different transmission lines and loads

3.1 INFINITELY LONG TRANSMISSION LINES

In the case of DC currents or low-frequency (60 Hz) AC currents, we have previously thought of energy as being propagated by the movement of conduction electrons throughout the volume of a conductor. We found that those charges and currents produced electromagnetic fields inside and outside of the conductors so that we could draw equivalence to the transport of energy to the transport of the fields they produce. In developing electromagnetic concepts by this historical path, we established a cause and effect that began with the creation of currents by internal electric fields created by an external source of electric potential, for example, a battery or a generator across a conductor. In these concepts, propagation times were limited to the speed of conduction electrons at the Fermi level (the electrons available for the conduction of charge) and, thus, of the fields they produced in the external nonconducting medium.

At higher frequencies, we found that external electromagnetic fields in the neighborhood of flat conductors could penetrate the surface layer of conductors with an exponential envelope, so that currents flowed mostly on a surface layer of conductors. In these problems, the source of external fields was a high-frequency source like a microwave generator or a light source. We viewed surface currents in

The Foundations of Signal Integrity, by Paul G. Huray

conductors as being induced by external sources of electromagnetic waves. The propagation speed in those concepts was the speed of light in a nonconducting medium such as air or a resin.

In high-speed circuits, it is sometimes more convenient to view electromagnetic waves as propagating from one end of a transmission line to another, thereby inducing surface currents or charges in the surrounding materials (even nonconducting materials like a fiber-optic cable). In these problems, we will have to be sure that the boundary conditions satisfy Maxwell's equations just like we did for DC and low-frequency AC currents. In a sense, we are creating an equivalence principle that states, "It does not matter which we consider as the source and effect—conduction currents that induce magnetic fields or external magnetic fields that induce currents." But, as we will see, the speed of the transmission of energy will be far higher if we can consider the currents in a surface layer of a conductor as not being restricted by the speed of conduction electrons at the Fermi energy.

Sometimes, the conducting boundaries of transmission lines are referred to as "waveguides" for the electromagnetic waves at high frequencies. Except in the case of perfectly electrically conducting boundaries, electromagnetic waves associated with the transmission lines are ***not*** pure TEM waves, but it will be convenient to treat them that way as a first approximation in the case of very good conductors.

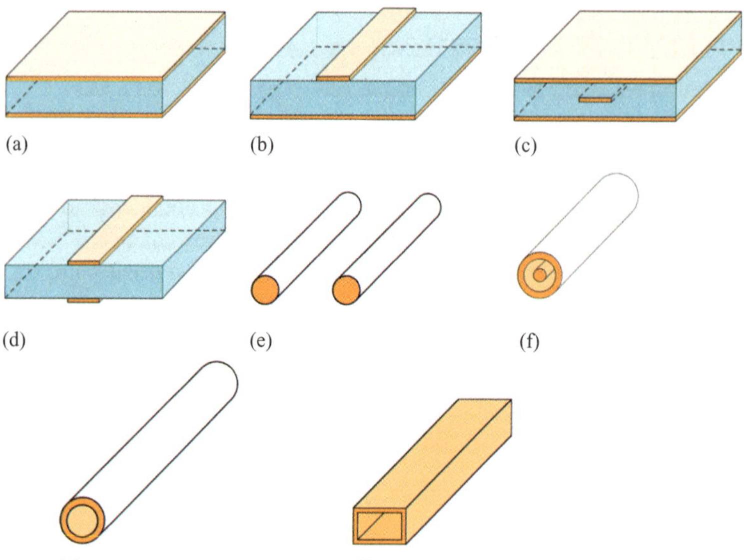

Figure 3.1 Typical geometric forms of transmission lines for the propagation of electromagnetic waves: (a) semi-infinite parallel-plates; (b) finite parallel "trace" with a semi-infinite ground plane (microstrip); (c) semi-infinite parallel-plates with a finite parallel "trace" (stripline); (d) finite parallel "traces"; (e) two-wire parallel (or twisted pair) cylindrical conductors; (f) coaxial cable; (g) cylindrical wave guide; (h) rectangular wave guide.

The geometric forms of transmission lines or waveguides may differ, but the equations that govern the relationship of the transport of voltage and current in the lines follow similar mathematical forms. Some common geometric forms of transmission lines are shown in Figure 3.1.

3.2 GOVERNING EQUATIONS

Assumptions

We will first consider the conductors to be almost perfectly conducting so that we can make the assumption that the waves can be guided in one direction (usually the z-direction) as TEM^z waves in a dielectric medium. In addition we will assume the following:

- The distance between any two conductors is small compared with the wavelength, λ, of the propagating electromagnetic waves in the dielectric medium.
- The length of the conductors in the z-direction is long compared with the wavelength λ.
- An external source causes a cosinusoidal electric/magnetic field in time [e.g., $\cos \omega t$ or $\mathrm{Re}(e^{-j\omega t})$] across one end of the transmission line.
- Four parameters will be sufficient to describe the characteristic behavior of the lines:

  R = resistance per unit length (in all conductors) in Ω/m,
  L = inductance per unit length (in all conductors) in H/m,
  G = conductance per unit length of the dielectric medium between conducting surfaces in S/m (or $\Omega^{-1}\mathrm{m}^{-1}$),
  C = capacitance per unit length between conducting surfaces in F/m.
- In the transmission lines, scalar potentials or voltages, $v(z, t)$, and corresponding surface currents, $i(z, t)$ exist at an instant in time.

We will then use Figure 3.2a to represent voltages and currents as a function of propagation distance, z, in any of the transmission lines of Figure 3.1.

We may apply Kirchoff's voltage law to Figure 3.2a to obtain $v(z, t) - R\Delta z i(z, t) - L\Delta z \partial i(z, t)/\partial t = v(z + \Delta z, t)$ or $[v(z + \Delta z, t) - v(z, t)]/\Delta z = -Ri(z, t) - L\partial i(z, t)/\partial t$ and, taking the limit as $\Delta z \to 0$,

$$\partial v(z,t)/\partial z = -Ri(z,t) - L\partial i(z,t)/\partial t \tag{3.1a}$$

We may apply Kirchoff's current law in Figure 3.2a at the point N to obtain

$$i(z,t) - G\Delta z v(z+\Delta z,t) - C\Delta z \partial v(z,t)/\partial t = i(z+\Delta z,t)$$

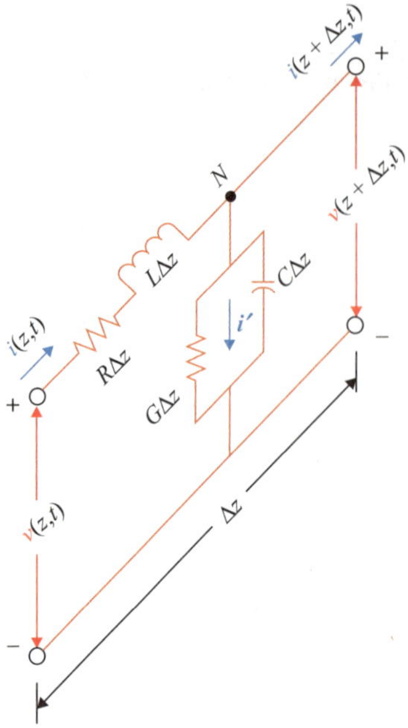

Figure 3.2a Schematic representation of the surface voltage, $v(z, t)$, and the surface current, $i(z, t)$ as a function of propagation distance, z, and as a function of the characteristic parameters R, L, G, and C.

or

$$[i(z+\Delta z,t)-i(z,t)]/\Delta z=-Gv(z,t)-C\partial v(z,t)/\partial t$$

and taking the limit as $\Delta z \to 0$,

$$\partial i(z,t)/\partial z=-Gv(z,t)-C\partial v(z,t)/\partial t \tag{3.1b}$$

Equations 3.1a and 3.1b are known as the **transmission line equations**.

Harmonic Inputs

If the voltage and current to a transmission line are driven by an external source that is harmonic, we can write the voltage and current as

$$v(z,t)=\text{Re}\left[V(z)e^{j\omega t}\right] \tag{3.2a}$$

$$i(z,t)=\text{Re}\left[I(z)e^{j\omega t}\right] \tag{3.2b}$$

in which case, Equations 3.1a and 3.1b are reduced to

$$dV(z)/dz = -RI(z) - j\omega LI(z) \tag{3.3a}$$

$$dI(z)/dz = -GV(z) - j\omega CV(z) \tag{3.3b}$$

Equations 3.3a and 3.3b are two coupled, first-order, linear ordinary differential equations for the phasor voltage $V(z)$ and phasor current $I(z)$. These equations may also be solved by taking another spatial derivative, so that

$$d^2V(z)/dz^2 = -(R + j\omega L)\,dI(z)/dz = (R + j\omega L)(G + j\omega C)V(z) \tag{3.4a}$$

$$d^2I(z)/dz^2 = -(G + j\omega C)\,dV(z)/dz = (G + j\omega C)(R + j\omega L)I(z) \tag{3.4b}$$

Equations 3.4a and 3.4b are of the same form:

$$d^2V(z)/dz^2 = \gamma^2 V(z) \quad \text{with } \gamma = \alpha + j\beta = \sqrt{(R + j\omega L)(G + j\omega C)} \tag{3.5a}$$

$$d^2I(z)/dz^2 = \gamma^2 I(z) \quad \text{with } \gamma = \alpha + j\beta = \sqrt{(R + j\omega L)(G + j\omega C)} \tag{3.5b}$$

where γ is called the propagation constant. The solution to these ordinary differential equations is of the form

$$V(z) = V_0^+ e^{-\gamma z} + V_0^- e^{\gamma z} \tag{3.6a}$$

$$I(z) = I_0^+ e^{-\gamma z} + I_0^- e^{\gamma z} \tag{3.6b}$$

The arbitrary constant coefficients V_0^+, V_0^-, I_0^+, and I_0^- have been labeled to indicate that the time-dependent solutions

$$v(z,t) = \mathrm{Re}\left[V(z)e^{j\omega t}\right] = \mathrm{Re}\left[V_0^+ e^{-\alpha z} e^{-j(\beta z - \omega t)} + V_0^- e^{\alpha z} e^{j(\beta z + \omega t)}\right] \tag{3.7a}$$

$$i(z,t) = \mathrm{Re}\left[I(z)e^{j\omega t}\right] = \mathrm{Re}\left[I_0^+ e^{-\alpha z} e^{-j(\beta z - \omega t)} + I_0^- e^{\alpha z} e^{j(\beta z + \omega t)}\right] \tag{3.7b}$$

are in the form of the sum of traveling waves that propagate in the positive z-direction, with amplitudes V_0^+ and I_0^+ and in the negative z-direction, with amplitudes V_0^- and I_0^-.

If we put solutions 3.6a and 3.6b back into Equation 3.3a, we obtain

$$V_0^+(-\gamma)e^{-\gamma z} + V_0^-(\gamma)e^{\gamma z} = -(R + j\omega L)\left[I_0^+ e^{-\gamma z} + I_0^- e^{\gamma z}\right],$$

and, equating the coefficients of the exponential terms individually, we see

$$\frac{V_0^+}{I_0^+} = \frac{(R + j\omega L)}{\gamma} = Z_0 \quad \text{and} \quad \frac{V_0^-}{I_0^-} = -\frac{(R + j\omega L)}{\gamma} = -Z_0 \tag{3.8}$$

The ratio of voltage to current is usually called impedance Z_0, and, from Equation 3.8, we see that this quantity is a complex number. Furthermore, Equation 3.8 gives us the impedance of the wave that propagates in the +z-direction as well as the impedance of the wave that propagates in the −z-direction. The quantity Z_0 is called the **characteristic impedance of the transmission line** and has units of ohms. In general,

$$Z_0 = \frac{(R + j\omega L)}{\sqrt{(R + j\omega L)(G + j\omega C)}} = \sqrt{\frac{(R + j\omega L)}{(G + j\omega C)}} \tag{3.9}$$

3.3 SPECIAL CASES

Semi-infinite Line

For a semi-infinite transmission line that begins at $z = 0$ and extends to $z = \infty$, we can see from Equations 3.7a and 3.7b that the coefficients V_0^- and I_0^- must be zero if the quantities $v(z, t)$ and $i(z, t)$ are to be finite in the limit as $z \to \infty$. Thus, for an external source that imposes a voltage and current, V_0^+ and I_0^+, across the transmission line at $z = 0$, we can see from Equations 3.7a and 3.7b that the voltage falls off exponentially with z with an attenuation constant, α, and a phase constant β. In these equations, the attenuation constant and the phase constant must have units of m^{-1}. The dimensionless name Neper (*Np*) is sometimes added to the attenuation constant, and the dimensionless name radian (*rad*) is sometimes added to the phase constant to remind us that we are not working in degrees.

Lossless Line

We say that a transmission line is *lossless* if both the resistance $R = 0$ and the conductance $G = 0$. For this special case, Equation 3.5 yields the propagation constant $\gamma = \alpha + j\beta = \sqrt{(R + j\omega L)(G + j\omega C)} = \sqrt{(-\omega^2 LC)} = j\omega\sqrt{LC}$. This quantity is purely imaginary so the attenuation constant, $\alpha = 0$ (confirming the name *lossless*), and the phase constant, $\beta = \omega\sqrt{LC}$ (a linear function of ω). From Equation 3.7, we can see that the phase velocity $u_p = \omega/\beta = 1/\sqrt{LC}$ is a constant. If we compare this phase velocity with the velocity of a TEM^z wave in a medium, $u_p = 1/\sqrt{\mu\varepsilon}$, we can use the equivalence $LC = \mu\varepsilon$. Using this equivalence, we can find L in terms of C (or vice versa) for a *lossless* line. In addition, using Equation 3.9, we can see that the characteristic impedance of a *lossless* line is $Z_0 = R_0 + jX_0 = \sqrt{L/C}$, that is, $R_0 = \sqrt{L/C}$ (constant) and $X_0 = 0$.

Conclusions

- The phase velocity of *lossless* lines is not dependent on frequency, so a pulse that is made up of a combination of different frequency components (e.g., as in its Fourier representation) will propagate without dispersion in time and space because each of its components travel at the same phase velocity, $u_p = 1/\sqrt{\mu\varepsilon}$.

- The amplitude of each component of a pulse does not attenuate with time or space for *lossless* lines because $\alpha = 0$.

Distortionless Line

We say a transmission line is *distortionless* if $R/L = G/C$. For this special case, Equation 3.5 yields the propagation constant $\gamma = \alpha + j\beta = \sqrt{(R+j\omega L)(RC/L+j\omega C)} = \sqrt{C/L}\,(R+j\omega L)$. The real and imaginary parts of the propagation constant, $\alpha = \sqrt{C/L}R$ and $\beta = \omega\sqrt{LC}$, for this special case show that voltages and currents attenuate with distance along the z-direction and that the phase velocity, $u_p = \omega/\beta = 1/\sqrt{LC}$, is a constant and is the same as the special case of a *lossless* line. Furthermore, we can see from equation 3.9 that $Z_0 = R_0 + jX_0 = \sqrt{(R+j\omega L)/(RC/L+j\omega C)} = \sqrt{L/C}$ that is, $R_0 = \sqrt{L/C}$ (constant) and $X_0 = 0$ as in the case of a *lossless* transmission line.

Conclusions

- Except for the attenuation constant, α, the characteristics of a *distortionless* line are the same as a *lossless* line.
- For a *distortionless* line $u_p = \omega/\beta = 1/\sqrt{LC}$ is a constant for all frequencies, ω. A pulse (which can be described by a Fourier transform in frequency space) will not broaden or distort its shape (except for attenuation) as it propagates down the transmission line in the positive z-direction; hence the name *distortionless line*.

Low Loss Line

If the transmission line is *low loss*, we can expand Equation 3.5

$$\gamma = \alpha + j\beta = \sqrt{(R+j\omega L)(G+j\omega C)}$$

$$\gamma = \sqrt{(j\omega L)(j\omega C)(1+R/j\omega L)(1+G/j\omega C)} = j\omega\sqrt{LC}\,(1+R/j\omega L)^{1/2}(1+G/j\omega C)^{1/2}$$

and, if we use the definition of *low loss* to be $R \le \omega L$ and $G \le \omega C$,

$$\gamma = j\omega\sqrt{LC}\left(1+\frac{1}{2}\frac{R}{j\omega L}+\frac{1}{8}\frac{R^2}{\omega^2 L^2}+\cdots\right)\left(1+\frac{1}{2}\frac{G}{j\omega C}+\frac{1}{8}\frac{G^2}{\omega^2 C^2}+\cdots\right)$$

$$\gamma_{LL} \approx j\omega\sqrt{LC}\left(1+R/2j\omega L+R^2/8\omega^2 L^2\right)\left(1+G/2j\omega C+G^2/8\omega^2 C^2\right)$$

so

$$\alpha_{LL} \approx 1/2\left(R/\sqrt{L/C} + G/\sqrt{C/L}\right), \beta_{LL} \approx \omega\sqrt{LC}\left[1+(1/8\omega^2)(R/L-G/C)^2\right] \quad (3.10)$$

From Equation 3.10, we can thus see that *low loss* (α_{LL} = small) implies that R must be small compared with $\sqrt{L/C}$ and that G must be small compared with $\sqrt{C/L}$. We see that the more general expression for *low loss* phase velocity is

$$u_{p,LL} = \omega/\beta_{LL} \approx \left(1/\sqrt{LC}\right)\left[1+(1/8\omega^2)(R/L-G/C)^2\right]^{-1} \quad (3.11)$$

which will lead to signal dispersion. Because time is measured to high precision with a Vector Network Analyzer (VNA), researchers often use the time delay per inch, TD_{in}, which would be given as

$$TD_{in,LL} = 1\,\text{in}/u_p = (1\,\text{in})\sqrt{LC}\left[1+(1/8\omega^2)(R/L-G/C)^2\right] \quad (3.12)$$

We can make use of the equivalence $\sqrt{LC} = \sqrt{\mu\varepsilon} = \sqrt{\mu_r\varepsilon_r}/c$ for nonmagnetic materials ($\mu_r = 1$) to see that

$$TD_{in,LL} = (1\,\text{in})(0.0254\,\text{m/in})\left(\sqrt{\varepsilon_{r,eff}}/c\right)\left[1+(1/8\omega^2)(R/L-G/C)^2\right], \quad (3.13)$$

where we have let the term $\sqrt{\varepsilon_r} \Rightarrow \sqrt{\varepsilon_{r,eff}}$ to permit an effective permittivity in microstrip applications where the electric field lines are partly in the propagating medium (e.g., fire retardant with $\varepsilon_r = 4$ FR4) and partly in air.

PROBLEM

3.1 Balanis provides a way to calculate $\varepsilon_{r,\,eff}$ for a microstrip of certain dimensions. For a trace width $w = 0.010''$, height $h = 0.006''$, thickness $t = 0.001''$ and a nominal $\varepsilon_r = 4$, the resulting value of $\varepsilon_{r,\,eff} \sim 3.055$. In this case, show that the constants in front of the square bracket to be 148 ps. Thus, for a microstrip

$$TD_{in,LL}(\text{microstrip}) \approx 148\text{ps}\left[1+(1/8\omega^2)(R/L-G/C)^2\right] \quad (3.14)$$

We can also expand the impedance of a transmission line to show

$$Z_0 = \sqrt{\frac{(R+j\omega L)}{(G+j\omega C)}} = \sqrt{\frac{j\omega L(1+R/j\omega L)}{j\omega C(1+G/j\omega C)}} = \sqrt{\frac{L}{C}}\left(1+\frac{R}{j\omega L}\right)^{1/2}\left(1+\frac{G}{j\omega C}\right)^{-1/2}$$

$$Z_0 = \sqrt{L/C}\,[1+(1/2)R/j\omega L+(1/8)R^2/\omega^2L^2+\cdots]$$
$$[1-(1/2)G/j\omega C-(1/8)G^2/\omega^2C^2-\cdots]$$

and, if we again use the definition of *low loss* to be $R \leq \omega L$ and $G \leq \omega C$,

$$Z_{0,LL} \approx \sqrt{L/C}\,\{1+(1/8\omega^2)(R/L-G/C)[(R/L+3G/C)-(j/2\omega)]\} \quad (3.15)$$

Conclusions

- For a *low loss* transmission line, the propagation constant $\gamma = \alpha + j\beta = \sqrt{(R + j\omega L)(G + j\omega C)}$ will (in general) depend upon ω, and this will lead to a frequency-dependent phase velocity, u_p.
- As different frequency components of a signal pulse propagate along the transmission line (with different velocities), the signal pulse will attenuate and broaden (i.e., it will suffer distortion and dispersion). A *low loss* transmission line (with nonzero R and/or G) is therefore generally dispersive.

Low Loss and *Distortionless* Line

From Equation 3.15, we can see that the *distortionless* condition, $R/L = G/C$, gives us $Z_0 \approx \sqrt{L/C}$ so the impedance of a *low loss*, *distortionless* line will have the same impedance as a *lossless* line. From Equation 3.11, we can also see that, for the *distortionless* condition, $u_{p,LL} = \omega/\beta_{LL} \approx 1/\sqrt{LC}$ is the same as the frequency-independent *lossless* line.

Conclusion

In the real world, there is no such thing as a *lossless* transmission line (especially at high frequencies), but, if we could make *low loss* lines *distortionless*, we could obtain a phase velocity that is still independent of frequency. This would be a great advantage for the transmission of pulses of information because each of the Fourier components would travel at the same phase velocity and the pulse would retain its shape (albeit attenuated) as it propagates.

3.4 POWER TRANSMISSION

For a semi-infinite transmission line, the power, $P(z, t)$, propagating in the z-direction is given by

$$P(z,t) = \mathrm{Re}\left[V_0^+ e^{-\alpha z} e^{-j(\beta z - \omega t)}\right] \mathrm{Re}\left[I_0^+ e^{-\alpha z} e^{-j(\beta z - \omega t)}\right] \tag{3.16}$$

In this equation, $I_0^+ = V_0^+/Z_0 = V_0^+/(R_0 + jX_0) = V_0^+ R_0/|Z_0|^2 - jV_0^+ X_0/|Z_0|^2$, so

$$P(z,t) = \left[V_0^+ e^{-\alpha z} \cos(\beta z - \omega t)\right]\left(V_0^+/|Z_0|^2\right) e^{-\alpha z}[R_0 \cos(\beta z - \omega t) - X_0 \sin(\beta z - \omega t)]$$

or

$$P(z,t) = \left[\left(V_0^+\right)^2/|Z_0|^2\right] e^{-2\alpha z}[R_0 \cos^2(\beta z - \omega t) - X_0 \cos(\beta z - \omega t) \sin(\beta z - \omega t)] \tag{3.17}$$

Equation 3.17 tells us that, at a particular point z, the power propagating in the z-direction has one term that varies like $\cos^2 \omega t$ and another term that varies like cos(ωt)sin(ωt). If we time-average the first term over one period, we will obtain ½, and, if we time average the second term over one period, we will obtain 0. Thus,

$$P_{Avg}(z) = (1/2)\left[\left(V_0^+\right)^2 R_0 / |Z_0|^2\right] e^{-2\alpha z} \tag{3.18}$$

From Equation 3.18, we can see that the time-averaged power being propagated along the positive z-axis is decreasing with an attenuation coefficient, 2α. Thus, power is being lost with distance z from the origin in a semi-infinite transmission line. The amount of power loss is

$$P_{loss}(z) = -\partial P_{Avg}(z)/\partial z = 2\alpha P_{Avg}(z) \tag{3.19}$$

So we can measure the attenuation constant α by measuring the power loss relative to the average power as

$$\alpha = P_{loss}(z)/2P_{Avg}(z) \tag{3.20}$$

Power is lost in resistive heating, either along the z-direction by the "conductor" as $I_{Avg}^2 R$ or across the "insulating" medium as $V_{Avg}^2 G$; that is, if we take the time average of

$$P_{loss}(z,t) = \left\{\text{Re}\left[I_0^+(z) e^{-\gamma z} e^{j\omega t}\right]\right\}^2 R + \left\{\text{Re}\left[V_0^+ e^{-\gamma z} e^{j\omega t}\right]\right\}^2 G \tag{3.21}$$

we will get

$$P_{loss,Avg}(z) = \frac{\left(V_0^+\right)^2 (R_0)^2}{2|Z_0|^4} e^{-2\alpha z} R + \frac{\left(V_0^+\right)^2 (X_0)^2}{2|Z_0|^4} e^{-2\alpha z} R + \frac{\left(V_0^+\right)^2}{2} e^{-2\alpha z} G \tag{3.22}$$

$$\text{or}\quad P_{loss,Avg}(z) = \left[\left(V_0^+\right)^2 / 2|Z_0|^2\right] e^{-2\alpha z} R + \left[\left(V_0^+\right)^2 / 2|Z_0|^2\right] e^{-2\alpha z} G |Z_0|^2 \tag{3.23}$$

$$\text{or}\quad P_{loss,Avg}(z) = \left[\left(V_0^+\right)^2 / 2|Z_0|^2\right]\left[R + G|Z_0|^2\right] e^{-2\alpha z} \tag{3.24}$$

and, using Equation 3.20,

$$\alpha = P_{loss}(z)/2P_{Avg}(z) = (1/2)\left[R + G|Z_0|^2\right]/R_0 \tag{3.25}$$

3.5 FINITE TRANSMISSION LINES

The mathematical analysis for finite transmission lines is essentially the same as that for infinite transmission lines, but we typically expect a load with impedance, Z_L, on one end and a source or generator of electromagnetic potential, V_g, on the other

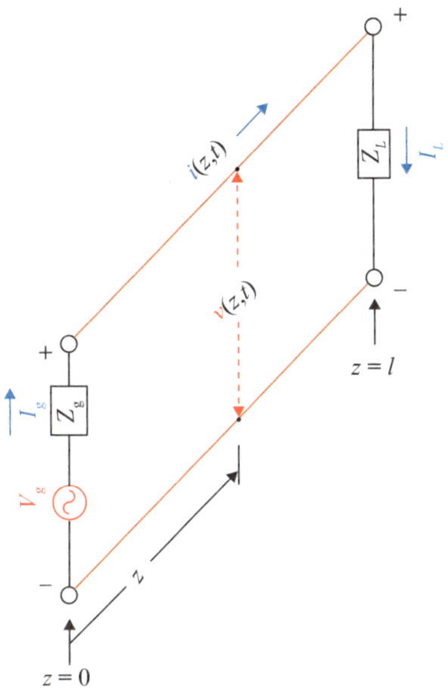

Figure 3.2b Variables used in the analysis of a finite transmission line that extends from $z = 0$ to $z = l$.

end. These lines may have any one of the cross-sectional shapes that were shown in Figure 3.1, but the schematic representation is shown in Figure 3.2b.

In Figure 3.2b, we have assumed the generator has some internal impedance, Z_g, and that the current through the generator is I_g. The current through the load of impedance, Z_L, is I_L.

In this analysis, we will assume that a generator voltage is cosinusoidal, so that the voltage across the transmission line at $z = 0$ is

$$v(0,t) = \mathrm{Re}\left[V(0)e^{j\omega t}\right] \tag{3.26a}$$

The voltage in the transmission line will then be a function of z so that

$$v(z,t) = \mathrm{Re}\left[V(z)e^{j\omega t}\right] \tag{3.26b}$$

and, at $z = l$, the voltage will be

$$v(l,t) = \mathrm{Re}\left[V(l)e^{j\omega t}\right] = \mathrm{Re}\left[V_L e^{j\omega t}\right] \tag{3.26c}$$

The current in the transmission line will also be a function of z so that

$$i(0,t) = \mathrm{Re}\left[I(0)e^{j\omega t}\right] \tag{3.27a}$$

$$i(z,t) = \mathrm{Re}\left[I(z)e^{j\omega t}\right] \tag{3.27b}$$

and, at $z = l$, the current will be

$$i(l,t) = \mathrm{Re}\left[I(l)e^{j\omega t}\right] = \mathrm{Re}\left[I_L e^{j\omega t}\right] \tag{3.27c}$$

In the previous section, we found that the solution to the spatial parts of the voltage and current were $V(z) = V_0^+e^{-\gamma z} + V_0^-e^{-\gamma z}$ and $I(z) = I_0^+e^{-\gamma z} + I_0^-e^{-\gamma z}$, where $\gamma = \alpha + j\beta = \sqrt{(R+j\omega L)(G+j\omega C)}$ and $V_0^+/I_0^+ = Z_0 = \sqrt{\dfrac{(R+j\omega L)}{(G+j\omega C)}}$ and $V_0^-/I_0^- = -Z_0$, where Z_0 is the transmission line characteristic impedance.

Applying the boundary conditions at $z = l$

$$V(l) = V_0^+e^{-\gamma l} + V_0^-e^{\gamma l} = V_L \tag{3.28a}$$

and

$$I(l) = \left(V_0^+/Z_0\right)e^{-\gamma l} - \left(V_0^-/Z_0\right)e^{\gamma l} = I_L \tag{3.28b}$$

Solving Equations 3.28a and 3.28b for V_0^+ and V_0^-,

$$V_0^+ = (1/2)(V_L + I_L Z_0)e^{-\gamma l} = (I_L/2)(Z_L + Z_0)e^{-\gamma l} \tag{3.29a}$$

and

$$V_0^- = (1/2)(V_L - I_L Z_0)e^{-\gamma l} = (I_L/2)(Z_L - Z_0)e^{-\gamma l} \tag{3.29b}$$

Putting these coefficients back into the transmission line equations,

$$V(z) = (I_L/2)\left[(Z_L + Z_0)e^{\gamma(l-z)} + (Z_L - Z_0)e^{-\gamma(l-z)}\right] \tag{3.30a}$$

and

$$I(z) = (I_L/2Z_0)\left[(Z_L + Z_0)e^{\gamma(l-z)} - (Z_L - Z_0)e^{-\gamma(l-z)}\right] \tag{3.30b}$$

Now, if we let $z' = (l - z)$ be the distance from the load back to the point z,

$$V(z') = (I_L/2)\left[(Z_L + Z_0)e^{\gamma z'} + (Z_L - Z_0)e^{-\gamma z'}\right] \tag{3.31a}$$

and

$$I(z') = (I_L/2Z_0)\left[(Z_L + Z_0)e^{\gamma z'} - (Z_L - Z_0)e^{-\gamma z'}\right] \tag{3.31b}$$

But $e^{\gamma z'} + e^{-\gamma z'} = 2\cosh\gamma z'$ and $e^{\gamma z'} - e^{-\gamma z'} = 2\sinh\gamma z'$ so

$$V(z') = (I_L/2)[Z_L \cosh \gamma z' + Z_0 \sinh \gamma z'] \tag{3.32a}$$

and

$$I(z') = (I_L/Z_0)[Z_L \sinh \gamma z' + Z_0 \cosh \gamma z'] \tag{3.32b}$$

Now, if we define $Z(z') = V(z')/I(z')$ as the line impedance at z',

$$Z(z') = Z_0 \frac{Z_L + Z_0 \tanh \gamma z'}{Z_0 + Z_L \tanh \gamma z'} \tag{3.33}$$

If we evaluate Equation 3.33 at the point $z = 0$ or ($z' = l$),

$$Z(z' = l) = Z(z = 0) = Z_0 \frac{Z_L + Z_0 \tanh \gamma l}{Z_0 + Z_L \tanh \gamma l} \tag{3.34}$$

This is the combined impedance that the harmonic generator "sees" as a result of the transmission line impedance and the load impedance. Equation 3.34 is one of the most famous equations in electrical engineering and is called the **finite transmission line equation**.

Conclusions

We can see from Equation 3.34 that, if $Z_L = Z_0$, the impedance seen by the generator is Z_0 regardless of the length, l, of the transmission line; that is, it is as if the transmission line were not present. Under the condition that $Z_L = Z_0$, we say the line has a **matched** load.

Lossless Lines

For *lossless* transmission lines in which $R = 0$ and $G = 0$, we have shown

$$\gamma = \alpha + j\beta = \sqrt{(R + j\omega L)(G + j\omega C)} = \sqrt{(-\omega^2 LC)} = j\omega\sqrt{LC}$$

so Equation 3.34 becomes

$$Z(z = 0) = Z_0 \frac{Z_L + Z_0 \tanh j\omega\sqrt{LC}l}{Z_0 + Z_L \tanh j\omega\sqrt{LC}l} = Z_0 \frac{Z_L + Z_0 j \tan\left(\omega\sqrt{LC}l\right)}{Z_0 + Z_L j \tan\left(\omega\sqrt{LC}l\right)} \tag{3.35}$$

and, because $Z_0 = R_0 = \sqrt{L/C}$ for a *lossless* line,

$$Z(z = 0) = R_0 \frac{Z_L + jR_0 \tan\left(\omega\sqrt{LC}l\right)}{R_0 + jZ_L \tan\left(\omega\sqrt{LC}l\right)} \text{ with } R_0 = \sqrt{L/C} \tag{3.36}$$

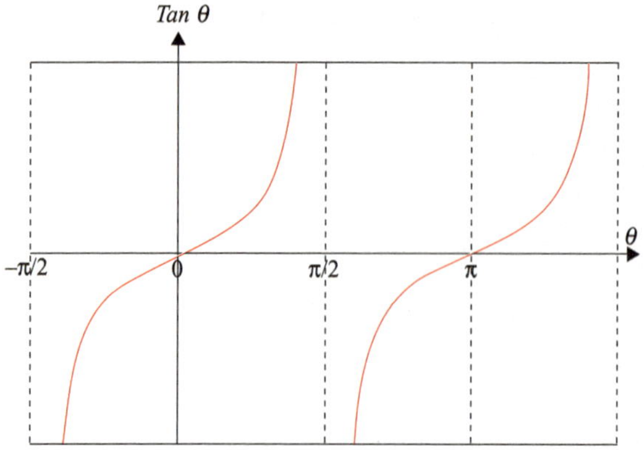

Figure 3.3 Functional form of tan θ vs. θ.

Special Case #1

If $Z_L \rightarrow \infty$ (an open circuit line), then $Z_{open}(z=0) = -j\sqrt{L/C}/\tan\left(\omega\sqrt{LC}l\right)$. Here, we find that the impedance seen by the generator is purely reactive and we can conclude from Figure 3.3 how the magnitude of Z varies with the argument of the tangent function.

We cconclude that Z_{open} can be either + or − and that it varies between −∞ and +∞. Additionally, if we write $u_p = 1/\sqrt{LC}$, we can express

$$\omega\sqrt{LC}l = 2\pi f(1/u_p)l = (2\pi/T)(1/u_p)l = 2\pi l/\lambda$$

so

$$Z_{open}(z=0) = -j\sqrt{L/C}/\tan(2\pi l/\lambda)$$

Very Special Case #1

If $Z_L \rightarrow \infty$ and if $l << \lambda$:

$$\begin{aligned} Z_{open}(z=0) &= -j\sqrt{L/C}/\tan(2\pi l/\lambda) \approx -j\sqrt{L/C}/2\pi l/\lambda \\ &= -j\sqrt{L/C}/\omega\sqrt{LC}l = -j/\omega Cl \end{aligned}$$

which is the impedance of a capacitor of Cl farads (C = capacitance per unit length).

Special Case #2

If $Z_L \rightarrow 0$ (a short circuit), then $Z_{short}(z=0) = jR_0 \tan\left(\omega\sqrt{LC}l\right)$, which is purely reactive and can be + or −.

Very Special Case #2

If $Z_L \to 0$ and if $l \approx \lambda/4$ then $Z_{short}(z = 0) = jR_0 \tan(2\pi l/\lambda) = jR_0 \tan(2\pi/2) = \infty$. For quarter wavelength transmission lines (or multiples), a short circuit acts as if there were no cable present.

Other Very Special Case #2

If $Z_L \to 0$ and if $l << \lambda$, $Z_{short}(z=0) = jR_0 \tan\left(\omega\sqrt{LC}l\right) \approx jR_0\left(\omega\sqrt{LC}l\right) = j\sqrt{L/C}\omega\sqrt{LC}l = j\omega Ll$, which is the impedance of a pure inductor because L = inductance per unit length.

Open and Closed Circuit Measurements

We may use the open and closed circuit measurements to find values of an unknown transmission line by recognizing in Equation 3.34 that

$$Z_{open}(z=0) = Z_0/\tanh \gamma l \tag{3.37a}$$

and

$$Z_{short}(z=0) = Z_0 \tanh \gamma l \tag{3.37b}$$

So we note that $Z_{open}\, Z_{short} = Z_0^2$ and $\gamma = (1/l) \tanh^{-1} (Z_{short}(z = 0)/Z_0)$, so

$$Z_0 = \sqrt{Z_{short}(z=0) \cdot Z_{open}(z=0)} \tag{3.38a}$$

and

$$\gamma = (1/l) \tanh^{-1} \sqrt{Z_{short}(z=0)/Z_{open}(z=0)} \tag{3.38b}$$

Thus, we can measure the values of $Z_{open}(z = 0)$ and $Z_{short}(z = 0)$ to find the values of Z_0 and γ.

Lumped-Circuit Approximation Model of Vias

In the very special case above, we have seen that electrically short ($l << \lambda$) transmission lines act as if they were lumped inductors or capacitors with $Z_{short}(z = 0) = j\omega Ll$ and $Z_{open}(z = 0) = j/\omega Cl$. This condition commonly occurs in the use of a via that connects a transmission line in the form of a microstrip to a transmission line in the form of a stripline on an imbedded layer of a printed circuit board (PCB), as shown in Figure 3.4.

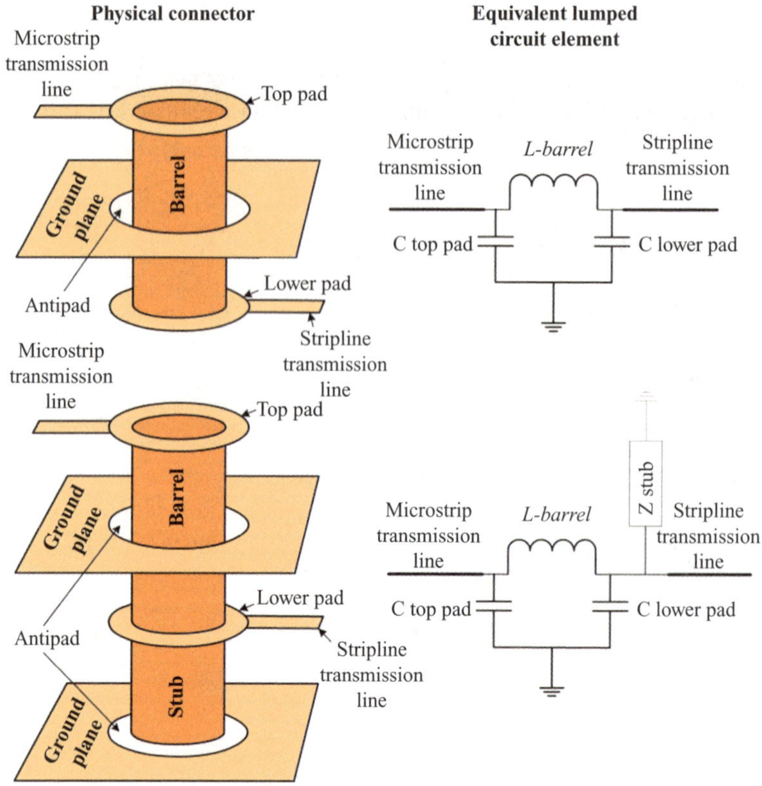

Figure 3.4 (Top) "Blind Via" barrel and via pads on a multiple layer PCB. (Bottom) "Through-hole Via" barrel and pads on a multiple layer PCB.

Because the pads for a "blind via" are on either side of the barrel, the equivalent circuit element is often shown as a lumped inductor and two capacitors in the middle of two transmission lines. The "through-hole via" is more common because it is easier to manufacture but it produces a compound equivalent circuit element that consists of two lumped circuits: one like the blind via to electrically connect two transmission lines and another that serves no constructive purpose but can act as an antenna to radiate electromagnetic fields that can interact with neighboring circuit components. As we will see later in this chapter, the effects of such lumped elements can also cause voltage reflections that interfere with the incident voltage from a transmitter.

3.6 HARMONIC WAVES IN FINITE TRANSMISSION LINES

In Equations 3.31a and 3.31b $V(z') = (I_L/2)[(Z_L + Z_0)e^{\gamma z'} + (Z_L - Z_0)e^{-\gamma z'}]$ and $I(z') = (I_L/2Z_0)[(Z_L + Z_0)e^{\gamma z'} - (Z_L - Z_0)e^{-\gamma z'}]$. These phasor terms may be developed for any harmonic quantity, even those which are the components of a Fourier transform, so that:

$$V(z,t)=(I_L/2)\left[(Z_L+Z_0)e^{(\alpha+j\beta)(l-z)+j\omega t}+(Z_L-Z_0)e^{-(\alpha+j\beta)(l-z)+j\omega t}\right] \quad (3.39a)$$

and $$I(z,t)=(I_L/2Z_0)\left[(Z_L+Z_0)e^{(\alpha+j\beta)(l-z)+j\omega t}-(Z_L-Z_0)e^{-(\alpha+j\beta)(l-z)+j\omega t}\right] \quad (3.39b)$$

The first term on the right-hand side of these equations is a wave that propagates in the positive z-direction, and the second term on the right-hand side is a wave that propagates in the negative z-direction. If we evaluate these terms at the point $z = l$, we can thus interpret the first term on the right-hand side as the magnitude of the net incident voltage (or current) to the load impedance and the second term as the net reflected voltage (or current) from the load impedance. The ratio of these terms is called the net voltage (or current) coefficient, Γ_V (Γ_I), where

$$\Gamma_V(z=l)=\left.\frac{V_0^-}{V_0^+}\right|_{z=l}=\frac{Z_L-Z_0}{Z_L+Z_0} \quad (3.40a)$$

and $$\Gamma_I(z=l)=\left.\frac{I_0^-}{I_0^+}\right|_{z=l}=-\frac{Z_L-Z_0}{Z_L+Z_0} \quad (3.40b)$$

Equation 3.40a is of the same form as Equation 2.23b $\Gamma_{23}\equiv\frac{E_0^-}{E_0^+}=\left(\frac{\eta_3-\eta_2}{\eta_3+\eta_2}\right)$ (except that here we are comparing the net voltage propagating in the negative z-direction with the net voltage propagating in the positive z-direction) and is in general a complex number (i.e., the voltage ratio coefficient has a magnitude, and phase; $|\Gamma_V|$ and θ_Γ).

Conclusions

1. Voltages propagating in the positive z-direction in a finite transmission line will **not** be reflected from the load impedance if the load is matched to the characteristic line impedance; that is, if $Z_L = Z_0$.
2. Voltages propagating in the positive z-direction in a finite transmission line will be reflected from the load impedance with a magnitude and phase $|\Gamma_V|$ and θ_Γ.
3. For $Z_L \neq Z_0$, $|\Gamma_V| \neq 0$ and the voltage equations produce standing waves with maxima and minima. For this case, $S=\frac{|V_{max}|}{|V_{min}|}=\frac{1+|\Gamma_V|}{1-|\Gamma_V|}$ or $|\Gamma_V|=\frac{S-1}{S+1}$. This equation implies that we can measure the standing wave ratio to determine the magnitude of the impedance of a finite transmission line.

Harmonic Generator Perspective

The time-harmonic solution of the finite transmission line has been found to be equivalent to the time-harmonic solution of the plane wave propagating in a material

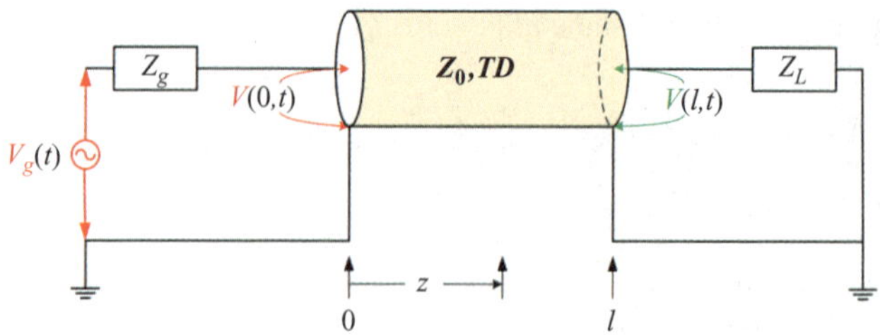

Figure 3.5 Schematic representation of the voltage across a generic transmission line with characteristic impedance Z_0 and delay time (propagation time) $TD = l/u_p = l\sqrt{LC}$.

as it orthogonally interacts with two boundaries in section 2.3. In the mentioned case, the boundaries exist at $z = 0$ and at $z = l$, and we have found the ratio of the voltage and current components at the load, $z = l$. The voltage and current components at $z = 0$ are those "seen" by the generator that we can write from Equations 3.39a and 3.39b as

$$V(0,t) = (I_L/2)\left[(Z_L + Z_0)e^{(\alpha+j\beta)l+j\omega t} + (Z_L - Z_0)e^{-(\alpha+j\beta)l+j\omega t}\right] \tag{3.41a}$$

and

$$I(0,t) = (I_L/2Z_0)\left[(Z_L + Z_0)e^{(\alpha+j\beta)l+j\omega t} - (Z_L - Z_0)e^{-(\alpha+j\beta)l+j\omega t}\right] \tag{3.41b}$$

We can thus picture the voltage across a generic transmission line (representing any one of the transmission line cross sections shown in Figure 3.1) as a schematic shown in Figure 3.5.

In Figure 3.5, we have included the delay time, *TD*, for a wave propagating at velocity $u_p = 1/\sqrt{LC}$ to move from $z = 0$ to $z = l$, which (along with Z_0) is a typical way to state the properties (characteristic impedance and length) of a transmission line.

Example Problem

A stripline trace on a printed wiring board with FR4 ($\varepsilon_r = 4$) has a *lossless* transmission line with characteristic impedance of 50 Ω and a length of TD = 3.333 ns, as shown in Figure 3.6. The generator voltage is 2V, ∠30° at 100 MHz and has an internal resistance of 50 Ω, with an internal capacitance of 31.8 pF. The load impedance is a 50-Ω resistor in series with a 159-nH inductor.

SOLUTION Let us find everything we know. Because the line is lossless, $\alpha = 0$. Because $\varepsilon_r = 4$, the phase propagation speed is $u_p = 1/\sqrt{\mu\varepsilon} = c/\sqrt{\mu_r\varepsilon_r} = 1.5\times10^8$ m/s so the transmission line is $l\ u_pTD$ = 0.5 m long, and the propagation constant at 100 MHz is $\beta = \omega/u_p = (2\pi\times10^8\ \text{s}^{-1})/(1.5\times10^8\ \text{m/s}) = 4.188\ \text{m}^{-1}$ and $\beta l = 2.094$.

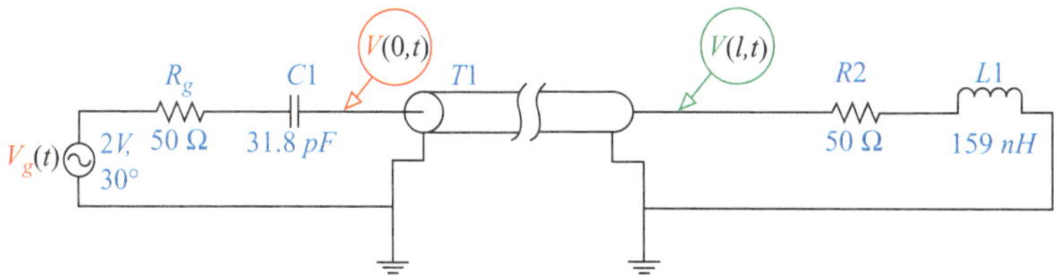

Figure 3.6 A 0.5-m long, 50-Ω stripline is in series with a 2-V AC generator operating at frequency f = 100 MHz. The internal resistance of the generator is $R1$ = 50 Ω, and its internal capacitance is 31.8 pF. The load resistance is $R2$ = 50 Ω, and its inductance is 159 nH.

We know that the characteristic impedance is related to the transmission line inductance, L, and capacitance, C, per unit length as $Z_0 = \sqrt{L/C} = 50\,\Omega$ and that the time delay is related to those same quantities as $TD = l\sqrt{LC} = 3.333\,\text{ns}$ so we can multiply these two equations to find lL = (3.333 × 10^{-9} s)(50 Ω) = 166.7 nH or L = 333.33 nH/m and we can divide the two equations to find lC = 3.333 × 10^{-9} s/50 Ω = 66.67 pF) or C = 133.33 pF/m.

We can find the voltage $V(0, t)$ and $V(l, t)$ as a function of the frequency f, as shown at the two points on the ends of the transmis sion line by solving Equations 3.33a and 3.33b for those respective values of z with the help of Equation 3.27:

$$Z(z') = Z_0 \frac{Z_L + Z_0 \tanh \gamma z'}{Z_0 + Z_L \tanh \gamma z'}$$

so for a lossless line in which $\gamma = j\omega\sqrt{LC} = j\beta$

$$Z(z = l) = \sqrt{L/C}\, \frac{Z_L + j\sqrt{L/C}\tan \omega\sqrt{LC}\,0}{\sqrt{L/C} + jZ_L \tan \omega\sqrt{LC}\,0} = Z_L = 50\,\Omega + j100\,\Omega$$

and from Equation 3.34a,

$$\Gamma_V(z = l) = \frac{Z_L - Z_0}{Z_L + Z_0} = \frac{(50\,\Omega + j100\,\Omega) - (50\,\Omega)}{(50\,\Omega + j100\,\Omega) + (50\,\Omega)} = 0.5 + 0.5j = 0.707e^{j0.785}.$$

Likewise,

$$\begin{aligned} Z(z = 0) &= Z_0 \frac{Z_L + jZ_0 \tan \beta l}{Z_0 + jZ_L \tan \beta l} \\ &= 50\,\Omega \frac{(50\,\Omega + j100\,\Omega) + j50\,\Omega(-1.7336)}{50\,\Omega + j(50\,\Omega + j100\,\Omega)(-1.7336)} = 10.8\,\Omega e^{j(0.622)} \end{aligned}$$

$$\Gamma_V(z = 0) = \frac{Z(z = 0) - Z_0}{Z(z = 0) + Z_0} = \frac{\left(10.8\,\Omega e^{j(0.622)}\right) - (50\,\Omega)}{\left(10.8\,\Omega e^{j(0.622)}\right) + (50\,\Omega)} = -0.705e^{-j0.259107}$$

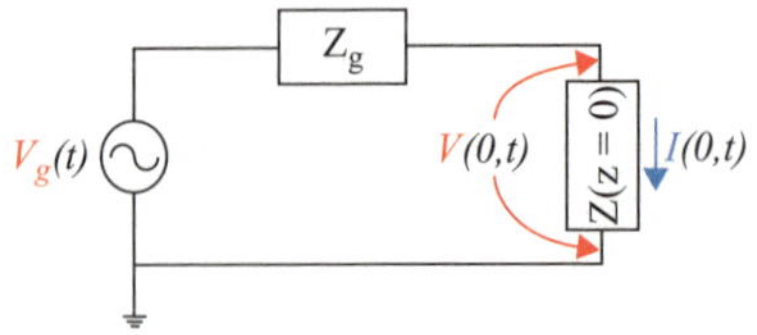

Figure 3.7 Voltage and current at the beginning of the transmission line as seen by the generator.

To get $V(0, t)$ these answers in terms of the generator voltage, $V_g(t)$, we use Figure 3.7.

Using the potentiometer drop across the transmission line impedance $Z\ (z = 0)$ seen by the generator, we can see from Figure 3.7

$$V(0,t) = \frac{Z(z=0)}{Z_g + Z(z=0)} V_g(t)$$

and using

$$V_g(t) = V_g e^{j\theta_g} e^{j\omega t} = 2Ve^{j\pi/6}$$

$$V(0,t) = V_g \frac{Z(z=0)}{Z_g + Z(z=0)}$$
$$= \frac{(2Ve^{j\pi/6})(10.8\,\Omega e^{j(0.622)})}{(50\,\Omega - j50\,\Omega) + (8.776\,\Omega + j6.294\,\Omega)} = 0.295\text{V}e^{j1.785} e^{j\omega t}$$

This voltage is thus imposed on the transmission line at $z = 0$ so, from Equation 3.6a,

$$V(z)|_{z=0} = \left(V_0^+ e^{-\gamma z} + V_0^- e^{\gamma z}\right)\Big|_{z=0} \text{ or } V(0) = V_0^+[1 + \Gamma_V(z=0)]$$

so

$$V_0^+ = 0.295\text{V}e^{j1.785} / [1 + \Gamma_V(z=0)]$$

so because $1 + \Gamma_V(z = 0) = 0.366e^{j(0.516)}$,

$$V_0^+ = 0.805\text{V}e^{j1.269}$$

we may now use this in

$$V(l) = V_0^+\left[e^{-j\beta l} + \Gamma_V(z=l)e^{j\beta l}\right] = 0.805\text{V}e^{j1.269}\left[e^{-j2.094} + \left(0.707e^{j0.785}\right)e^{j2.094}\right]$$

$$V(l) = 0.805\text{V}e^{j1.269}\left[1.365e^{j3.665}\right] = 1.099\text{V}e^{j(4.934)}$$

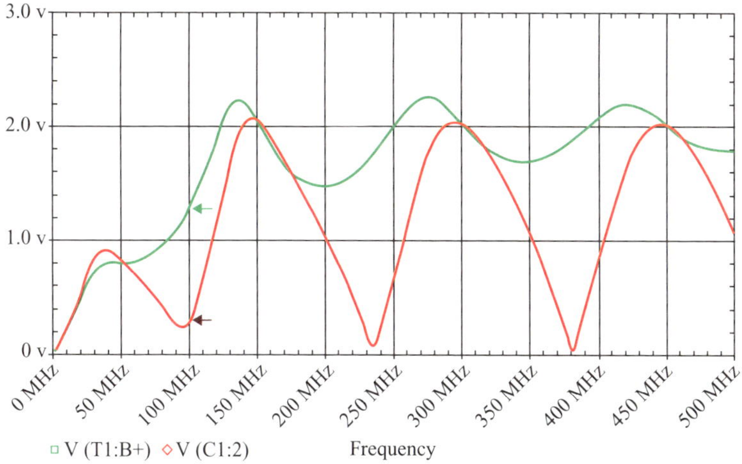

Figure 3.8 Voltage magnitude $V(0, t)$ and $V(l, t)$ at the two ends of the transmission line shown in Figure 3.6 plotted as a function of frequency. The red and green arrows show the value of the magnitude of the voltage at 100 MHz.

ANSWER $V_{100\ MHz}(l, t) = 1.099\,\mathrm{V} e^{j(4.934)} e^{j(2\pi\times10^8)t}$, $V_{100MHz}(0, t) = 0.295\mathrm{V} e^{j1.785} e^{j(2\pi\times10^8)t}$

3.7 USING AC SPICE MODELS

This problem can also be solved numerically with the use of the PSpice program and the net list below to yield the voltage (voltage), as shown in Figure 3.8.

Net List:	VS	1	0	AC	2	30
	RS	1	2	50		
	CS	2	3	31.8p		
	T	3	0	4	0	Z0 = 50 *TD* = 3.333n
	RL	4	5	50		
	LL	5	0	159n		
	.AC	LIN	1	1E6	0.5E9	
	.PRINT	AC	VM(3)	VM(4)		
	.END					

3.8 TRANSIENT WAVES IN FINITE TRANSMISSION LINES

Today, transmission lines mainly support the propagation of digital information in the form of 1s and 0s, as seen in Figure 3.9.

As shown in Figure 3.9a, the digital representation of 010000 is a voltage pulse that is approximately a square pulse in time. However, because of their finite power,

Effect of different loads

Arbitrary Load

For the general case of an arbitrary load (the most common case), we see from Equation 3.42 that there is a "reflected" wave equal to a fraction of the incident wave that propagates back along the negative z-axis. In general, this wave is also shifted in phase so that $\Gamma_L = |\Gamma| e^{j\theta_r}$.

Matched Load

For the special case of a matched load in which $Z_L = Z_0$, we see from Equation 3.42 that there is no "reflected" wave so the incident pulse propagates entirely into the load since $\Gamma_{matched} = 0$.

Zero Load

For the special case of a zero load in which $Z_L = 0$ (a short circuit), we see from Equation 3.42 that there is a negative "reflected" wave so the incident pulse interferes destructively with the reflected wave, $\Gamma_{short} = -1$.

Infinite Load

For the special case of an infinite load (an open circuit), Equation 3.42 yields a "reflected" wave equal to the incident wave that propagates back along the negative z-axis ($\Gamma_{open} = +1$). This is the special case shown in Figure 3.11.

In Figure 3.11, we see that the two waves interfere constructively to the left of the load and produce a voltage, as shown by the upper solid red curve.

There is no voltage wave for points $z > l$. The curves in this region are shown only for modeling purposes to indicate that the incident and reflected waves move in opposite directions.

Had the reflection coefficient Γ_L been less than 1 but greater than 0, the reflected wave in this sequence would be reduced in amplitude at every point in space.

For $t = TD + \Delta t$ and $t = TD + 2\Delta t$, we can see that the voltage at the load, $z = l$, exceeds the initial voltage produced by the generator.

Idealistic Digital Examples with Real Impedances

Heavyside Function

As a simple case, let us consider a DC transition in source voltage for the case of a real internal generator impedances, a transmission line with a real characteristic impedance, and a real load, as shown in Figure 3.12.

The voltage as a function of time at point *1* is thus shown in Figure 3.13a.

Here, we cannot use the effective impedance for a harmonic generator. as developed in section 3.5 because this is an ideal digital pulse. For this very fast snapshot in time, the heavyside function is imposed across the 25-Ω internal imped-

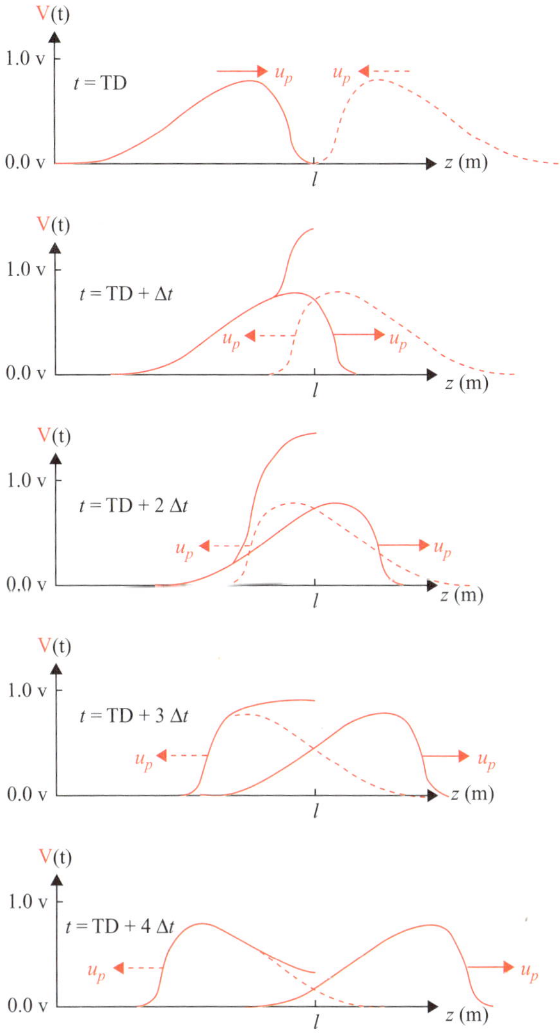

Figure 3.11 Special case of the interference (addition) of incident and reflected waves near an infinite load (open circuit) at a sequence of points in time $t = TD + n\,\Delta t$ for $n = 0, 1, 2, 3, 4$.

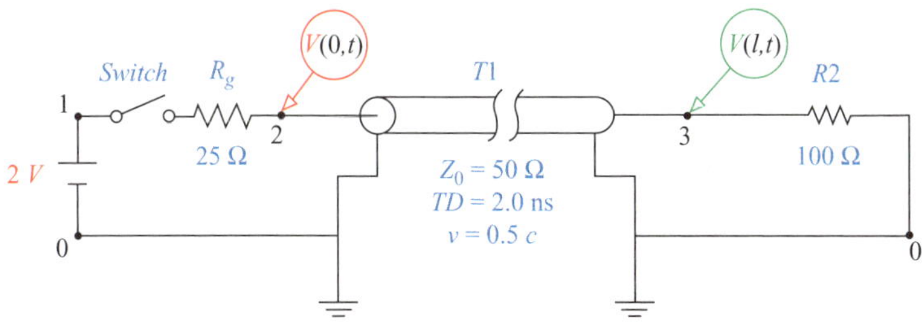

Figure 3.12 Special case of a DC source voltage switched at $t = 0$.

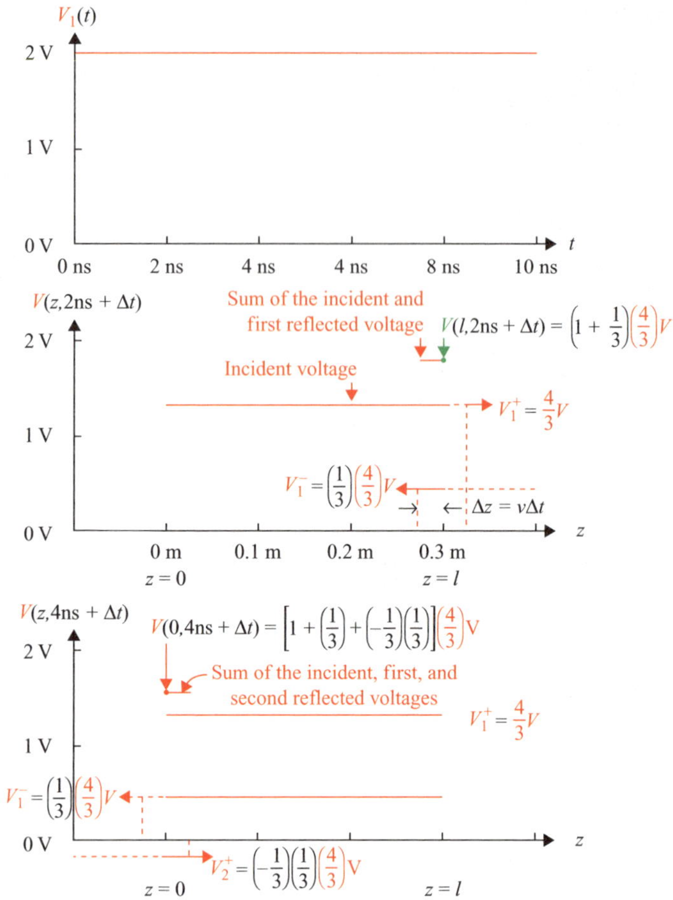

Figure 3.13 (a) Voltage at point 1 as a function of time; (b) snapshot of the propagating wave on the transmission line at time $t = 2.167$ ns and the virtual reflected wave at $z = l$ propagating in the negative z-zone; (c) snapshot of the propagating wave on the transmission line at time $t = 4.167$ ns and the virtual reflected wave at $z = 0$ propagating in the positive z-zone.

ance in series with the characteristic impedance of the 50-Ω transmission line. One might argue that, at this initial time, the propagating voltage has no knowledge of what load exists on the other end of the transmission line. We can thus consider the potential divider for a digital pulse (similar to that for Figure 3.7) but as shown in Figure 3.14.

Here, we see that the voltage drop, $V(0, t)$, across the transmission line is

$$V(0,t) = 2\text{V}\frac{(50\,\Omega)}{(25\,\Omega + 50\,\Omega)} = \frac{4}{3}\text{V} = 1.333\text{V}$$

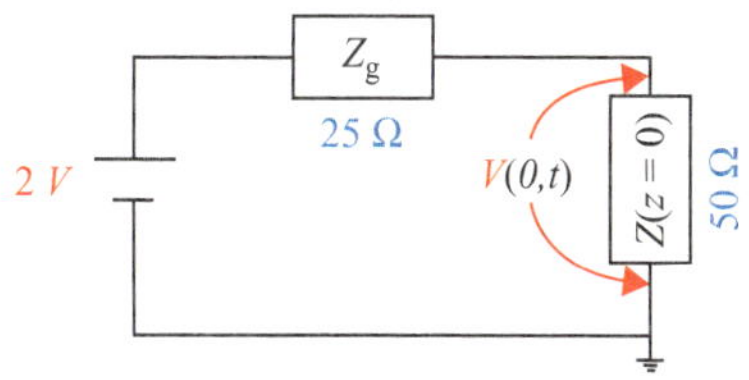

Figure 3.14 Potential division of the source voltage across the 50 Ω characteristic impedance of the transmission line.

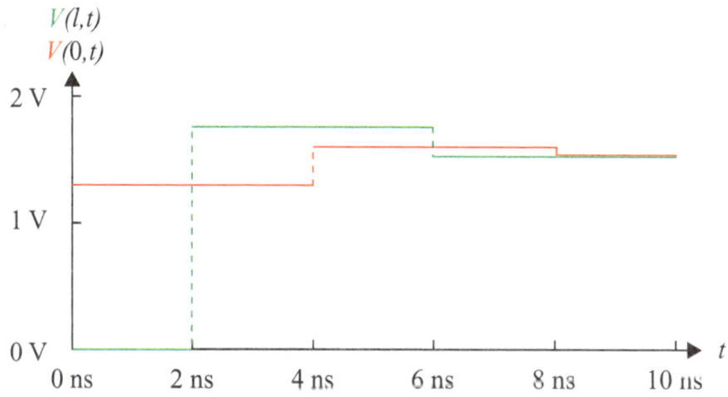

Figure 3.15 Voltage (voltage) at the ends of the transmission line in Figure 10.12 as a function of time up to 6 ns.

We can see from Figure 3.13a that the voltage at point 1 (in Figure 3.12) is steady 2 V after the switch has been closed (t = 0). We can see from Figure 3.14 that the voltage applied across the transmission line after the switch is closed is $V(0, t)$ = 1.333 V. This applied voltage will remain applied to point 2 by the battery, but a wave front will propagate down the transmission line at speed v = 0.5 c until ($t = TD$ = 2 ns) it reaches the end of the transmission line ($z = l$ = 0.3 m).

The voltage, $V(l, t)$, at point 3 in Figure 3.12 (or at 0.3 m in Figure 3.13b)) will be zero until the time $t = TD$ = 2 ns when it experiences both the incident voltage V_0^+ = 1.333 V and the reflected voltage V_0^- = Γ_{load} (1.333 V) = (1/3)(1.333 V) = 0.444 V (i.e., it will jump up to 1.777 V).

The reflected wave will then propagate back toward the battery until ($t = TD$ = 4 ns) it reaches the beginning of the transmission line (z = 0 m). At this time, the voltage will jump up from the applied voltage of 1.333 V to a higher value caused by the sum of the reflected voltage 0.444 V and the re-reflected voltage (Γ_g) (0.444 V) = (−1/3)(0.444 V) = −0.148 V from the beginning of the transmission line (i.e., 1.333 V + 0.444 V − 0.148 V = 1.629 V).

We can now plot the voltage (voltage) at the beginning (end) of the transmission line (at points 2 and 3) as a function of time up to 6 ns as shown in Figure 3.15.

We can see that, at time t = 6 ns the re-reflected wave of voltage −0.148 V will reach the end of the transmission line where it will cause a jump in the voltage of −0.148 V − (*1/3*)(−0.148 V) = −0.197 V. This amount will be added to the 1.777 V already at that point.

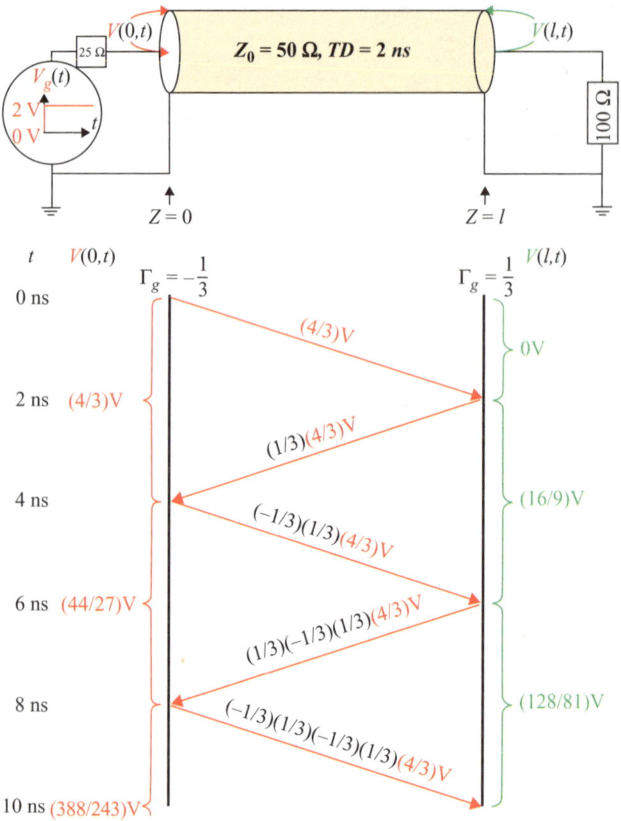

Figure 3.16 Voltage reflection diagram (ladder diagram) corresponding to Figure 3.15.

The process will continue with a new evaluation at each end every *2TD*. It is surprising how complex it is to keep track of all of the reflected, re-reflected, and re-re-reflected voltages at each end of the transmission line. Fortunately, a tool called the ladder diagram has been developed to assist us in keeping track of all the components. This is shown in Figure 3.16.

In this analysis, we follow the incident wave as it first reflects from the load with a magnitude $V_1^- = \Gamma_L V_1^+ = (1/3)V_1^+$ propagating in the negative z-direction. As we see in Figure 3.16, the incident wave at the first reflection is $V_1^+ = (4/3)\text{V}$ and it strikes the load end of the transmission line at $t = 2$ ns.

Until this voltage wave hits the load end, the voltage on the load end is zero, as shown on the right-hand side of the ladder diagram. The voltage on the generator end of the transmission line is (4/3)V, as shown on the left-hand side of the ladder diagram, and it remains at this value until the reflected wave returns to the generator end at $t = 4$ ns.

The reflected wave from the load has a magnitude of $V_1^- = \Gamma_L V_1^+ = (1/3)V_1^+$ and strikes the generator end of the transmission line at $t = 4$ ns, where it is re-reflected

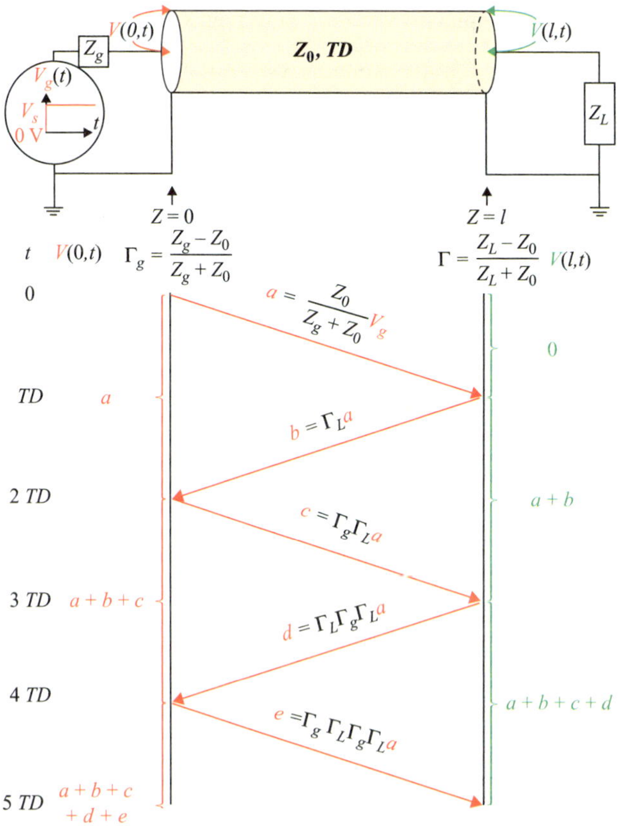

Figure 3.17 General case of reflection from each end of a transmission line of impedance Z_0 and time delay *TD*.

with an amplitude of $V_2^+ = \Gamma_g V_1^- = \Gamma_g \Gamma_L V_1^+ = (-1/3)(1/3)V_1^+$. The voltage at the generator end of the transmission line is, thus, the sum of (4/3)V as a result of the generator potential division plus the incoming wave amplitude $V_1^- = (1/3)V_1^+$ plus the re-reflected wave amplitude of $V_2^+ = (-1/3)(1/3)V_1^+$ (i.e., $V(0, t) = [1 + 1/3 - 1/9]$ (4/3)V = (44/27)V), as shown on the left-hand side of the ladder diagram. This voltage remains on the generator end until the pulse has time to travel $2TD = 4\ ns$ to the load end and back.

We can continue to follow the wave as it is reflected for the third time at the load end to produce a negative traveling wave voltage of $V_3^- = \Gamma_L V_2^+ = (1/3)(-1/3)$ (1/3)(4/3)V. The sum of all incident and reflected waves is, thus, $V(l, t) =$ (16/9)V − (4/27)V − (4/81)V = (128/81)V.

We continue to follow the reflected waves at each boundary to find the total voltage (voltage) at each end as a function of time. The general reflection diagram is shown in Figure 3.17.

In the reflection diagram, we emphasize the following points:

- Voltage represented by the first propagation is listed as a and is seen to be a potential division of the source voltage by the factor $Z_0/(Z_g + Z_0)$.
- Voltage at the load end is zero until *TD* at which time it jumps to $a + \Gamma_L a$ as a consequence of the incident voltage and the first reflected voltage.
- Reflected voltage propagates back toward the generator whose voltage has remained at value a until time *2TD* at which time it jumps to $a + \Gamma_L a + \Gamma_g \Gamma_L a$ as a consequence of the incident amount of reflected voltage and the subsequent second reflected voltage.
- The process continues at each end of the transmission line as time increases to jump at intervals *2TD* as the reflections propagate one round trip.

QUESTION What would be different about the values of voltage (voltage) measured at each end of the transmission line if there were an amplitude reduction of $e^{-\alpha l}$ during each transit of the reflected voltages? Could you modify the values shown in Figure 3.17 to account for this attenuation?

Propagation of a Digital Pulse of Voltage

Suppose the voltage source is a transient pulse that rises to 2 V at time $t = 0$ and then returns to 0 V after a period of time Δt_g. In this case, the transient pulse would remain at 2 V at the generator end for only the time Δt_g and would drop back to zero. Note that it does not matter how long the period Δt_g is relative to *TD* but, for clarity of the answers, let us consider the *Special Case* #1 of $\Delta t_g < TD$. Here, we will also assume that there is no attenuation of the pulses as they propagate and that all of the Fourier components propagate with the same phase velocity, $v = u_p$. The resulting voltages (voltages) at the two ends of the transmission line are shown in Figure 3.18.

Special Case #1 would apply for a *lossless* transmission line, but we could take attenuation into account at each end if the transmission line were *distortionless*. However, we can see that the pulse shape at each end would be distorted from a digital pulse if the transmission line were not *distortionless*.

PSpice Idealistic Digital Examples with Real Impedances

Figure 3.18 was created by copying the output from PSpice with the Vpulse generator function in which the pulse width was set to 1 ns. Let us consider *Special Case #2* in which the pulse width is set to 3 ns. The results are shown in Figure 3.19.

In principle, we could have produced this plot analytically with the help of a ladder diagram. However, as is shown in Figure 3.19, the waveform is surprisingly complex when the pulse width, Δt_g, is greater than *TD*. It takes patience to keep track of the various parts to be summed as a function of time, so PSpice becomes a valuable tool for such special cases.

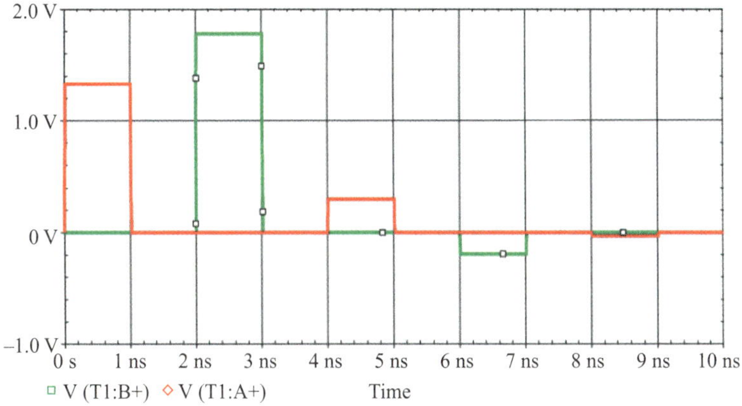

Figure 3.18 Voltage (voltage) at the ends of a transmission line for *Special Case #1* in which the generator creates a digital pulse of voltage across the transmission line of Figure 3.16 for a time $\Delta t_g = 1$ ns.

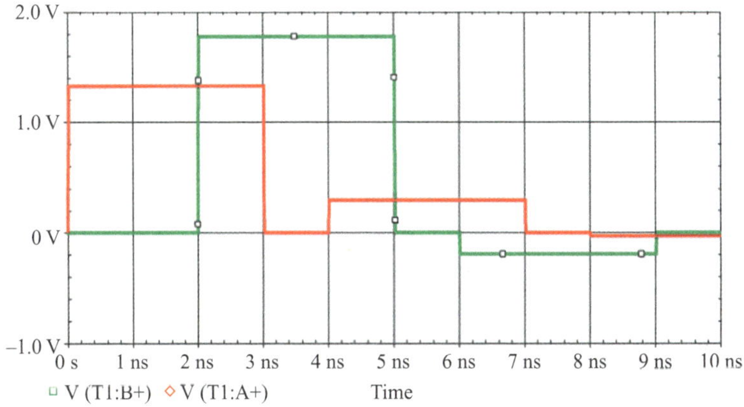

Figure 3.19 Voltage (voltage) at the ends of a transmission line for *Special Case #2* in which the generator creates a digital pulse of voltage across the transmission line of Figure 3.16 for a time $\Delta t_g = 3$ ns.

With PSpice, we can also consider more realistic waveforms such as a trapezoidal pulse with a rising and falling edge. In the output shown in Figure 3.20, the generator pulse has been modified to have a rising and falling edge of 0.2 ns.

Distortionless Lines

If the transmission line is lossy but *distortionless*, then the basic shape of the pulse will change only in amplitude, by an amount of $e^{-\alpha l}$ each time it propagates across

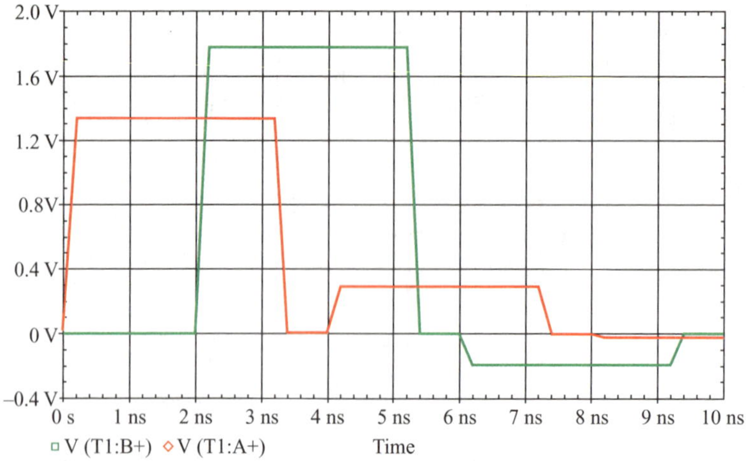

Figure 3.20 Voltage (voltage) at the ends of a transmission line for *Special Case #3* in which the generator creates a trapezoidal digital pulse of voltage with rising and falling edges of 0.3 ns across the transmission line of Figure 3.16 for a time $\Delta t_g = 2$ ns.

the transmission line. In this case, we can modify the ladder diagram in Figure 3.17 to account for each transit, as shown in Figure 3.21.

As in the reflection diagram in Figure 3.17, we emphasize the following points:

- Voltage represented by the first propagation is listed as a_g and is seen to be a potential division of the source voltage by the factor $Z_0/(Z_g + Z_0)$ but has also been attenuated by $e^{-\alpha l}$ after time *TD*.
- Voltage at the load end is zero until *TD* at which time it jumps to $a_g e^{-\alpha l} + \Gamma_L a_g e^{-\alpha l}$ as a consequence of the incident voltage and the first reflected voltage.
- Reflected voltage propagates back toward the generator whose voltage has remained at value a_g until time *2TD* at which time it jumps to $a_g + \Gamma_L a_g e^{-2\alpha l} + \Gamma_g \Gamma_L a_g e^{-2\alpha l}$ as a consequence of the incident amount of reflected voltage and the subsequent second reflected voltage.
- The process continues at each end of the transmission line as time increases to jump at intervals *2TD* as the reflections propagate one round trip, which also causes them to attenuate in amplitude by $e^{-2\alpha l}$.

The PSpice model does not have the capability to include attenuation except through the use of personal macros written by the user nor does it have the capability to include dispersion effects caused by the frequency dependence of the permittivity. It does have the capability to include current driven sources or voltage-regulated impedances such as those found in diodes and transistors. This makes the tool powerful for almost any initial analysis, and it virtually never loses track of components of the answer like a hand solution using a ladder diagram is likely to do. Several additional features of PSpice and some references to assistance are noted below.

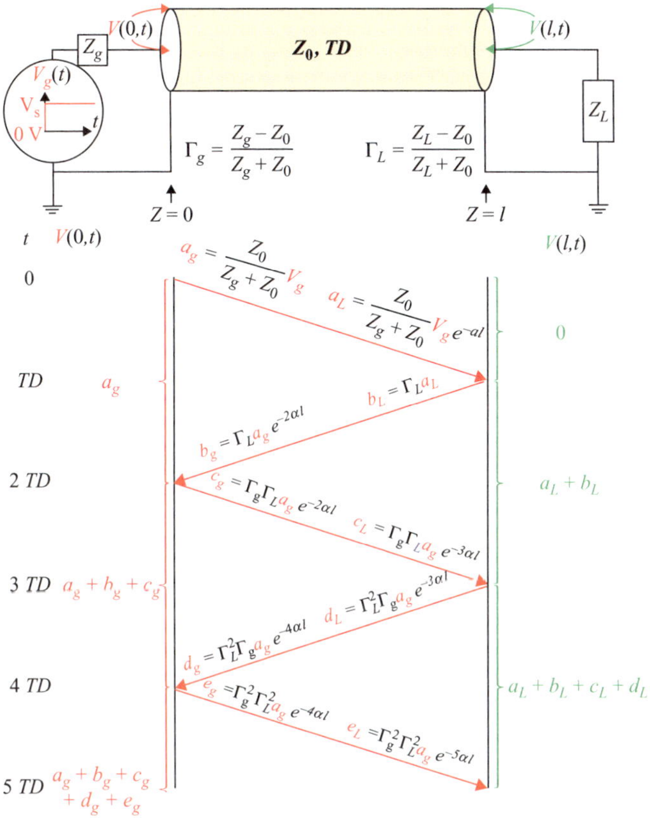

Figure 3.21 Special case of reflection from each end of a *distortionless* transmission line of impedance Z_0 and time delay *TD* with an amplitude attenuation of $e^{-\alpha l}$ for each transit.

Quantities You Can Find with PSpice

- Fourier analysis (on the trace window menu bar, traceàFourier).
- Mathematical interpretation of signals (on the trace window, traceà add traceà functions or Macros.
- Node voltages and branch currents over a time interval (Transient analysis), and over a range of frequencies (AC analysis).

PSpice References

1. http://www.eng.auburn.edu/~tdenney/ENGR1110/Tutor91_2.pdf
2. http://rock.uta.edu/dillon/pspice/
3. http://www.glue.umd.edu/~oramahi/PSPICE-TUTORIAL.pdf
   PSPICE 9.2 Version
4. http://www.ee.unlv.edu/kevin/Images/PSpice.pdf

This text was designed to focus on the fundamentals of signal integrity that are based on the solutions to Maxwell's equations, not on the use of specialty tools. For that reason, a decision has been made not to extend the tool analysis further or to include the use of other very powerful graphical tools such as the Smith Chart or the Bergeron diagram. Those tools and the use of more powerful commercial tools, such as HSpice provided by the Synopsys Corporation, Mountain View, CA, are left for the many descriptive texts on the subject to which this text can be a supplement. In essence, all of these tools are methods for examining the numerical solutions to Maxwell's equations. However, as an example of the power of a suite of such tools, examples of applications for a suite of tools by Ansoft (now owned by the Ansys Corporation, Pittsburg, PA) are included in the last chapter of this text.

Other comparable numerical tools are Advanced Design System (ADS) provided by the Agilent Corporation (Santa Clara, CA), Microwave Studio (MWS), provided by the CST Corporation of America (Framingham, MA), and Sonnet Suites, provided by the Sonnet Corporation (Syracuse, NY). The reader should be forewarned that all numerical tools provide approximate solutions to Maxwell's equations and, in some cases, do not provide the capability to include such effects as field penetration into conductors and frequency dependence of the propagating medium, or they do not easily include real-world effects like surface roughness. The good news is that these suites of tools often interoperate and are constantly being improved to grow in capability with time. Understanding the limitations of the numerical codes is important to the student so that results are not misinterpreted.

Chapter 4

Ideal Models vs. Real-World Systems

LEARNING OBJECTIVES

- Estimate the magnitude of time-dependent scalar voltage, surface charge density, surface current, electric field intensity, and magnetic field intensity in an ideal model for a printed circuit board (PCB) trace and ground plane
- Identify and justify low- or high-profile surface irregularities of a real-world PCB at the microscopic level
- Know what impurities or voids are likely to occur in real-world printed wiring boards PWBs and how to change their character in an electrodeposition process
- Understand resistance measures of purity compared with bulk Cu for voids, impurities, and magnetic content
- Identify and justify the dominant inclusions in the electromagnetic propagation region of a PCB
- Understand the basis for Cu grain boundaries as they relate to surface roughness and know how annealing effects them
- Identify real-world boundary conditions that exist because of the complexity of a typical PCB and understand why numerical solutions are needed for a real-world analysis of signal integrity

INTRODUCTION

Various different communities have been involved in the study of material properties for the propagation of electromagnetic waves in real-world systems. Most of these communities did not overlap with the others, and, thus, the theory of electromagnetic propagation has evolved in various stages (some of which overlapped in time).

Solid-state physicists, like Paul Drude, at the turn of the twentieth century were interested in the explanation of conductivity of metals and alloys as a function of temperature and the components that were involved in the impurity content of a material versus its thermal lattice vibrations (phonons). This community formulated the microscopic form of Ohm's law in terms of the scattering of conduction electrons

The Foundations of Signal Integrity, by Paul G. Huray

that were assumed to exist as a free electron gas. Solid-state physicists later applied the Fermi exclusion principle to electrons in a reciprocal (frequency) space, and this led Sommerfeld and others to explain the mechanisms of scattering of electrons from lattice imperfections and phonons in a quantum mechanical formalism. For example, we showed[1] that a free electron Fermi gas has a maximum velocity of $v_F = \hbar k_F/\mathrm{m} = (\hbar/\mathrm{m})(3\pi^2 N/V)^{1/3}$ for the conduction electrons. Furthermore, we listed[2] the values of v_F for monovalent metals and found that, for copper, $v_F = 1.56 \times 10^6$ m/s. It was natural to include those constitutive material properties in *Maxwell's Equations* by considering the induced current density of metals in the presence of time-harmonic electric field intensity to be a term related to the electric permittivity of the medium. Both of these treatments were approximations that checked relatively well with each other and with measurement because we could rely on time averages of large ensembles of atomic structures that ignored details. However, both theories were constructed by making an idealistic assumption that atomic cores could be specified in a more or less periodic structure and that we could consider large volumes of atoms to occur in a single crystal structure. As we shall see, these are gross approximations that do not exist in a microscopic analysis of real-world systems.

In the 1920s and 1930s, induced dipoles from otherwise neutral atoms or molecules were studied by infrared or ultraviolet frequency stimulation that displaced the center of mass of atomic electrons from their positive nuclei. The degree to which this displacement happens was considered by Linus Pauling, William Shockley, and others. This was called the electronic polarizability of the atoms and expressed the degree to which the separation would occur for different atoms. The amount of separation of the electrons from their binding nucleus was proportional to the applied electric field intensity, but there was a restoring force brought about by the nuclear electron separation that was proportional to the amount of separation (for small separations). Because the resulting vibrations were of higher frequency, radiation was expected to lead to a loss mechanism as were interactions with nearest neighbor atoms. Thus, a drag term was included in the mechanical analog models.

Later, optical and electron-scattering Raman studies led to a formalism of solids that were considered to be an electrically neutral gas of electrons surrounding a lattice of positively charged ion cores, that is, a plasma. The excitation of these two charged (intermeshed) materials at very high frequencies (often at optical frequencies) showed modes of oscillation of the two media by the application of a periodic electric field intensity. The resulting permittivity of the plasma medium exhibited macroscopic properties that explained the transmission of light through the materials at frequencies above a threshold plasma frequency and the reflection of that light below the plasma frequency.

In the 1950s and 1960s, developers of microwave ovens sought to use the properties of permanent dipoles for heating by rotating the polar molecules (particularly water molecules) at gigahertz frequencies to cause the breaking of bonds between adjacent molecules. In the 1980s and 1990s, the engineering community was interested in making capacitors that would hold more charge for a given scalar voltage than a set of empty conducting plates, so they formulated molecular materials with permanent electric dipole moments that would yield greater capacitance by increasing the permittivity of the medium. Recent advances in engineered nanoscale materials of carbon and titanium dioxide have permitted the construction of supercapacitors that hold very large quantities of charge per unit volume. In 2008, a solid-state version of the memristor that depends on nonlinear losses in titanium dioxide thin film-layered structures was reported.[3] There is some debate about whether or not this device represents a fourth kind of fundamental device (along with resistors, inductors, and capacitors), but, in any case, our simple theory of resistance and capacitance involving reversible polarization will not suffice to describe the phenomenon.[4]

For the use in real-world transmission lines used in the PWB industry, we will need to consider all of these types of interactions from DC to optical frequencies that will effect electric conductivity, permittivity, and permeability. Some of those studies are considered in this chapter, and their consequences for electromagnetic propagation, as well as the means for including the effects into numerical models that will be presented in simulations Chapter 8, are developed.

4.1 IDEAL TRANSMISSION LINES

In Chapter 3, we found that finite transmission lines operating at high frequency yield complicated voltage and current waveforms because of the reflection of incident waves at a load or a source. These waveforms interfere with one another in confusing ways so that it is often advantageous to use numerical codes such as open source PSpice or commercial HSpice provided by the Synopsys Corporation (Mountain View, CA) for their analysis. A more complete numerical analysis using Ansoft tools provided by the Ansys Corporation (Pittsburgh, PA) for large structures is shown in Chapter 8. After conducting the analysis by hand for some simple cases, we found that it is easy to make a mistake or to forget a term so that a preanalysis by a computation code assists the analysis of outcomes for practical transmission lines. However, the material of Chapter 3 was mainly limited to the special case of *lossless* transmission lines, and we know that real transmission lines have losses provided from mechanisms that depend on frequency and that the phase velocity of the various Fourier components of a pulse is a function of frequency that leads to signal dispersion in real-world systems.

In Chapter 5, we will show that various authors have studied propagation of electromagnetic waves in materials that had conductivity, molecules with permanent dipole moments and atoms that could be induced to produce a dipole moment, and acted like a plasma of electrons and ions at high frequency. These materials often have a loss mechanism that leads to the decay of electromagnetic power and energy in a source wave front or harmonic oscillation. Thus, real transmission lines used in printed wiring board (PWB) materials must include terms that account for delay, attenuation, and dispersion. We will show how these effects change pulses in the time domain in Chapters 5 and 7.

We also found,[5] at high frequencies, that external electromagnetic fields in the neighborhood of conductors predominantly penetrate the surface layer of conductors

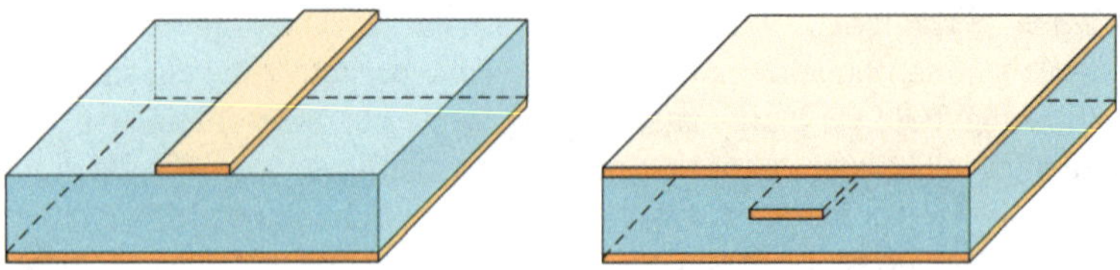

(a) (b)

Figure 4.1 (a) Finite microstrip geometry; (b) finite stripline geometry.

and that surface charges are induced on the surface layer to match boundary conditions required by Gauss's law at a conducting interface.

Finally, in Chapter 3 we treated practical transmission lines like strip lines and microstrips in an ideal manner as if they could be represented by a coaxial cable or symmetric waveguide. We know that this will not be the case because electric field lines will readjust to meet real-world boundary conditions (BC) and that "return paths" for ground planes will distribute charges and currents over large surfaces. In the case of transmission lines with traces and ground planes made of very good (but not perfect) conductors, the electromagnetic waves will not be pure TEM waves.

Two practical lines useful in printed wiring boards, the microstrip (a) and the stripline (b), are shown in Figure 4.1.

4.2 IDEAL MODEL TRANSMISSION LINE INPUT AND OUTPUT

In the finite microstrip transmission line, an input scalar potential function is often imposed externally on one end (called Port 1 in a two-port junction model), and a subsequent output scalar potential function is measured at the other end (called Port 2 in a two-port junction model), as shown in Figure 4.2.

We call the propagation direction the z-direction so that the input voltage is the externally imposed voltage at $z = 0$ and the output voltage is the measured voltage at $z = l$, which is consistent with out notation in Chapter 3 (where we colored the voltage at $z = l$ green). The details of how the input voltage is provided are important because they give us the stimulating voltage distribution and resulting electric field distribution as a function of time. For example, if the end of the trace of Figure 4.2 is considered to be a perfect electric conductor (PEC) and the voltage is measured relative to another PEC of identical shape a distance h below it on the ground plane, the initial electric field intensity can be approximated as a uniform sheet pointing in the x-direction at the $z = 0$ plane. Such a stimulus might be provided by a soldered tab on each of the conductors. We show this "lumped port" distribution in Figure 4.3a.

By comparison, we show in Figure 4.3b that the electric field intensity lines at the $z = l$ plane after the charge distribution have spread out to the sides of the trace and over a great horizontal distance on the ground plane. These "wave port" distri-

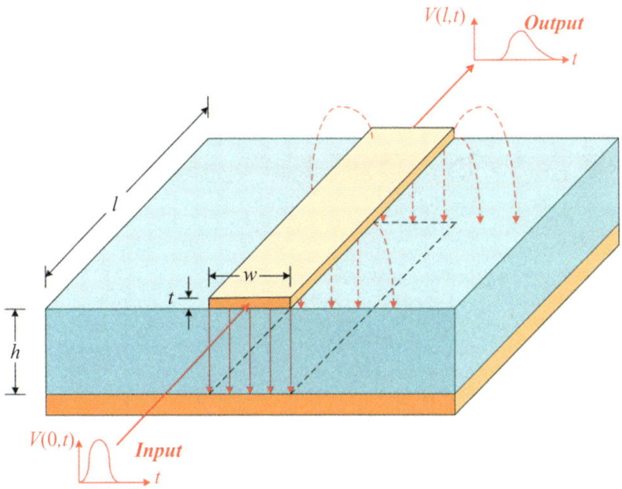

Figure 4.2 Hypothetical output voltage at $z = l$ (Port 2) as a function of time t and input voltage at $z = 0$ (Port 1) as a function of time t.

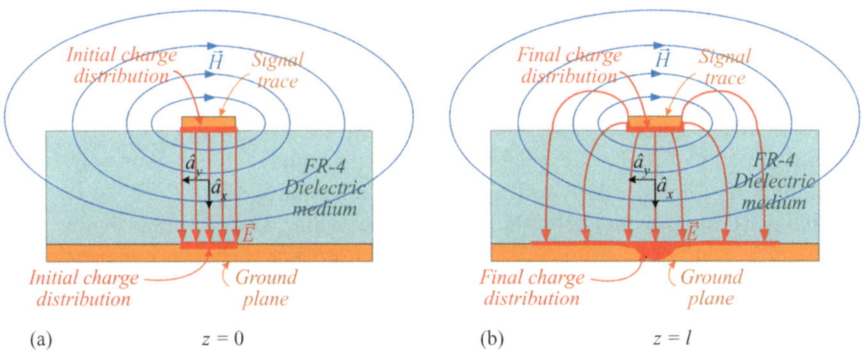

Figure 4.3 Electric field intensity lines and magnetic field intensity lines on (a) the $z = 0$ plane and (b) on the $z = l$ plane.

butions will be found by a solution of Maxwell's equations for the real-world boundary conditions shown in which the electric field intensity lines and potentials at infinity (in the x–y plane) go to zero.

The solution to this boundary value problem is too difficult to perform analytically so we will have to resort to a numerical solution to Maxwell's equations for these boundary conditions. Three types of numerical codes are typically employed to solve these problems: the finite difference time domain (FDTD), method of moments (MOM), and finite element method (FEM). These three types of solutions are described in Chapter 8 on numerical simulations. However, we will note here

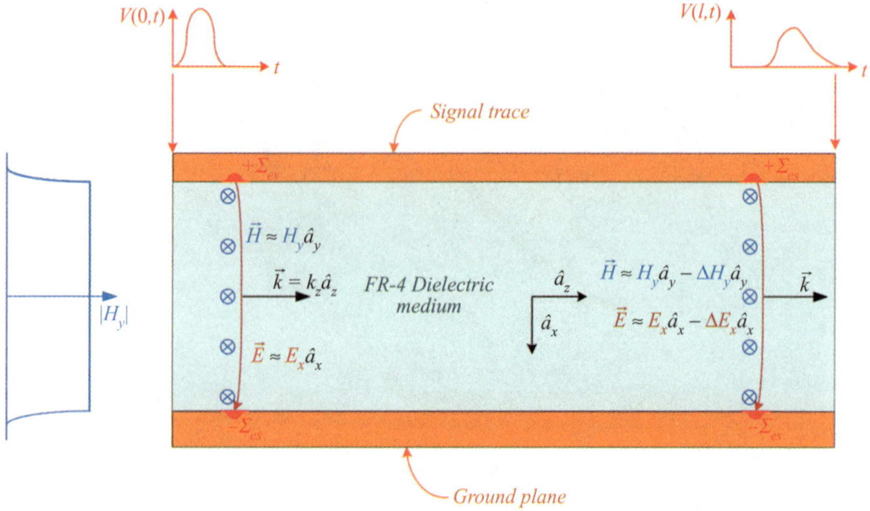

Figure 4.4 Electric field intensity (in the x-direction) and magnetic field intensity lines (in the y-direction) for an electromagnetic wavefront in FR-4 as it propagates from $z = 0$ to $z = l$.

that a solution to a geometric problem that includes points at $x = \pm\infty$ and $y = \pm\infty$ will require an infinite number of mesh points at which we will solve for potential and fields, and this makes the boundary value problem impossible to solve numerically. We must therefore make an approximation that the fields go to zero on the surface of a box of finite extent but large compared with the dimensions of the trace width and FR-4 thickness. As a rule of thumb, a so-called air box will be taken to be at least 10 times the width w shown in all dimensions. Only a series of numerical calculations can tell us if this is large enough to produce a change in electric field intensity values to a precision of our choice (e.g. $\leq 10^{-6}$ change for each doubling of spatial extent).

To examine the propagation of an electromagnetic wave as it moves from Port 1 toward Port 2 in these numerical models, we will view the cross section of the trace and ground plane, as shown in Figure 4.4.

In Figure 4.4, we show that the magnitude of the magnetic field intensity, H_y, is approximately constant for different values of x but that it penetrates exponentially into the trace and ground conductors; the magnetic field intensity from a harmonic signal actually penetrates into a flat conductor as an exponentially damped cosine function.[6] In Figure 4.4, the electric field intensity is shown as being approximately in the x-direction (a TE^z mode), but, because of the surface impedance, it has a small component in the z-direction to make it non-TE^z. We further note that the electric and magnetic field intensities at the $z = 0$ plane are decreasing as they move in the z-direction (having lost an amount ΔE_x and ΔH_y by the time they arrive at Port 2. We will describe the mechanisms for these losses in the subsequent sections. We also note in Figure 4.4 that there is a surface charge density $+\Sigma_{es}$ on the trace and

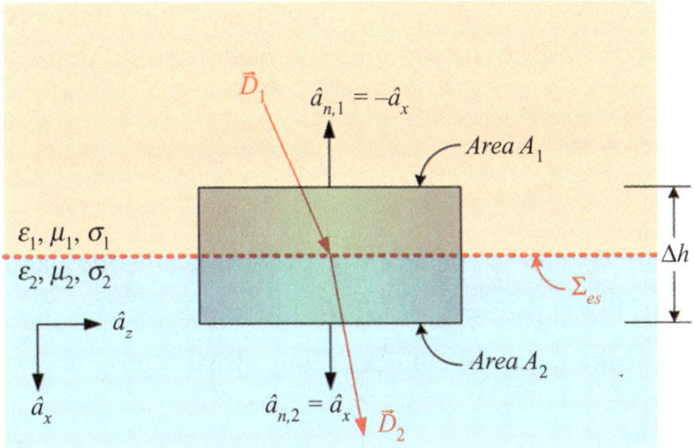

Figure 4.5 Gaussian surface (pill box) of height Δh and cross-sectional area A.

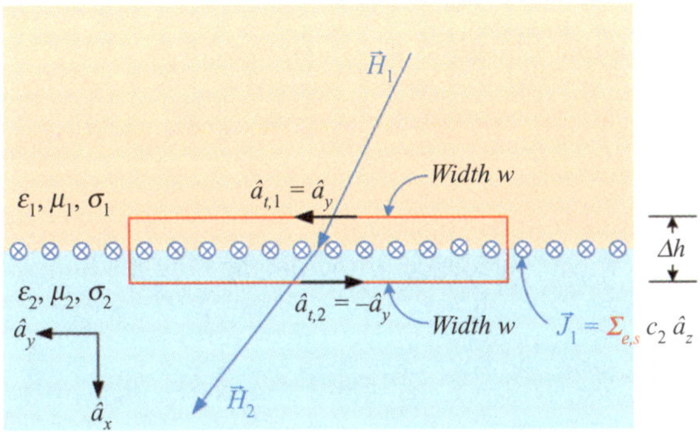

Figure 4.6 Closed path (w Δh) for the evaluation of the linear current density, $\vec{J}_l = \Sigma_{e,s} c_2 \hat{a}_z$.

$-\Sigma_{es}$ on the ground plane, as required by Gauss's law to make the discontinuous value of E_x match on the surface of the conductor, as shown in Figure 4.5:

We can use the divergence theorem $\oint_C \vec{D} \cdot d\vec{s} = \int_V \vec{\nabla} \cdot \vec{D} d^3 v$ with Maxwell's equation $\vec{\nabla} \cdot \vec{D} = \rho_v$ to obtain Gauss's law that $\oint_C \vec{D} \cdot d\vec{s} = Q_{enclosed}$ so that $\int_{area\ A_1} \vec{D}_1 \cdot \hat{a}_{n,1} \rho d\varphi d\rho + \int_{cylinder} \vec{D} \cdot \hat{a}_\rho \rho d\varphi dz + \int_{area\ A_2} \vec{D}_2 \cdot \hat{a}_{n,2} \rho d\varphi d\rho = (D_{2,n} - D_{1,n}) A = \Sigma_{es} A$ in the limit as $\Delta h \to 0$. For a perfect electric conductor with $\sigma_l = \infty$, we thus conclude that $D_{2,n} = \Sigma_{e,S}$ or

$$\Sigma_{e,S} = \varepsilon_2 E_{2,x} \tag{4.1}$$

We can further use Stokes's theorem $\int_S \left(\vec{\nabla} \times \vec{H} \right) \cdot d\vec{s} = \oint_C \vec{H} \cdot d\vec{l}$ with Ampere's law $\vec{\nabla} \times \vec{H} = \vec{J}$ for the surface current, as shown in Figure 4.6.

Here, we have used the phase velocity of the propagating electric field intensity wave, $u_p\hat{a}_z$, in the limit as $\Delta h \to 0$ for a PEC to compute the linear current density (A/m):

$$\vec{J}_l = \Sigma_{e,s} u_p \hat{a}_z \tag{4.2}$$

We note that this linear current density is consistent with Stokes's theorem and Ampere's law above because $\int_S (\vec{\nabla} \times \vec{H}) \cdot d\vec{s} = \oint_C \vec{H} \cdot d\vec{l}$ is the same as $\vec{J}_l \cdot w\hat{a}_z = (H_{2,t} - H_{1,t})w$ because the height of the path, Δh, can be infinitesimal with a purely surface current.

For a perfect electric conductor with $\sigma_l = \infty$, we thus conclude that $H_{l,t} = 0$ so that

$$\vec{H}_2 = \Sigma_{e,s} u_p \hat{a}_y \tag{4.3}$$

Using $u_p = c_2 = 1/\sqrt{\mu_2 \varepsilon_2}$, we can see that Equations 4.1 and 4.3 yield

$$H_{2,y} = \varepsilon_2 E_{2,x} / \sqrt{\mu_2 \varepsilon_2} = E_{2,x} / \eta_2, \tag{4.4}$$

which is consistent with the magnitude of the magnetic field intensity being related to the electric field intensity by the wave impedance of the propagating medium.

Conclusions

- The induced surface charge density needed to satisfy Gauss's law at a perfectly flat, perfect conductin metal boundary is $\Sigma_{e,s} = \varepsilon_2 E_{2,x}$.
- The surface charge density moves at the phase velocity of the propagating wave to form a linear current density $\vec{J}_l = \Sigma_{e,s} u_p \hat{a}_z$.
- The phase velocity of the propagating surface current is much larger than the velocity of bulk metal conduction electrons at the Fermi level so the surface charge density is **not** being transported by conduction electrons but by transverse induction of the conduction electrons by the propagating electromagnetic waves.

EXAMPLES

4.1 Suppose a time-dependent source is imposed on the microstrip shown in Figure 4.2 with $h = 0.006''$, $w = 0.010''$, and $l = 1.00''$. If the applied voltage reaches a maximum at $1\ V$, (a) What is the magnitude of a TE_{10}^z electric field intensity that propagates down the trace waveguide? (b) What local electric charge density is induced on the top and bottom conductors? (c) What magnetic field intensity accompanies the propagating electric field intensity?

SOLUTION In order to support a discontinuity in the electric field intensity at the trace conductor/propagating medium interface in Figure 4.2, a surface charge density, $\Sigma_{e,s} = \varepsilon_2 E_{2,x}$ is required to exist by Equation 4.1. Similarly, at the ground plane/propagating medium interface, a surface charge density, $-\Sigma_{e,s} = -\varepsilon_2 E_{2,x}$ is required.

a. For a pure TE^z_{10} wave at 1 V, $E_x \approx 1\,V/0.006\,in = 6.7 \times 10^3\,V/m$ so

b. the surface charge density is

$$\Sigma_{e,s} = \varepsilon_r \varepsilon_0 E_x = (4)(8.85\times10^{-12}\,C^2/Nm^2)(6.7\times10^3\,V/m) = 0.237\times10^{-6}\,C/m^2$$

c. $\vec{J}_{l,\,trace} = \Sigma_{e,s} u_p \hat{a}_z = (0.237\times10^{-6}\,C/m^2)(3\times10^8\,m/s/\sqrt{4})\hat{a}_z = 35.5\,A/m\,\hat{a}_z,$

$\vec{I}_{trace} = (35.5\,A/m)(0.010\,in)\hat{a}_z = 8.9\,mA\hat{a}_z$

and

$\hat{a}_x \times (\vec{H}_2 - \vec{H}_1) = \vec{J}_{trace} - \vec{J}_{ground\ plane} = 71\,A/m\hat{a}_z$

so if $\vec{H}_1 = 0$,

$\vec{H}_2 = 71\,A/m\hat{a}_y$

NOTE The electric field intensity that results from the application of a 1 V signal on a PCB of 6.7 kV/m is about 1% of the breakdown (local lightening) electric field intensity of humid air, so we should not be surprised that PCBs sometimes fail upon exposure to humidity. With such a large electric field intensity, the response of the material in the propagating medium plays a large role in the propagating field loss. Correspondingly, $H = 71$ A/m (or $B = 0.89$ G) is about the magnitude of the earth's magnetic flux density.

4.2 (a) Evaluate the total current that propagates down the transmission line and (b) compare this current to the conduction current caused by the applied potential for a 50 Ω transmission line from a circuit model analysis.

SOLUTION From Example 4.1,

a. $\vec{I}_{total} = \vec{I}_{trace} - \vec{I}_{ground\ plane} = 17.8\text{ mA}\hat{a}_z$ and

b. $I_c = \dfrac{V}{Z_0} = \dfrac{1V}{50\Omega} = 20\,mA$

NOTE This is a modest average current, but, if we have a very fast pulse that has a small duty cycle, instantaneous currents can be thousands of amps at an instant in time. Is it a surprise that current can cause electromigration[7] or that thermal spikes cause heat damage over time?

4.3 Assuming a perfectly flat copper trace and ground plane surface, (a) find the number of atoms per unit area on the surface, (b) find the distance, x_0, that the electrons in the first atomic layer would need to be displaced to produce

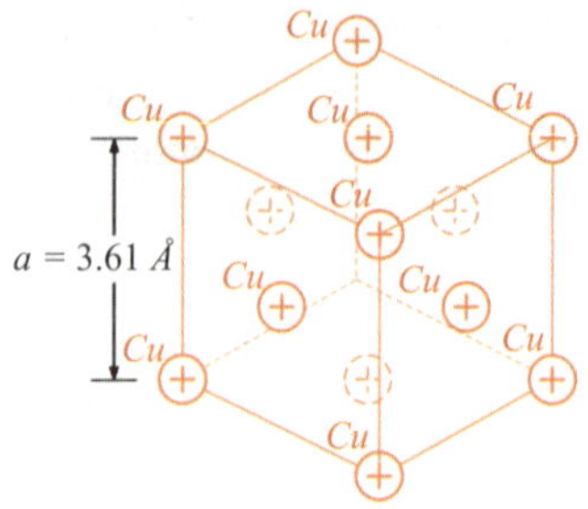

Figure 4.7 Copper face-centered cubic crystal structure.

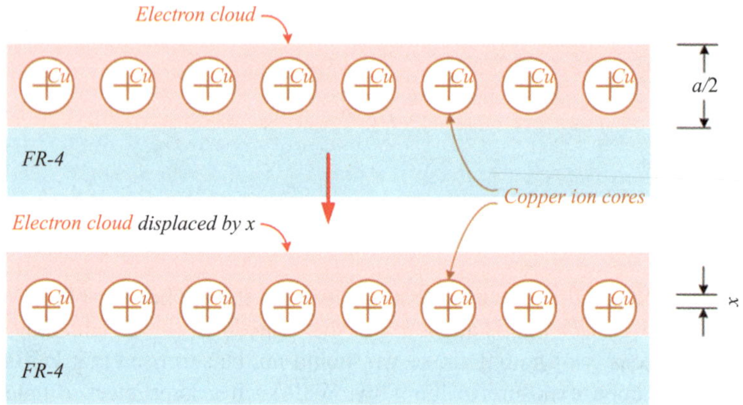

Figure 4.8 (Top) Immersion of positive copper ion cores in a cloud of negative electrons to produce an electrically neutral atomic layer; (Bottom) displacement of the electron cloud by an amount ***x*** to produce an effective positive charge density at the trace/propagating medium interface.

the charge density of Example 4.2, and (c) put this in the context of known distances.

SOLUTION Copper atoms are arranged in a face-centered cubic lattice with lattice parameter $a = 3.61$ Å, as shown in Figure 4.7.

There are $N_V = N/V = 4\,\text{atoms}/(3.61 \times 10^{-10}\ \text{m})^3 = 8.45 \times 10^{28}\,\text{atoms/m}^3$ and $N_A = N/A = 2\,\text{atoms}/(3.61 \times 10^{-10}\,\text{m})^2 = 0.154 \times 10^{20}\,\text{atoms/m}^2$ in an atomic layer.

If we assume that there is one "free" electron per copper atom, then we can view the atomic layer adjacent to the propagating medium, as shown in Figure 4.8.

a. By displacing the uniform electron cloud by an amount x, a charge density $\Sigma_{e,s} = (1.6 \times 10^{-19}\,C/\text{e}^-)(0.154 \times 10^{20}\,\text{e}^-/\text{m}^2)[x/(a/2)]$ is produced on the bottom surface.

b. In order to produce $\Sigma_{e,s} = 0.237 \times 10^{-6}\ C/\text{m}^2$, we thus need $x/a = 4.81 \times 10^{-8}$.

c. A displacement of $x = (3.61 \times 10^{-10}\ \text{m})(4.81 \times 10^{-8}) = 17.4 \times 10^{-18}$ m is a fraction of a nuclear dimension!

PROBLEM

4.1 In Chapter 3, electric field intensity below a transmission line trace was treated as the *source* of a surface charge density (causal duality). The necessary surface charge density, $\Sigma_{e,s}$, and the magnitude of displacement, x, of the electron cloud were calculated above. Use the restoring force (hint, see Chapter 5) for transverse plasma electron charge density displacement in copper to find the work required per unit area to displace the charge by x for one cycle of a harmonic wave. What fraction of the wave energy per unit length of propagation would be lost to the work of induction of this surface charge density?

NOTE The ideal models for a transmission line above assume that the trace and ground plane are perfectly flat and parallel to one another and that the propagating medium for electromagnetic waves is homogeneous.

4.3 REAL-WORLD TRANSMISSION LINES

The manufacturing process does not produce perfect rectangular traces, as shown in the above ideal models, that are flat to microscopic precision and that are perfectly uniform in cross section. For this fact to be demonstrated, a scanning electron microscope (SEM) photograph[8] of a typical copper trace used in a PCB transmission line is shown in Figure 4.9 for four magnifications from (a) to (d).

Several features of the real-world transmission line are worthy of comparison to the ideal transmission line model:

- Cross sections of copper traces are not perfect rectangles, as shown in the ideal model drawing of section 4.2 but are more accurately described as trapezoids with a rectangle on top with rounded edges and a very rough surface on the bottom. This cross section occurs because of the mask and etch process of making traces on a surface and an intentional roughening procedure on the bottom to improve adhesion between the trace and the dielectric propagating material. This process will be discussed in a subsequent section.
- The copper ground plane below the trace is not perfectly flat nor is it exactly parallel to the trace. The copper surface facing the trace is smoother than that on the bottom and is also smoother than the bottom of the trace. We will try to account for the scattering of electromagnetic waves between these two real-world conductors with a more accurate model in chapter 7 on advanced signal integrity.

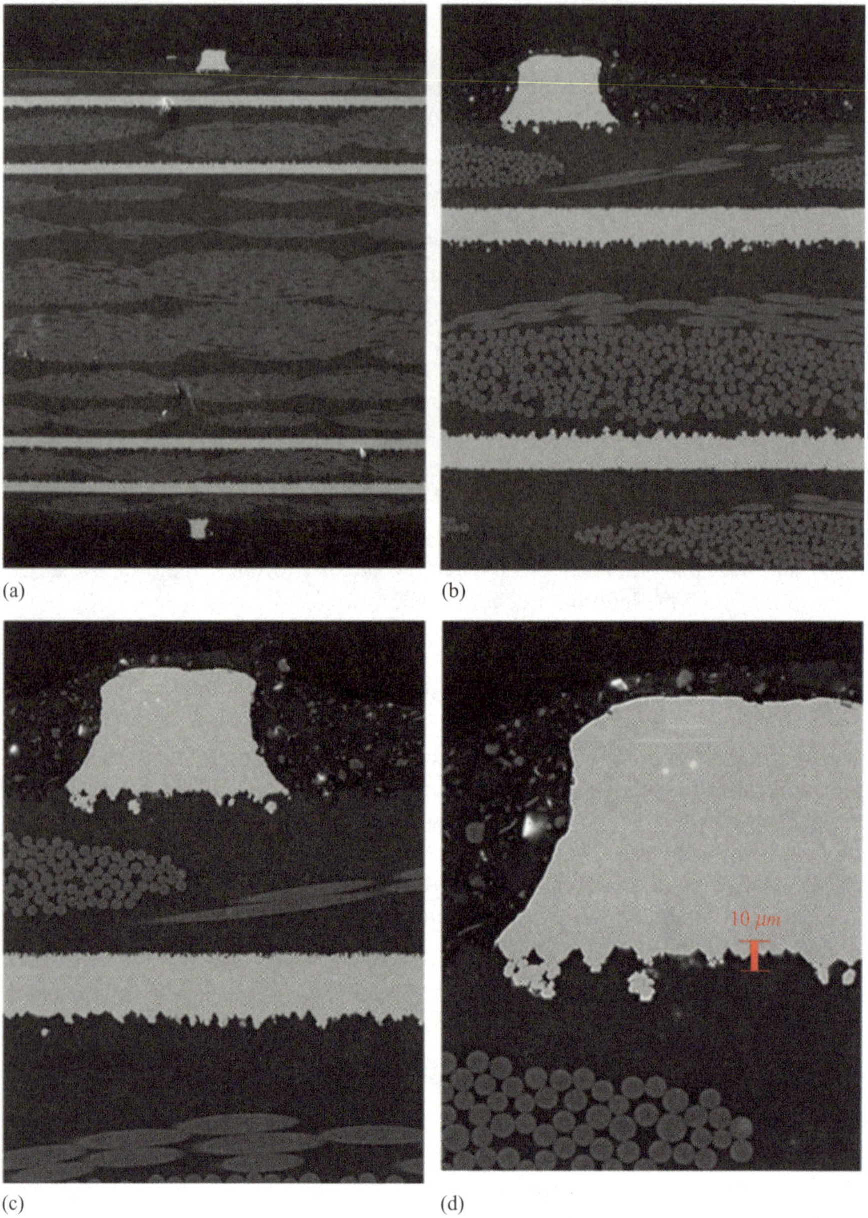

Figure 4.9 Increasing magnifications of a scanning electron microscope photograph of a cross section through a six-layer printed circuit board with a copper microstrip trace on the top and bottom (shown in white) and four copper ground or power planes in the middle. A red bar is shown in part (d) to give a sense of the scale of the photographs. Images were captured by using a backscattered electron detector.

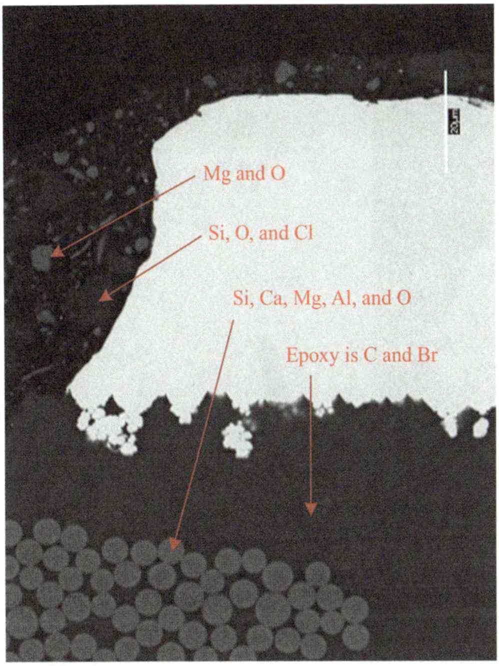

Figure 4.10 Cross section of part of a copper trace (covered by solder mask varnish) over an FR-4 propagating medium showing the Auger measurements of the elemental composition of various parts. Hydrogen is not observed in Auger measurements, and stoichiometry is not obtained.

- As shown in Figure 4.10, bundles of fiberglass (SiO_2, CaO, MgO, and Al_2O_3) are imbedded in the FR-4 material (C, O, H, and Br) between the trace and the ground plane at a random location (as seen on other similar cross sections). Some of the fibers in the glass weave are seen to be approximately parallel to the trace in the propagation direction, and some are approximately perpendicular, as shown in the elliptical cross sections of the lighter materials.

  NOTE Hydrogen is not observed in the Auger process, but we know that it is present from the original composition.

- The trace is not exposed to air, as indicated in the ideal trace of section 4.2, but is covered by a "solder mask" of varnish that has a permittivity close to that of FR-4. This is observed by eye on real motherboards as a green coating on top and bottom of a PCB. The solder mask is not homogeneous but is filled with small particulates that we have identified as bits of material composed of Mg, Si, O, and Cl.

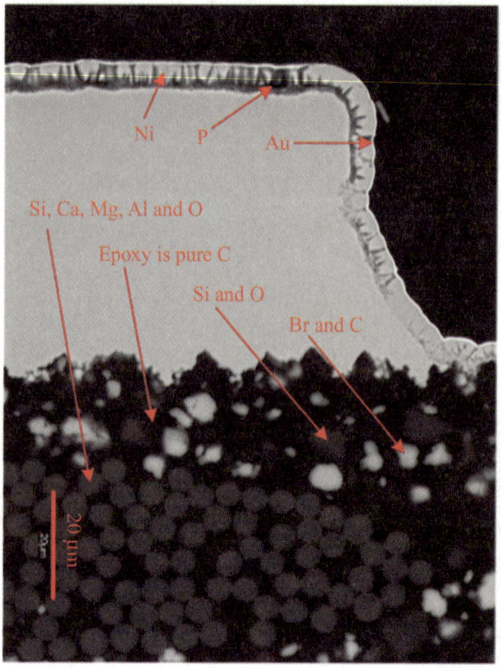

Figure 4.11 Scanning electron microscope photograph of a copper trace above a copper ground plane (off scale below) with Isola 620 http://www.isola-group.com/en/products/name/detail.shtml?28 (Chandler, AZ) in the electromagnetic propagation region. This trace is covered by a NiP coating for protection rather than the solder mask varnish shown in Figures 4.9 and 4.10. A thin coat of pure Au is applied to the NiP surface.

In many cases, PCBs are constructed with a low loss material in the electromagnetic propagation region. Rogers Corporation (Rogers, CT), Nissho Electronics Corporation (Santa Clara, CA), and Isola Group (Chandler, AZ) are three corporations that manufacture materials with a lower tan δ than FR-4). A SEM photograph of a different PCB constructed with Isola 620 is shown in Figure 4.11.

We see in Figure 4.11 that small particles of Si and O and of Br and C have been interspersed in the C-based epoxy to reduce the value of tan δ. This figure is shown again in a wide-angle photograph in Figure 4.12 to emphasize the impurities in the top intervening layers, as contrasted with the more homogeneous lower core layer of FR-4.

Note that the Au layer in Figure 4.12 extends horizontally well beyond the copper trace for this coating probably because of the lower etch characteristics of Au than Cu. This much wider conductor will contrast even more than the solder mask-covered copper trace in its electromagnetic properties compared with the ideal model used in section 4.2.

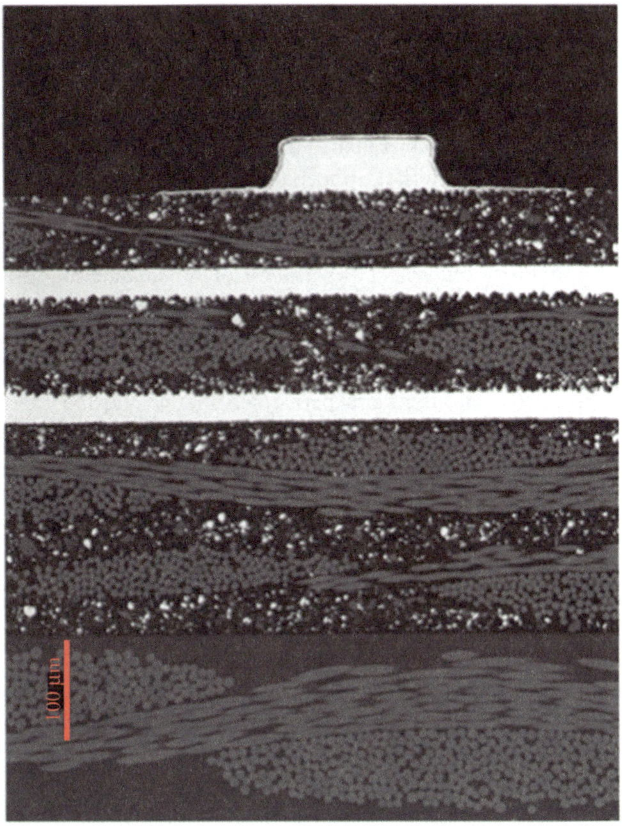

Figure 4.12 Scanning electron microscope photograph of a NiP-covered copper trace with a thin film of gold.

4.4 EFFECTS OF SURFACE ROUGHNESS

Copper foil manufacturers make copper surfaces rough so that they will adhere to the underlying dielectric epoxy. Formed under pressure and heat, the stack-up of dielectric and copper layers, with connecting vias, should be able to withstand cycles of thermal contractions, electric and magnetic forces, and the effects of adsorbed humidity and subsequent desiccation. Furthermore, the stack-up must be able to withstand a solder process that produces large thermal stress. Even if there is local heating from high currents in electronic components, the nearby material must maintain its integrity for the useful lifetime of the product. These threats to physical integrity are addressed by many additives to a copper surface that are applied shortly after the electrodeposition of copper foil onto a smooth titanium drum.

Each manufacturer has its own proprietary process for creating a rough surface, but most use the electrodeposition of pure copper onto a relatively smooth surface from an acid bath with controlled pH, temperature, and deposition current. The thickness is determined by the deposition current and the rate of rotation of the drum. Each manufacturer strives to make the highest-purity and highest-quality copper surfaces that will withstand long-term chemical interactions with air, water, and the corrosive effects of halogens in the epoxy. Finally, the PCB manufacturer must achieve the prescribed values of characteristic impedance within some limits of variation as required by the circuit designer.

Not only must the absolute characteristic impedance of transmission lines meet a certain margin of variability but also the variation of impedance along a continuous trace as long as 20–30 in. must be relatively constant. Measures of the product variations are often said to lie within the "corners" of variation (the worst case deviation from a specification) on a multiple axis chart of design performance. Measures of the product's variations are performed first inside the PCB manufacturer and later at the facilities of the circuit designers as they assess the prototypes, and then the production runs for consistency of performance relative to a given target.

At the integrator corporation, there are often subsequent "shake and bake" tests of the PCBs populated with the circuit components to ensure that they will meet acceleration and thermal spikes to some level often well beyond the specifications; these variations are typically company secrets for they determine the number of products per million that will statistically be permitted to fail. Also tested are performance measures under variations in power supply voltages and currents, variations in component input and output impedance, and electromagnetic radiation from nearby sources that act as cause of cross talk between transmission lines. Some of these must meet Federal Communications Commission (FCC) regulations, and some determine the perceived quality of the final product. The inability of any supplier or buyer of these products to meet design criteria may have serious or fatal financial consequences. Fortunately, the suppliers and buyers of PCBs often work closely together to design tests that will assure performance success before the product goes into mass production. This work falls within the realm of the Signal Integrity Engineer.

Unfortunately, some characteristic design features may be optimized to meet one criterion at the expense of another. This has been the case for surface roughness, where adhesion has been selected as the most important design criterion at the expense of transmission line signal losses. At low frequencies, the secondary effect can be overcome by good circuit design and large timing margins, but, as the frequency of PCBs increase to meet customer demand, the secondary effect can prove to be a fatal flaw. Such a circumstance exists in the case of surface roughness so it is given special emphasis. A typical electrodeposition process for producing copper foil is shown in Figure 4.13.

Many conferences and books have been written about the electrodeposition process.[9] Studies of the subsequent product show that impurities are removed from the copper wire stock in the purification process but are not completely eliminated. The deposition process causes a reasonably well-controlled material variation of the density and chemical composition of the surface atoms, as described below.

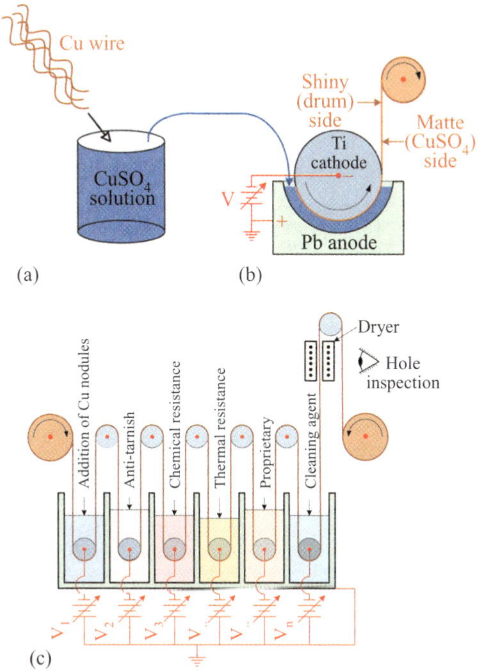

Figure 4.13 Schematic process for the production of copper foil used in printed circuit boards: (a) 99% pure copper wire is dissolved in a H_2SO_4 bath and then purified to better than 99.99%; (b) proprietary additives are made to the solution to adjust the pH, and then the solution is put into a temperature and potential gradient-controlled electrodeposition chamber where copper ions are deposited onto a rotating titanium drum; (c) the copper foil receives a series of different surface treatments to attach "anchor nodules" that make the surface rough for good adhesion, protect it from oxidation, separate it from the bromine in the epoxy, and give it thermal protection.

Electrodeposition of Cu from a $CuSO_4$ (0.1 M) + H_2SO_4 (0.5 M) Solution

If an external current greater than the limiting current, j_L, is forced through an electrode in an electrodeposition cell, the potential of the electrode will change until some other process (other than the deposition of Cu^{2+}) can occur. The limiting current density is of great practical importance in metal deposition because the type and quality of metal deposits depend on the relative values of the deposition current and the limiting current. An example[10] is shown in Figure 4.14.

Growth Mechanisms

The exact growth process on an atomic scale is probably chaotic and statistical in nature, but solid-state physicists have shown two possible growth scenarios, as implied in Figure 4.15.

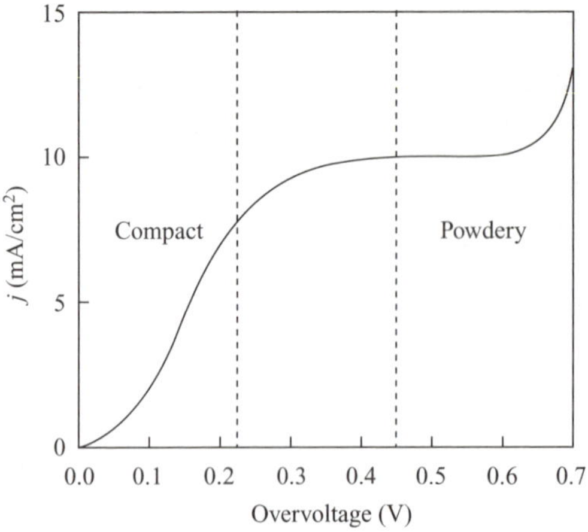

Figure 4.14 Overpotential characteristic of transition from a compact to a powdery deposit.

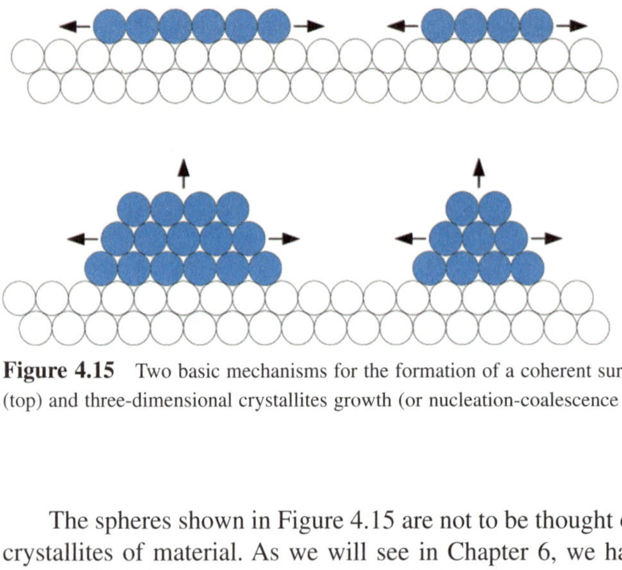

Figure 4.15 Two basic mechanisms for the formation of a coherent surface deposit: layer growth (top) and three-dimensional crystallites growth (or nucleation-coalescence growth) (bottom).

The spheres shown in Figure 4.15 are not to be thought of as atoms but as small crystallites of material. As we will see in Chapter 6, we have reason to call them "snowballs" that deposit out of the solution. In a sense, electrodeposition is the formation of a layer of copper "snowballs" on a relatively flat surface.

Electroless Cu Deposition

The deposition rate depends on several variables such as deposition current shown in Figure 4.14 and the pH of the chemical bath. For the effect of temperature to be seen by itself, Figure 4.16 shows a deposition process with no deposition current (electroless deposit).

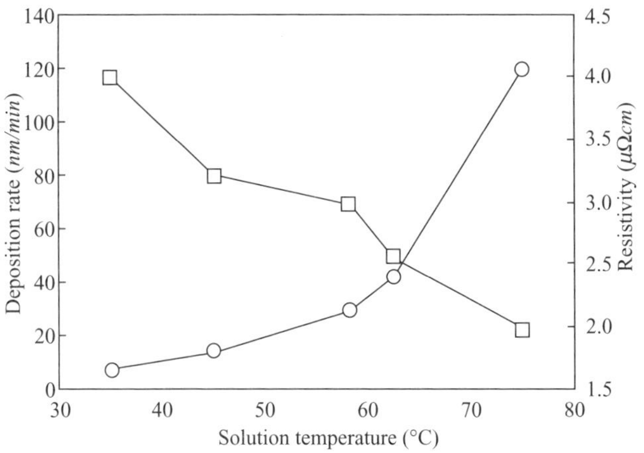

Figure 4.16 Depositon rate (○) and electrical resistivity (□) of a Cu layer of thickness 0.5 μm, deposited on titanium nitride.

From Figurc 4.16, we see that the electrical resistivity of a copper surface layer decreases from 4.0 to 2.0 μΩcm, with an increase of the deposition solution temperature from 35° to 75°C. The resistivity subsequently decreases down to 1.8 to 1.9 μΩcm after annealing at 200°C for 2 hours. Because the electrical resistivity of bulk copper is about 1.7 μΩcm, *we* conclude that electrodeposition from a low-temperature bath likely results in an amorphous surface layer with a large number of voids.

Cu Deposition Dependence on pH

In Figure 4.17, an electroless Cu deposition took place from a 0.05 M $CuSO_4$, 0.15 M ethylenedinitrilo-tetra-2-propanol 0.07 M paraformaldehyde with sufficient NaOH to givc a desired pH. The maximum rate of deposition was obtained at a pH value of 12.5 so the process engineer might choose to maximize the throughput by choosing this value for the solution pH.

Electrodeposited Cu Film Microporosity

From Figure 4.18, we can see that electrodeposited copper films contain a large number of microscopic voids (pores) with locally unfilled rcgions inside the lattice. Most microvoids are generated at the boundaries between 3-D faceted crystallites during their coalescence. This has been speculated to be a result of the incorporation of H_2 bubbles into (25–30 μm) thick Cu films in uniform concentrations, as shown. In these samples, large (2000 Å) bubbles were found to be trapped at grain boundaries. The density of electroless Cu films is found to be 8.56–8.76 g cm^{-3} compared with pure bulk Cu at 8.9331 g cm^{-3}.

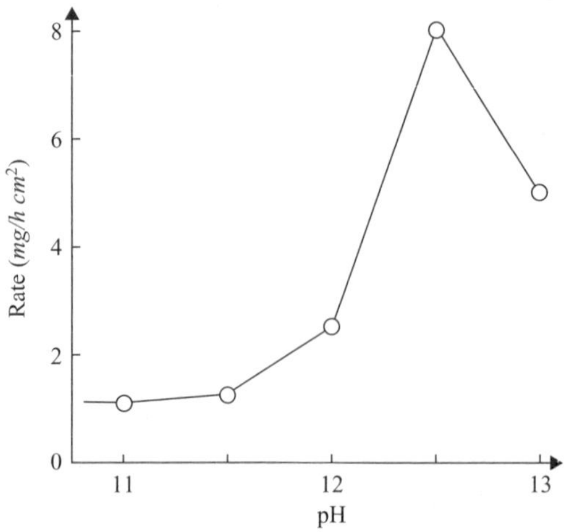

Figure 4.17 Rate of electroless Cu deposition as a function of pH.

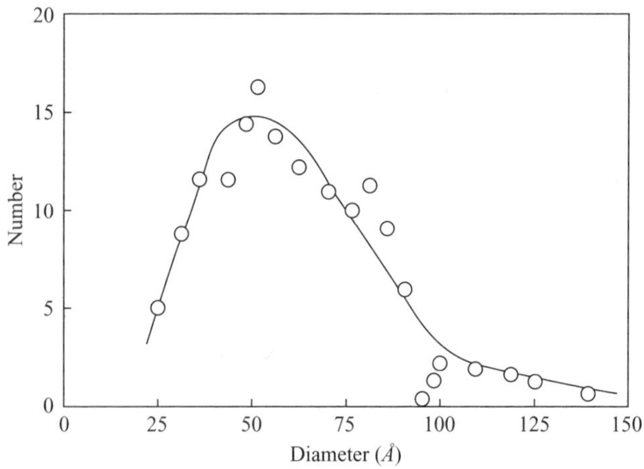

Figure 4.18 Number of voids per unit volume (10^{15} to 10^{16} cm^{-3}) as a function of average void diameter (25–100 Å) from various electrodeposited copper films[11].

SEM Measurements

To determine how the PCB surface looks at microscopic levels, SEM photographs have been made of the surfaces. One such photograph is shown in Figure 4.19.

In Figure 4.19, the photo magnified 1,000× indicates that the surface of a high-profile sample consists of mounds of crystallites about 10 μm high; a profilometer measurement of the same surface shows the mounds to average 11.6 μm in separa-

Figure 4.19 Scanning electron microphotograph of the rough side of a "high profile" electrodeposited surface in four magnifications taken at an angle of 32 degrees from normal. A dimension bar on each of the photographs is shown for comparison of a typical feature.

tion. The arrangement of the mounds is somewhat random but partially hexagonally close packed as seen in the photo magnified 5,000×. The photo magnified 10,000× shows "snowball" crystallites that have a distribution of sizes and shapes but could be roughly called spheres that average about 2 μm in diameter. The photo magnified 25,000× shows that the "snowballs" are probably constructed of smaller kernels that average below 0.1 μm in diameter. These features are consistent with the solid-state packing model of Figures 6.2 and 6.23 and the microporosity plot of Figure 4.18. The surface of Figure 4.19 was electrochemically designed to be of "high profile" so that the surface features would adhere well to the underlying epoxy medium.

It is common sense that such a high-profile surface will contribute more surface resistance than a low-profile surface, so some copper traces and ground planes are made with a low-profile electrodeposit process. Such a surface is shown in Figure 4.20.

In Figure 4.20, the photo magnified 1,000× indicates that the surface of a low-profile sample consists of scattered crystallites on a relatively flat plane. The arrangement of the mounds is very random, as seen in the photo magnified 5,000×. The photo magnified 10,000× shows "snowball" crystallites that have a distribution of sizes and shapes but could be roughly called spheres that average less than 2 μm in diameter. The photo magnified 25,000× shows that the "snowballs" are probably constructed of smaller kernels that average below 0.1 μm in diameter.

Figure 4.20 Scanning electron microphotograph of the rough side of a "low profile" electrodeposited surface in four magnifications taken at an angle of 32 degrees from normal. A dimension bar on each of the photographs is shown for comparison of a typical feature.

Surface Chemical Composition

As shown in Figure 4.13, copper foil manufacturers employ a post-electrodeposition treatment to protect the surface from oxidation and separate it from the bromine in the epoxy. One of the additives is typically a Zn deposition that creates a CuZn (brass) layer on the surface. Brass is any alloy of copper and zinc; the proportions of zinc and copper can be varied to create a range of brasses with varying properties. One of the brasses most resistant to oxidation is composed of 30% Zn and 1% Sn (called "Admiralty Brass") that inhibits leaching of Zn from brasses (*dezincification*) in halogen-rich environments (such as those with Br present). Dezincification is thought to dissolve copper and zinc simultaneously, with copper precipitating back from the solution with a remaining copper-rich sponge with poor mechanical properties and color changing from yellow to red. To combat this, arsenic can also be added to brass. For the chemical composition of the PCB surface to be understood, Auger measurements have been made of the surface to determine the chemical composition as a function of depth, as shown in Figure 4.21.

In Figure 4.21, it is seen that sputtering with 20 keV Ar ions wears away the surface layer to a depth of less than 1 μm. Measurements are made by sputtering the surface for 1 min so as to remove about a 20-nm layer followed by an Auger electron analysis. The elemental composition as a function of depth obtained in this manner is shown in Figure 4.22.

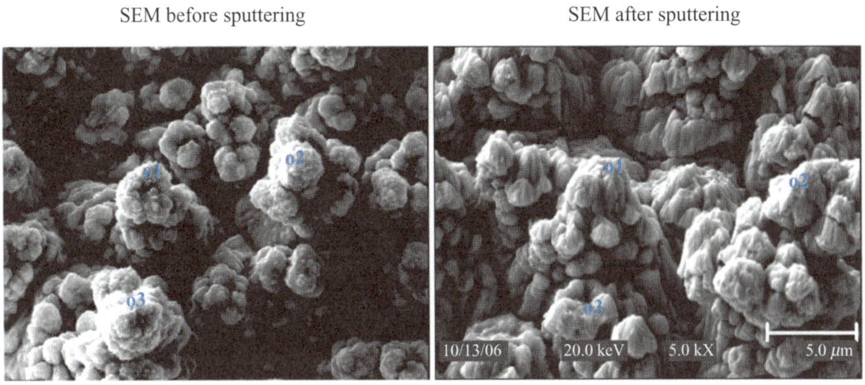

Atomic concentration table

Area	C	N	O	Si	Cu	Zn
1	44.45	6.96	19.89	20.80	1.94	5.96
2	43.67	7.61	20.84	19.44	1.68	6.76
3	42.81	7.66	22.41	19.95	1.34	5.82

Atomic concentration table

Area	C	Cu
1	26.37	73.63
2	13.89	86.11
3	20.65	79.35

Figure 4.21 Surface of a high-profile printed circuit board film prior to sputtering with 20 keV Ar ions.[12] In this figure magnified 5000×, three locations are identified from which Auger electron energy measurements are made, following a specific time of sputtering that wears down the surface. On the left, the elemental composition is shown in the table below the photo before sputtering. The scanning electron microscope photo to the right shows the surface material after a set of sputtering applications with an elemental composition measurement in the table below the photo after complete sputtering.

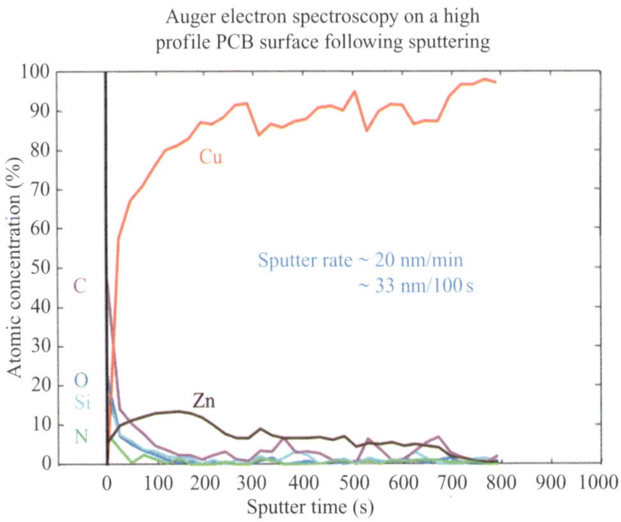

Figure 4.22 Elemental composition profile of a high-profile "copper" film that has been post-electrodeposition treated with proprietary additives.

Figure 4.22 shows that a CuZn (brass) layer of varying composition extends into the "snowball" surface about 0.2 μm deep. C, O, Si, and N elements of smaller concentration are found to be located within about 0.01 μm of the surface. The importance of the thickness of this surface layer is seen in how it compares with the "skin depth" of pure copper as a function of frequency. Consistent with calculations for Table 1.4 copper skin depth, given by $\delta = \sqrt{2/\mu\sigma\omega}$, has values of $\delta = 2.1\,\mu\text{m}$ at 1 GHz, 0.66 μm at 10 GHz and 0.21 μm at 100 GHz. Because the electrical conductivity of brass is about half that of copper, we can see that the brass layer is carrying a large fraction of the current at high frequencies and is thus producing more power loss than pure copper because of the zinc in the protection layer. An analysis of this loss has been simulated by David Aerne for his MS thesis at the University of South Carolina. Some of those results are shown in Chapter 7 on "Advanced Signal Integrity."

4.5 EFFECTS OF THE PROPAGATING MATERIAL

Conduction Electrons

Electrons in metals, semimetals, and semiconductors have some fraction of their electrons that produce a current density that is proportional to the applied electric field intensity.[13] Usually, the propagating material for a transmission line is chosen to be a good insulator, but there can be metal impurities included in the material or even some smaller fraction of electrons that obey microscopic Ohm's law. As is shown in Chapter 5, the constitutive properties of these "free" electrons can be treated as a part of the complex permittivity of the propagating medium so that they add to the normal permittivity as shown in Equations 5.79 and 5.80:

$$\varepsilon'_{r,\,total}(\omega) = \varepsilon'_r(\omega) - \frac{\sigma_e(0)}{\varepsilon_0}\left(\frac{1/\tau_s}{\omega^2 + 1/\tau_s^2}\right) \tag{5.79}$$

$$\varepsilon''_{r,\,total}(\omega) = \varepsilon''_r(\omega) + \frac{\sigma_e(0)}{\varepsilon_0\omega}\left(\frac{1/\tau_s^2}{\omega^2 + 1/\tau_s^2}\right) \tag{5.80}$$

The degree to which the real term change is significant depends upon the magnitude of the conductivity, $\sigma_e(0)$, relative to the permittivity of free space. The imaginary term clearly contains a singularity at zero frequency, but this has occurred because we have formed the ratio for our mathematical convenience by multiplying and then dividing by frequency. For a weakly conducting medium, we concluded in Equation 1.30b that $\sigma/\omega\varepsilon \ll 1$ so we are not permitted to evaluate this quantity for very small values of frequency that violate this requirement.

Permanent Dipoles

The properties of the propagating medium have an influence on the electric flux density in the space between the trace and the ground plane.[14] For example, if the

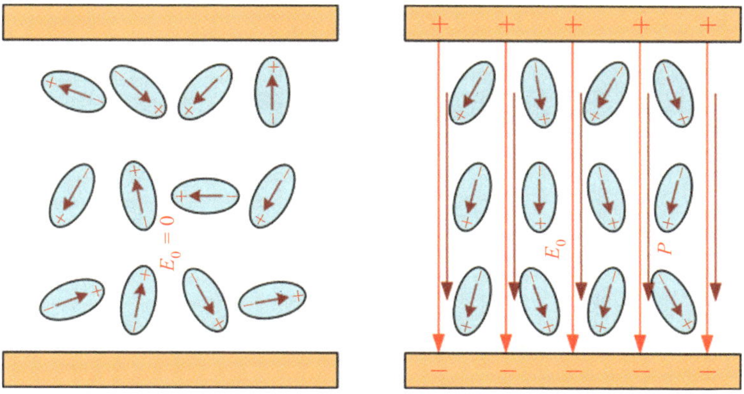

Figure 4.23 (Left) Random orientation of permanent electric dipoles in the absence of an external electric field intensity and (Right) partial alignment (in thermal equilibrium with their environment) of the permanent dipoles in the presence of an applied external field E_0.

material has molecules with a permanent electric dipole moment, they will be partially aligned by the application of the external electric field intensity between the trace and the ground plane, as shown in Figure 4.23.

In practical applications, the external electric field intensity is never strong enough to saturate the electric dipole moments (make them perfectly align) because of the finite temperature of the lattice in which they are bound. However, the thermal averaging leaves a net positive charge density of the dipoles near the ground plane and a net negative charge density near the trace, and this reduces the magnitude of the applied electric field intensity at any point in the propagating medium. The volume average of the dipole moments P has been evaluated in Chapter 5.

Induced Dipoles

In addition, the presence of applied electric field intensity can *induce* a dipole moment in neutral atoms by displacing the geometrically symmetric electron cloud relative to the positive nucleus of the atom. This effect is shown in Figure 4.24.

We can define the induced electric dipole moment, $\vec{P}_i$, of the ith atom, as shown in detail in Figure 4.25 to be the product of the displacement, $\vec{x}_i$, between the center of mass of the electrons and the nucleus, and the amount of charge that is displaced, q_i.

The degree to which the polarization occurs can be expressed in terms of the fraction of an electron that is displaced as

$$\vec{p}_i \equiv q_i \vec{x}_i = -e\alpha_i \vec{x}_i, \tag{4.5}$$

where α_i is called the polarizability of the atom of type i. We can also take the volume average polarization, $\vec{P}$, of the induced atomic moments to be

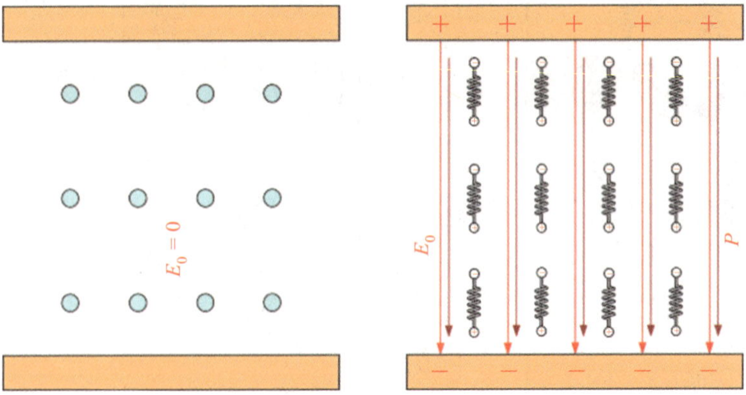

Figure 4.24 (Left) Electrically neutral atoms within a material in the absence of an external electric field intensity; (Right) induced dipole polarization, $\vec{P}$, within a material because of the application of an external electric field intensity, $\vec{E}_0$.

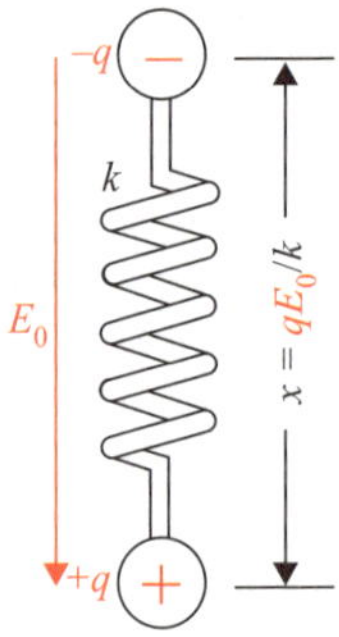

Figure 4.25 Distortion of the atomic electric charge cloud on an otherwise neutral atom from its atomic core.

$$\vec{P} = \frac{\sum_i \vec{p}_i}{\Delta v} = N_i \vec{p}_i = -N_i e \alpha_i \vec{x}_i, \tag{4.6}$$

where N_i is the number of atoms per unit volume.

We have shown that the electric flux density inside the propagating medium for an external electric field intensity is given by

$$\vec{D} = \varepsilon_0 \vec{E} + \vec{P} = \varepsilon_0 \left(1 + \frac{\vec{P}}{\varepsilon_0 \vec{E}} \right) \vec{E} = \varepsilon_0 \varepsilon_r \vec{E} = \varepsilon \vec{E} \tag{4.7}$$

so that the permittivity of the propagating medium is

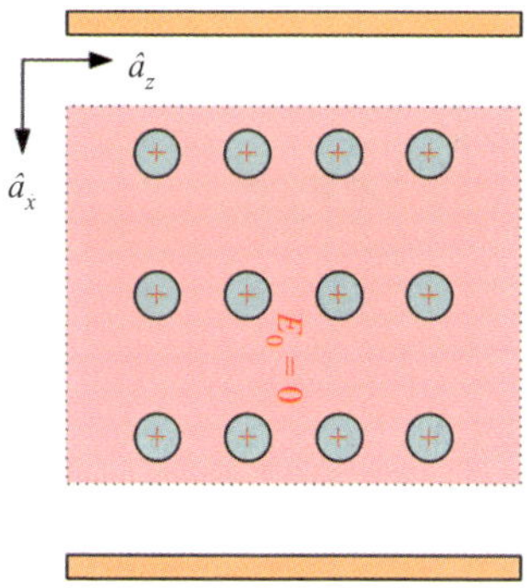

Figure 4.26 Neutral ion cores immersed in a conduction electron charge cloud in the *absence* of applied electric field intensity.

$$\varepsilon_r = \left(1 + \frac{\vec{P}}{\varepsilon_0 \vec{E}}\right) = 1 + \sum_i \frac{N_i \alpha_i (e^2/m_e)}{(k_i/m_e - \omega^2) - j(b_i/m_e)\omega}, \tag{4.8}$$

as is shown in Chapter 5. This can be written in terms of measured quantities as

$$\varepsilon_r(\omega) = \varepsilon_{r\infty} + \frac{(\varepsilon_{rs} - \varepsilon_{r\infty})}{1 - (\omega/\omega_1)^2 - j\tau_e\omega} \tag{4.9}$$

Plasmons

We can treat a conducting media (in which there are some "free" electrons) as a lattice of positive ions immersed in a uniform plasma cloud of conduction electrons, as shown in Figure 4.26.

When a harmonically oscillating electric field intensity acts on the electrons in either a longitudinal or transverse manner, the charge cloud acts in a collective manner to form a "plasma oscillation" at frequency $\omega_p = \sqrt{Ne^2/\varepsilon_0 m_e}$ as deduced in Chapter 5. These frequencies are very high for electrons in an alkali metal (1.93×10^{15} Hz for Li, 1.38×10^{15} Hz for Na, 0.94×10^{15} Hz for K) where the free electron charge density exceeds 10^{28} e^-/m^3 but can be much smaller (10^8 Hz) for semiconductors like Si, where the free electron charge density is just above 10^{14} e^-/m^3.

The permittivity of the plasma for a transverse oscillation (like those found in a transmission line) is found in Chapter 5 to behave as

$$\varepsilon_r'(\omega) + j\varepsilon_r''(\omega) = 1 + \left[1 - \omega^2/\omega_p^2\right]^{-1} \tag{4.10}$$

Equation 4.10 is a purely real quantity, but we note that the treatment of plasmons assumes no loss mechanism of the electron cloud as it oscillates relative to the lattice of ion cores and we regard this as an unrealistic assumption.

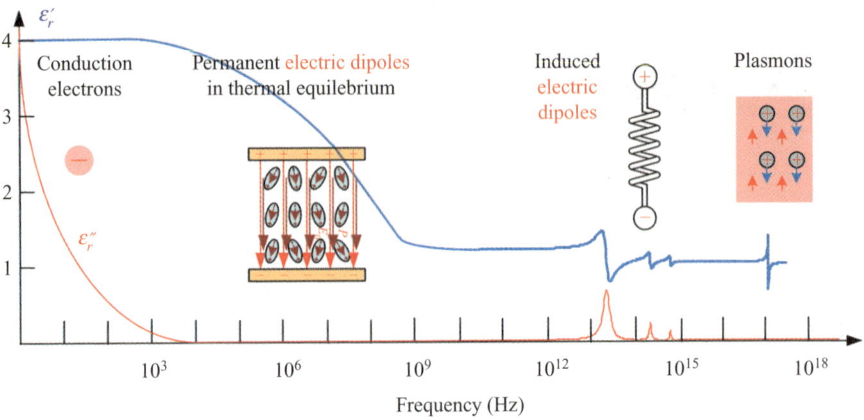

Figure 4.27 Total real, ε_r', and imaginary, ε_r'', parts of the relative electric permittivity for propagating materials from four basic mechanisms: conduction electrons, permanent electric dipoles in thermal equilibrium with their environment, induced electric dipoles with restoring and drag forces, and collective oscillations of the entire electron cloud (plasmons).

Conclusions

The electric permittivity of a propagating medium depends upon its conductivity, the permanent dipole moments of its constituent molecules, the degree to which its atoms will polarize (its polarizability), and the number of free electrons per core atom that are available for plasmon oscillations. The permittivity is additive in frequency space, as is shown in Figure 4.27.

4.6 EFFECTS OF GRAIN BOUNDARIES

The electrodeposition process shown in Figure 4.13b produces copper foils of various thickness and matte finish (initial surface profile roughness) as the speed of deposition and rotation rate of the titanium drum is changed. Depending on the PCB application, foils of thickness as small as 10 μm (0.5 oz of Cu per square foot) or as large as 80 μm (4.0 oz of Cu per square foot) can be produced by copper foil manufacturers with consistent reliability. The side of the foil adjacent to the drum is called the shiny side because it is as smooth as the titanium drum upon which it is deposited and is, thus, not a typical variable for roughness (except for postdeposition treatment, as shown in Figure 4.13c). The side toward the $CuSO_4$ solution is controllable by the electrodepositor by varying the pH, temperature, and over potential, as shown in section 4.4. A product of this initial deposition process is shown in Figure 4.28.

In Figure 4.28, the terms *HTE* (Standard), *Super-HTE*, and *VLP* are Oak Mitsui Corporate (Camden, SC) designations that refer to tensile strength, elongation properties and low-profile characteristics of the foils (http://www.oakmitsui.com/media/

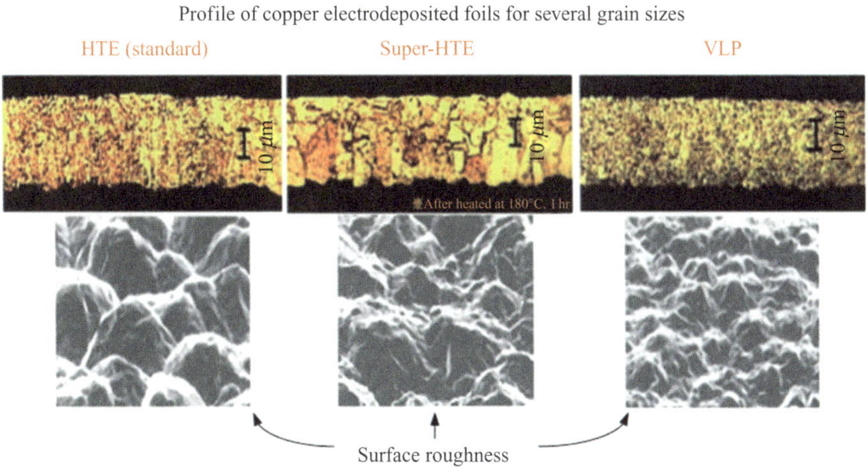

Figure 4.28 Scanning electron microscope profile of a copper foil following electrodeposition for several Oak Mitsui foils (Camden, SC). The top photo series is a cross section stained to emphasize the grain structure. The bottom photo shows the corresponding surface profile.

ML.pdf). From the top photos, we can observe that grain size is a function of process variables and subsequent heat treatment.

Studies of resistivity of copper also show[15] that grain boundary scattering can be a dominant factor in the overall resistance and subsequently power loss in copper foils. This study concludes that the mean free path of electrons in copper can be significantly reduced by grain-boundary scattering alone and these results are expressed in terms of a reflection coefficient, R, that varies with grain size in nm as follows: 0.57 (106 nm), 0.56 (122 nm), 0.59 (160 nm), 0.59 (180 nm), 0.62 (240 nm), and 0.62 (305 nm). This reflection coefficient was proposed[16] and changes the electrical conductivity, σ_e, as

$$\sigma_e/\sigma_0 = 1 - 1.5\alpha + 3\alpha^2 - 3\alpha^3 \ln(1 + 1/\alpha) \quad (4.11)$$

$$\text{where } \alpha = (l_0/d)(R/1 - R), \quad (4.12)$$

σ_0 is the conductivity in the absence of grain boundaries, l_0 is the background electron mean free path length, and d is the average random distance between grains.

4.7 EFFECTS OF PERMEABILITY

Most signal integrity engineers begin problems with the statement "Let us assume the materials all have a relative permeability of $\mu_r = 1$." For many problems, this is a valid assumption, but we have already seen in Figures 4.11 and 4.12 that some signal traces are protected by a layer of nickel and phosphorus, which can be at least partially magnetic. In addition, many ceramic capacitors were originally designed with very thin, high-melting-point, tarnish-free platinum sheets separated by a barium titanate dielectric medium to boost relative permittivity into the thousands. As the cost of platinum increased, those capacitor manufacturers choose to replace the platinum sheets with nickel because it could also withstand high temperatures and was relatively tarnish free once it was coated with copper or encased in a ceramic medium. But, unless the nickel capacitors undergo subsequent annealing, the capacitors remain strongly magnetic and can be picked up by a horseshoe magnet.

For example, the electrical resistivity of electrodeposited Ni is shown in Figure 4.29 as a function of anneal time at various temperatures.

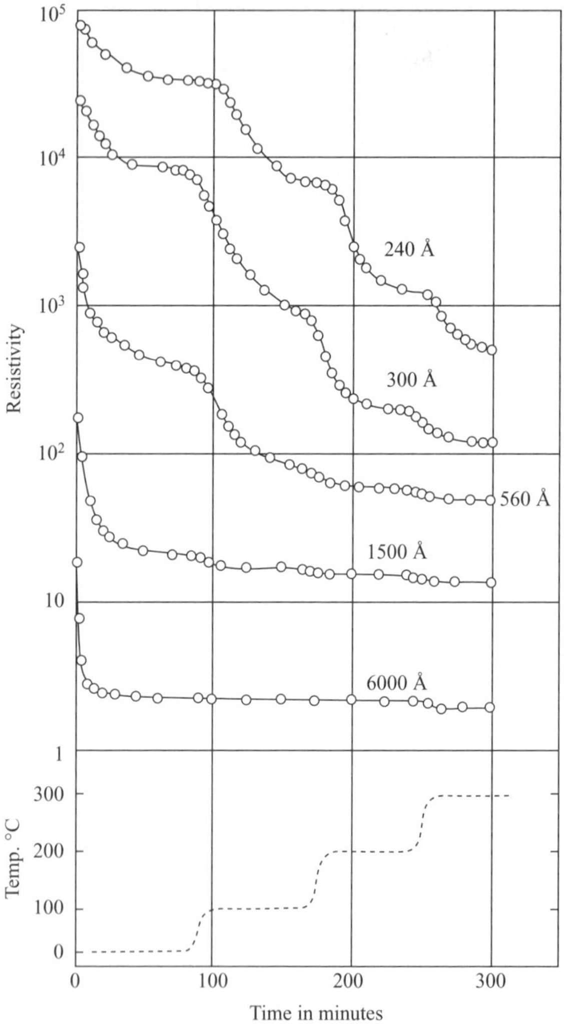

Figure 4.29 (Upper curves) Resistivity (in ohm/square) changes in time of Ni-P films of various film thicknesses in Å during heating. (Bottom [dashed] curve) Heat-treatment temperatures as a function of time.

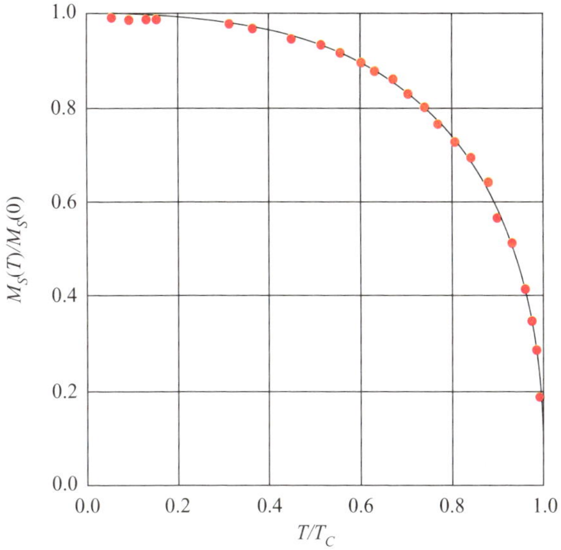

Figure 4.30 Saturation magnetization of ferromagnetic Ni as a function of temperature.

The decrease in electrical resistivity with annealing is an indication that non-magnetic Ni-P is changing from an amorphous solid solution state to separate metallic phases of pure Ni with P inclusions, as we saw in Figure 4.11. Pure Ni has a classic saturation magnetization curve as a function of temperature that is shown in Figure 4.30.

Parameters derived from this data give $M_s = 0.0485\,T$ at 300 K, and $\mu(0\,\text{K}) = 0.606\,\mu_B$, where $\mu_B = e\hbar/2m_e = 9.274 \times 10^{-24}\,\text{J}\,T^{-1}$, and the Curie temperature of Ni is $T_c = 627\,\text{K}$. If these data apply to the Ni in the material surrounding a copper trace or to Ni in a ceramic capacitor, then it is relatively magnetic at room temperature. Signals on such a trace yield fields below an ideal model that can be calculated by the application of HFSS by Ansoft and yield dispersion as calculated by HSpice by Synopsys. A field distribution and the signal dispersion caused by various levels of μ_r have been studied[17] by using an ideal model cross section, as shown in Figure 4.31.

The approximation of uniform material properties is a poor representation of the magnetic grain boundaries associated with the pockets of P interspersed in the Ni coating (top photo), and we have ignored the copper grain boundaries within the Cu trace. Perhaps the worst approximation is that we have modeled a smooth surface on the bottom of the trace that was manufactured to have a high profile roughness.

Figure 4.32 shows the simulated insertion loss of three different kinds of 0.040″ inserts in a 1″ transmission line as a function of frequency. We see that, at high frequencies, the high-permeability material has much greater insertion loss than

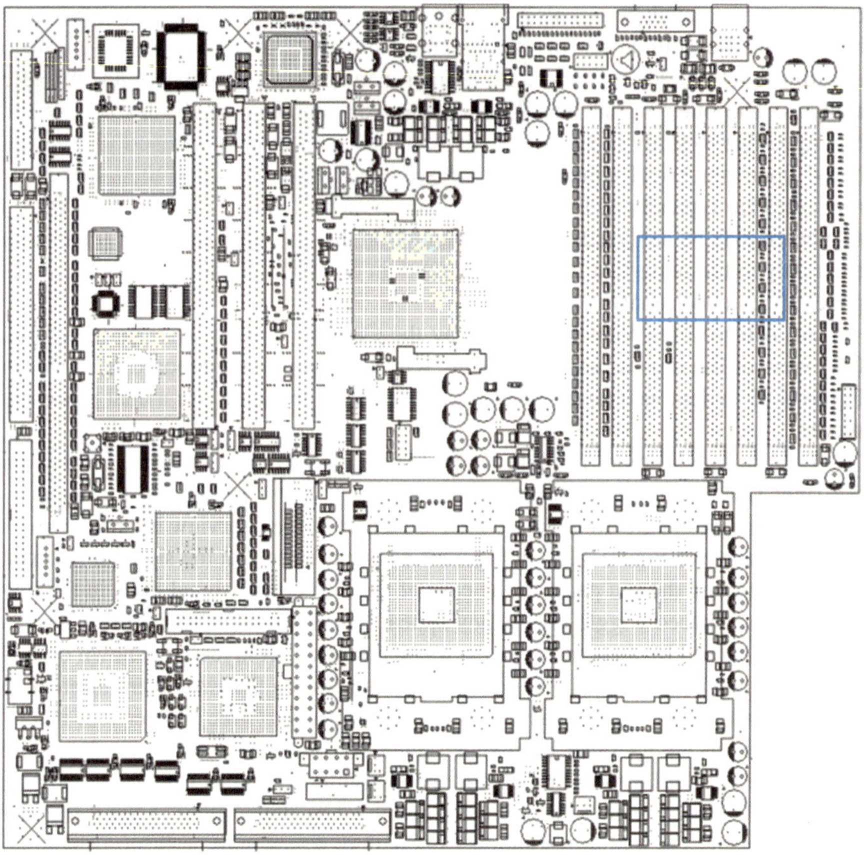

Figure 4.33 Computer-aided design layout of the major components of an Intel "Kings Canyon" server motherboard with the connection pins shown in black. In particular, note the complexity of the ball grid array for each of the central processing unit (CPU) and memory channel handler (MCH)-integrated circuit packages. Shown in blue is a segment of the board that is magnified in Figure 4.34.

research engineers have written specialty codes that improve upon some aspect of the commercial tools. However, most corporations require some continuity in the definition of signals so these "home-made" tools (even improved versions) are not often employed in the prototyping of circuit designs. The signal integrity engineer must have experience in the most popular tools available if he is to work in a major corporation. The USC curriculum provides most of those popular tools to its graduate students in their course of study courtesy of the Ansys, Agilent, CST of America, and Sonnet Corporations.

An example of the use of a suite of tools from the Ansys Corporation is presented in Chapter 8.

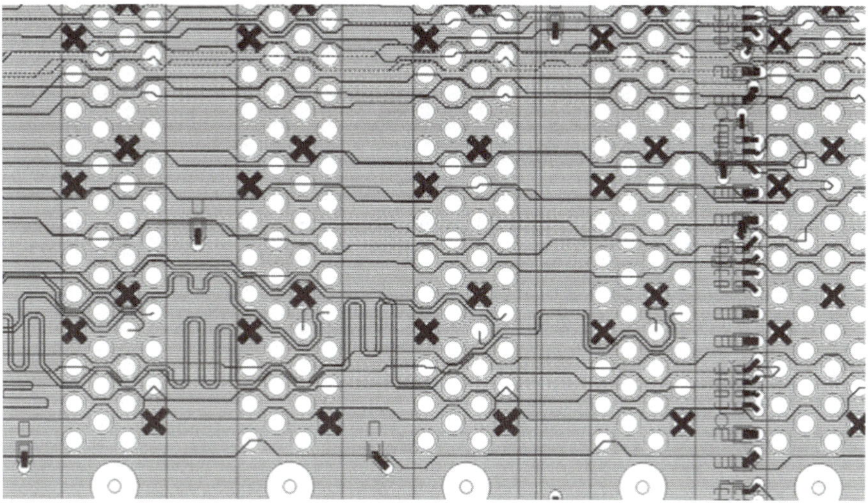

Figure 4.34 Computer-aided design layout of the top layer and first signal layer of the copper traces that connect the DIMM pins shown in the blue rectangle in Figure 4.33.

4.9 FINAL CONCLUSIONS FOR AN IDEAL VERSUS A REAL-WORLD TRANSMISSION LINE

- Copper can be electrodeposited as a compact or powdery material depending upon the solution temperature, pH, and overpotential.
- The density of electrodeposited Cu is lower than that of bulk Cu because of a uniform inclusion of microscopic voids per unit volume of 10^{15}–$10^{16}\,cm^{-3}$, with a distribution of crystallite diameters (copper "snowballs") between *25* and 100 Å.
- The surface layer of electrodeposited Cu consists of exponentially decreasing amounts of Si, Cr, and O, with depth, and a solid solution of Zn (brass) that decreases more slowly.
- The resistivity of electrodeposited Cu is higher than that of pure Cu (even at large depths) but approaches that of bulk Cu with annealing.
- Electrodeposited Ni is a solid solution of P in amorphous Ni that includes intermetallic Ni_3P interspersed with polycrystalline Ni metal and pure P. The resistivity of the amorphous Ni-P is much higher than that of pure Ni and decreases with temperature and time in an annealing process (reflecting a structural change). Heat treatment changes the amorphous films from non-magnetic to weakly ferromagnetic because of Ni crystallization.
- There are many inclusions (filled with Si, Br, Cl, Mg, Ca, and O) in the propagating medium and in the solder mask to scatter and absorb electromagnetic power.

- Grain boundaries in copper cause an increase in the resistivity of a trace and ground plane and thus have an influence on the power lost per length.
- Magnetic materials may not always be neglected in simulation or theoretical models (especially in Ni-coated transmission lines or ceramic capacitors).
- PCBs are complex systems with many trace segments on 8 to 40 different signal, power, or ground layers. Signal lines are in close geographic proximity to one another so that they will communicate (cross talk) via electromagnetic radiation. While fashion designers may want to produce the thinnest or lightest products for the consumer, the signal integrity engineer must take account of the cross talk and assure there are margins of the signal to sustain a low bit-error-rate of information transfer.

ENDNOTES

1. Paul G. Huray, *Maxwell's Equations* (Hoboken, NJ: John Wiley & Sons, 2009), Chapter 5.
2. Ibid., Table 5.2.
3. Dmitri B. Strukov, Gregory S. Snicer, Duncan R. Stewart, and R. Stanley Williams, "The Missing Memristor Found," *Nature* 453 (May 2008): 80.
4. Huray, *Maxwell's Equations*, Section 7.14.
5. Ibid., Chapter 7.
6. Ibid., Section 7.6.
7. Ibid., Section 5.2.
8. Research measurements at ORNL sponsored by the Assistant Secretary for Energy Efficiency and Renewable Energy, Office of FreedomCAR and Vehicle Technologies, as part of the High Temperature Materials Laboratory User Program, Oak Ridge National Laboratory, managed by UT-Battelle, LLC, for the U.S. Department of Energy under contract number DE-AC05-00OR22725.
9. Mordechay Schlesinger and Milan Paunovic, eds., *Modern Electroplating* (New York: John Wiley & Sons, 2000).
10. N. Ibl, *Advances in Electrochemistry and Electrochemical Engineering*, Vol. 2, C. S. Tobias, ed. (New York: Wiley, 1962) as reported by Milan Paunovic in the chapter "Electrochemical Aspects" of *Modern Electroplating*, p. 12.
11. S. Nakahara, "Properties and Barrier Material Interactions of Electroless Copper," *Acta Metall.* 36 no. 7 (1988): 1669–1681.
12. S. G. Pytel, P. G. Huray, S. H. Hall, R. I. Mellitz, G. Brist, H. M. Meyer III, L. Walker, and M. Garland, "Analysis of Copper Treatments and the Effects on Signal Propagation," In *IEEE-ECTC Conference Proceedings* (Lake Buena Vista, FL, May 27–30, 2008).
13. Huray, *Maxwell's Equations,* Chapter 5.
14. Ibid., Chapter 3.
15. W. Wu, M. Van Hove, and K. Maex, "Influence of Surface and Grain-Boundary Scattering on the Resistivity of Copper in Reduced Dimensions," *Applied Physics Letters* 84, no. 15 (April 2004): 2838–40.
16. A. F. Mayadas and M. Shatzkes, "Electrical-Resistivity Model for Polycrystalline Films: The Case of Arbitrary Reflection at External Surfaces," *Phys. Rev. B* 1, no. 4, (Feb. 1970): 1382–89.
17. Brandon Gore, "Effective Losses from Printed Wiring Board Components Containing Nickel," Master's thesis, University of South Carolina, 2007, pp. 1–49.

Chapter 5

Complex Permittivity of Propagating Media

LEARNING OBJECTIVES

- Describe and quantify the basic mechanisms that influence complex electric permittivity in propagating medium materials in the frequency domain
- Explain delay, attenuation, and dispersion of electromagnetic waves on a printed circuit board in the time domain via complex permittivity of propagating media
- Assure that models describing properties of the medium obey causal relationships
- Find the medium response function to an arbitrary signal

INTRODUCTION

In a laboratory setting or a practical electronic device, electromagnetic fields are bounded by a metal or insulator. Some of the boundaries, such as the boundary at infinity, are implied, but most are provided by some man-made structure, often metallic. Because the wavelength of E&M waves at low frequencies are long compared with the cross section boundaries, those boundaries may be approximately treated as smooth, but, at high frequencies, the propagation of E&M waves is strongly influenced by irregularities in the boundaries. The influence of surface roughness will be considered in Chapter 6.

Because of geometry of the boundary conditions, electromagnetic responses often resonate at distinct eigenfrequencies (geometric modes that correspond to them are called *eigenmodes*). Eigenmodes are sometimes desirable if very sharp (high Q) values of frequencies are desired (such as in a Klystron for a particle accelerator or in a cell phone signal generator/antenna receiver) but can present a problem when a pulsed signal (made up of a Fourier spectrum of various component frequencies) propagates from one device to another. This is an aspect of Signal Integrity that lacks a sound fundamental basis, (especially at high frequencies), but there is a limit to what can be considered in one book so the topic of mixed eigenmodes shall be reserved for future work.

The Foundations of Signal Integrity, by Paul G. Huray

In Chapter 5, the effect of Boundary Conditions on E&M waves at the interface between a perfectly insulating and a perfectly conducting boundary will be ignored to concentrate instead on the influence of the materials that make up the propagating medium (e.g., FR-4, Rogers Corporation, Rogers, CT; Nelco Advanced Circuitry Materials, Nelco, CA; Isola Group, Chandler, AZ; polytetraflourethylene (PFTE or Teflon, DuPont Chemical) dielectrics, glass fibers, semiconductors, bubbles, inclusions, voids, water, included atmosphere, and other impurities). We will attempt do this by assuming constant, perfect model boundaries discussed in Chapter 4. This approach will allow us to concentrate on one variable at a time as we change the propagating medium material properties and to consider the outcome on the electric and magnetic fields that propagate on a printed circuit board (PCB).

In a sense, Chapter 5 will consider the influence of the nineteenth-century "jelly" in which E&M waves propagate. We will find that it is not enough to describe propagation in the luminiferous ether because charged particles in the propagating medium have inertia, experience forces, and undergo acceleration. The energy that produces these accelerations comes from the propagating E&M wave and causes signals to attenuate and disperse.

5.1 BASIC MECHANISMS OF THE PROPAGATING MATERIAL

When an electric field intensity, $\vec{E}$, acts on a charge, q, in an atom or molecule immersed in a propagating medium, it produces a force, $\vec{F}_{external} = q\vec{E}$ proportional to the magnitude of the charge and to the magnitude of the electric field that causes the charge to move. Because electrons have much smaller mass, m_e, than the ions with which they are associated, they accelerate much more than their parent ions. There are many types of charges that can move from an equilibrium (or random) position to produce an electric dipole moment density and thereby change the electric permittivity of the medium. Some examples are (a) relatively isolated **permanent** electric polar moments in dielectric materials that partially align with the field (by rotation of polar molecules) or electric dipoles in an organic molecule that rotate or exhibit bending modes, (b) **induced** electric dipoles from electrons in unfilled shells in an atom, (c) **conduction electrons** in the partially filled band of a lossy dielectric, and (d) **plasma** electrons. In case (a), the permanent electric dipoles are in thermal equilibrium with their environment, practical fields are not strong enough to bring them into saturation (in contrast with magnetic interactions), and the polar molecules have an inertia that does not permit them to follow the phase of the applied field except at low frequencies. The molecules experience no other restoring force to bring them back into partial alignment and they experience losses as they interact with their neighbors. In case (b), there is a strong restoring force of induced electric dipoles in an atom that is proportional to the electron's displacement relative to its nucleus that tends to restore displaced charge to an equilibrium position. In addition, there is often a strong damping force that is proportional to the velocity of the charge because of radiation losses or frictional interactions with neighboring molecules. In case (c), a small fraction of the electrons (at the Fermi level of an unfilled band) are "free" to produce a current in proportion to the applied electric field but they lose energy as they move and experience no restoring force once they have been displaced. In case (d), there is a solid with equal concentration of positive charges from ion cores and negative charges from the conduction electrons; plasma electrons are usually treated as mobile and experience no drag force. Here, the displacement of the electron cloud from the ion lattice produces a restoring force for transverse fields but none for longitudinal fields. In a sense, cases (c) and (d) are special cases of (a) or (b) in that they either experience restoring forces or loss but not both.

5.2 PERMITTIVITY OF PERMANENT POLAR MOLECULES

At low frequencies, we have treated insulators as dielectrics with polar molecules in thermal equilibrium with their environment in the static limit $\lim_{\omega \to 0} \varepsilon(\omega)$ for which there was partial alignment in proportion to the direction of the applied electric field intensity. This was the topic of static electric field intensity for which we concluded[1] that the macroscopic electric flux density in a linear homogeneous medium could be approximately written:

$$\vec{D} = \varepsilon_0 \vec{E} + \vec{P} = \varepsilon_0 (1 + \chi_e) \vec{E} = \varepsilon_0 \varepsilon_r \vec{E} = \varepsilon \vec{E}$$

In an oscillating field, electric dipoles have inertia so the degree to which this expression is valid depends upon the ability of the dipoles to follow the changing electric field intensity. Each type of dipole will, in general, have a different inertia so it is difficult to predict the range of frequencies for which the behavior applies. In general, at some point $\vec{x}$ in space, we will thus expect the static equation of this expression to be frequency dependent and the relationship between $\vec{D}(\vec{x}, \omega)$ and $\vec{E}(\vec{x}, \omega)$ to be related by a complex quantity, $\varepsilon'(\omega) + j\varepsilon''(\omega)$, to permit the electric flux density to lag or lead the electric field intensity. Here, $\varepsilon'(\omega)$ is the real part of the permittivity, and $\varepsilon''(\omega)$ is the imaginary part of the permittivity so, in general, a more complete expression would be:

$$\vec{D}(\vec{x}, \omega) = [\varepsilon'(\omega) + i\varepsilon''(\omega)]\vec{E}(\vec{x}, \omega) \tag{5.1}$$

For a propagating medium in the electronics industry, great lengths are taken to minimize the polarization of the material and to reduce the permittivity's complex component so as to make the quantity in square brackets as close to a real constant, ε_0, as possible. For media in which predictable propagation of fields is important, materials are often chosen so that material electric dipole moments are small, and the constituent molecules are highly symmetric. The most common material is an FR-4-type epoxy resin in which the chemical structures of the brominated epoxy resin (the bromine being present to impart flame retardancy) and the curing agent are shown[2] in Figure 5.1a and Figure 5.1b.

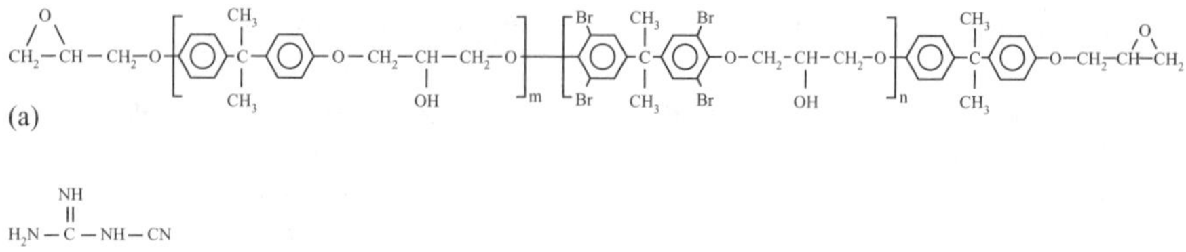

Figure 5.1 (a) Chemical structure of a brominated FR-4 type epoxy; (b) chemical structure of curing agent (dicyandiamide).

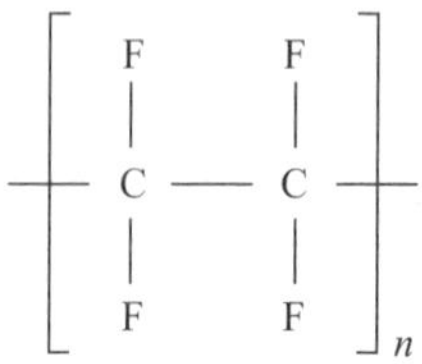

Figure 5.2 Structural repeat unit of polytetrafluoroethylene (Teflon).

The precise network structure produced by these reactants cannot be given because there are several unknowns. Values of the subscripts *m* and *n* vary, giving rise to both a distribution of chain lengths between cross-links and a variation in the bromine content. The number of epoxy groups that react with the dicyandiamide will also vary, although it is typically assumed to be about four. Ideally, the material should have a low real-part relative permittivity ($\varepsilon_r' \cong 3.6$ for FR-4), and it should have a very low imaginary part (loss tangent, $\tan\delta \cong 0.010$ for FR-4) at 1 GHz. There are several improvements possible in even lower permittivity and loss tangent that omit the hydroxyl groups, but their adhesion properties are impaired; they do not meet flame-retardant requirements or are too expensive. One high-quality (but expensive) material that fits the improvements and works predictably at microwave frequencies is Teflon or polytetrafluoroethylene (PTFE), with an $\varepsilon_r' \cong 2.0$ and $\tan\delta \cong 0.0002$ shown in Figure 5.2:

The European Union (EU) has mandated a halogen-free material for printed wiring boards (PWBs) of the future for health and environmental reasons, and many US firms are seeking materials or composites with improved dielectric properties and processing characteristics for the manufacture of reliable high-speed digital PWBs. Producers of PWBs are seeking low-loss laminates (at 10 GHz) that are less expensive, halogen- and lead-free, have higher temperature and better thermal cycling properties, are thinner because of embedded passive materials, exhibit poor water uptake, and have better thermal conductivity with a low coefficient of thermal expansion.

Water as an Example of a Polar Molecule

A molecule of water is composed of one oxygen atom (with eight protons and eight electrons) and two hydrogen atoms (with one proton and one electron each). The most common isotopic form of water has eight neutrons in the oxygen nucleus and zero neutrons associated with each of the hydrogen nuclei, but studies exist for water with the isotope of hydrogen with one neutron (deuterated water) and for the isotope of hydrogen with two neutrons (tritiated water).

Atomic Notation

One of the simplest forms of Schrödinger's equation results from the solution of the two-body problem of an electron in the presence a positive nucleus (one or more protons) in an otherwise empty universe. Quantum mechanics restricts the electron to an orbital in which there are Bohr quantum numbers that describe the energy levels of electron bonding in terms of an integer, n, and two other quantum numbers, l and s, that describe the orbital and spin angular momentum of the electron. For a given integer n, l is restricted to be an integer less than n, and s is restricted to be $\pm\frac{1}{2}$. The lowest energy state of the hydrogen atom is the $n = 1$, $l = 0$, $s = \frac{1}{2}$ state, but, in the absence of an external electric field intensity and ignoring the small interaction between the magnetic moment of the nucleus and the electron, the spin of the electron does not affect the energy of the atom so there are two possible states with different spins that can occupy the $n = 1$, $l = 0$ state without violating the **Pauli exclusion principle** that no two Fermions may have the same quantum state. Fermions are half-integer spin particles like electrons. The expression for the *ground state* (lowest energy state) is often written in a spectroscopic notation as the $1s^1$ state of the electron, and this term is shown on many charts of the periodic table, as indicated in Figure 5.3. In this notation, the first number $n = 1$ is the principal quantum number, the letters s, p, d, f, ... signify electrons having orbital angular momentum 0, 1, 2, 3, ... in units of $\hbar$, and the exponent 1 implies the number of electrons having this n and l. It is convention to leave off the number of electrons having this n and l unless there are more than one. The electron can be raised to a higher energy state by the addition of energy to the ground state in which case it can assume the next level of energy (the $2s$ state). This so-called *excited* hydrogen atom will reassume its ground state in an exponentially decaying period of time, called the life-time and will in turn emit an X-ray of energy equal to the difference between the first excited ($2s$) and the ground ($1s$) states (equal to about 10.2 eV). If hydrogen is excited by an energy >13.6 eV, the electron can escape from its proton nucleus and the atom is said to be ionized. By contrast, the ground state of oxygen has two $1s$ electrons, two $2s$ electrons, and four $2p$ electrons so its ground state is shown in the periodic table as ($1s^2 2s^2 2p^4$) or sometimes just ($2s^2 2p^4$), with the implied understanding that the $1s$ shell of electrons is full. Of course, we see that, for the ground state of oxygen, the $2s$ shell of electrons is also full but that the $2p$ states are not full because there can be $2(2l + 1) = 6$ of the $2p$ electrons; that is, two spins for each of the $m_l = -1$, $m_l = 0$ and $m_l = +1$ states.

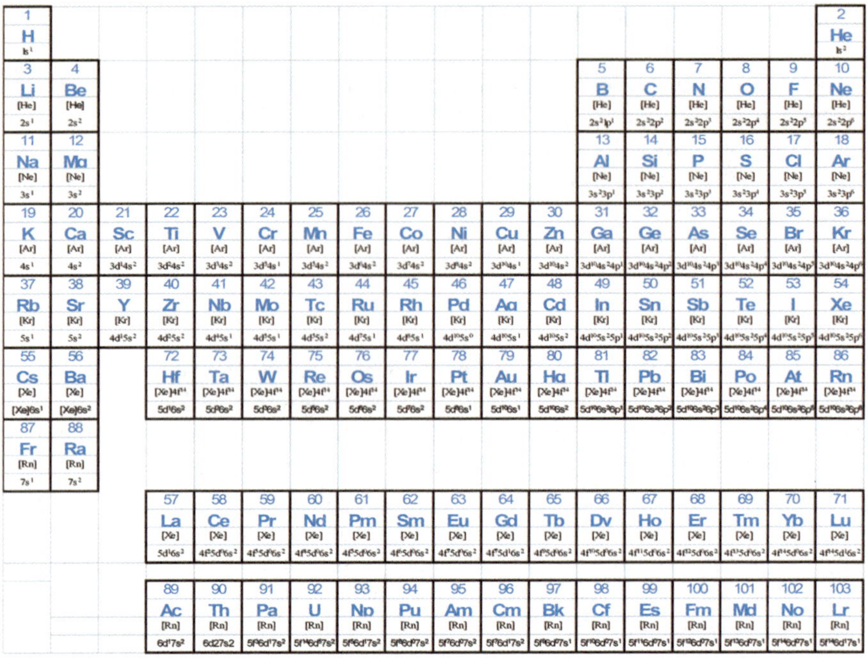

Figure 5.3 Periodic table of elements with their corresponding electron configurations.

Molecular Notation

For two hydrogen atoms in an otherwise empty universe, the lowest energy level occurs when the atoms are spaced at a specific distance (about 0.6 Å) apart, as shown in Figure 5.4. In this figure, we see that the separation is stable for antiparallel electron spins and not for parallel electron spins as long as their kinetic energy is less than about 3.1 eV. Furthermore, we see that the parallel orientation of electron spins does not produce a stable equilibrium because there is no minimum in the potential energy curve.

This arises because of the four-body interaction between the two positive nuclei and the two negative electrons. In this calculation, the electrons fall into energy states that are almost the same as the 1s states but are slightly modified by the nearby presence of the other hydrogen atom. These levels are called *molecular orbitals* and are described by chemists as having either covalent (σ) or ionic (π) bonds. Similar calculations for molecular oxygen yield similar results in which the lowest energy states occur for a diatomic configuration with a spacing of about 3 Å between the oxygen atoms.

Water Molecules

If we have two hydrogen atoms and one oxygen atom in an otherwise empty universe, the lowest energy state occurs when the hydrogen atoms bond to the oxygen

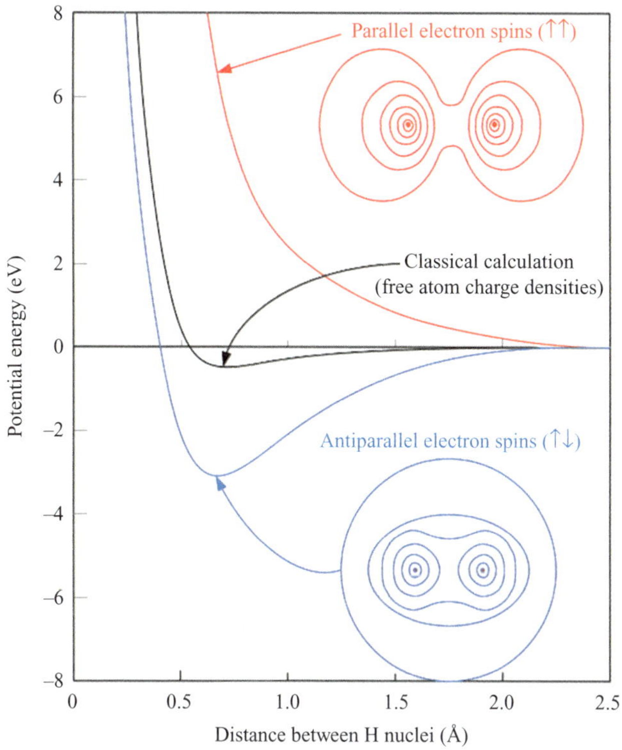

Figure 5.4 Potential energy of molecular hydrogen as a function of the separation of their nuclei for parallel and antiparallel electron spins.

atom in molecular orbitals that are similar to the atomic 1*s* and $2s2p^4$ states of the respective atoms but are slightly modified by the nearby presence of the other two atoms. Rather than form a symmetric molecule, the H_2O molecule is in its lowest energy state when the hydrogen atoms make an angle of about 105° with respect to one another, as shown in Figure 5.5. The length between the oxygen nucleus and the hydrogen nucleus is 0.958 Å. The covalent radius of H is 0.32 Å, and the covalent radius of O in a water molecule is 0.73 Å.

In Figure 5.5, the large blue spheres indicate the location of the electron associated with each hydrogen atom, and the red sphere indicates the location of the electrons associated with the oxygen atom. However, because the 10 electrons are in molecular rather than atomic orbitals, they interact with one another and with the neighbor nuclei and, in a sense, can lose knowledge of their original parent nucleus; the less tightly bound electrons are now in states of the molecule. The lowest energy state occurs when the angle subtended between the two hydrogen nuclei is 105°.

Figure 5.6 shows the electron density contours of the molecular orbits. Because oxygen is a relatively electronegative atom, it tends to attract more electrons on that side of the molecule at the expense of electrons associated with the hydrogen atoms.

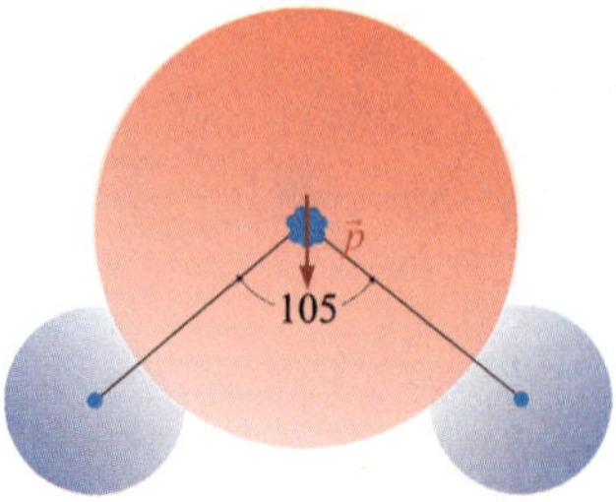

Figure 5.5 Graphic orientation of one oxygen and two hydrogen atoms in an H_2O molecule.

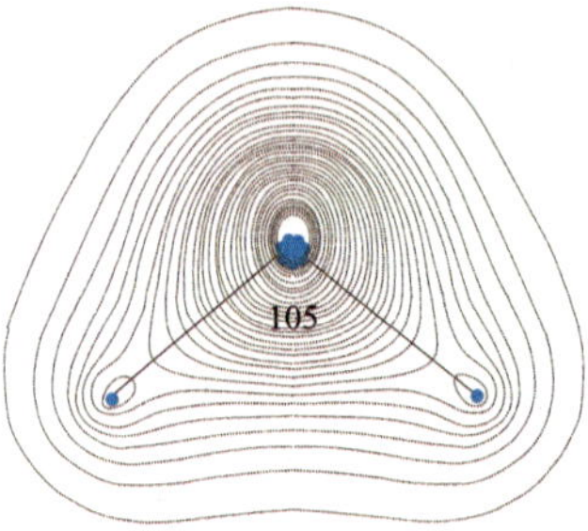

Figure 5.6 Electron density lines for an isolated water molecule.

This geometric shift of the electron cloud center of mass is about –0.7 e on the O atom with +0.15 e associated with each of the hydrogen atoms, which causes a permanent electric dipole moment, $\vec{p}$, of the molecule in its ground state as shown in Figure 5.6. A neutral water molecule in its vapor state has an electric dipole moment of 6.2×10^{-30} C m.

The moment of inertia of the water molecule about a vertical axis passing through the center of mass in Figure 5.5 or 5.6 permits a low-energy rotational mode about that axis, but there are two other perpendicular axes with different moments of inertia as possible rotational axes. A general rotation would have components of all three.

Water Dimers

In *liquid* water, molecules possess a distribution of electric dipole moments that range between $(6.2 - 10.3) \times 10^{-30}$ C m because of the various environments of the H_2O molecules. Water molecules in the liquid state do not stay together for more than a few milliseconds because of the exchange of hydrogen atoms between neighbors (a process known as protonation/deprotonation). However, this residence time is much greater than the time scale of a typical applied electric field so that we can usually treat the molecule as a permanent structure. With this model, the structure of water clusters has been shown to produce dimers in which an adjacent hydrogen atom is attracted toward the oxygen atom of a neighbor H_2O molecule, as shown in Figure 5.7.

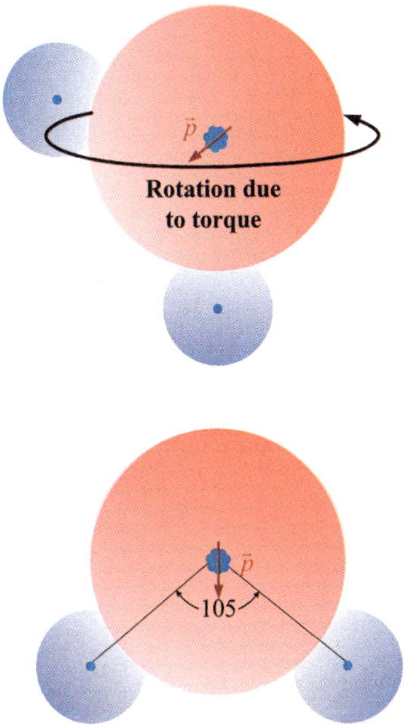

Figure 5.7 Relative orientation of two water molecules.

Pauling and others have shown that the upper water molecule has its lowest energy when the hydrogen atoms are oriented with one close to the other oxygen atom as shown. However, the thermal interactions between adjacent water molecules are large enough to make this subtle orientation only a statistical average. Furthermore, if there is a torque applied to the electric dipole moments, $\vec{p}$, by an external electric field intensity, either of the two molecules can rotate about an axis between the two oxygen atoms, but there will be a slight minimum in the dimer energy when the top molecule has its hydrogen atom rotated by 90° in or out of the page.

Water Trimers

If water molecules are clustered in a group of three or more, rotation of the middle molecule will possibly result in the breaking of a neighbor hydrogen/oxygen bond, as shown in Figure 5.8. This loss of energy produces internal heat in the liquid that is used in microwave ovens.

Water Molecules Near Other Hydrogen Rich Materials

Although not perceived as very reactive, water molecules can interact with neighboring materials that have a rich hydrogen concentration such as FR-4. The interactions should be dependent on water concentration, temperature, and the strength of the hydrogen bond. The attraction of the water molecule to neighbor atoms also causes a restoring force, hence, an effective spring constant and a change in the damping

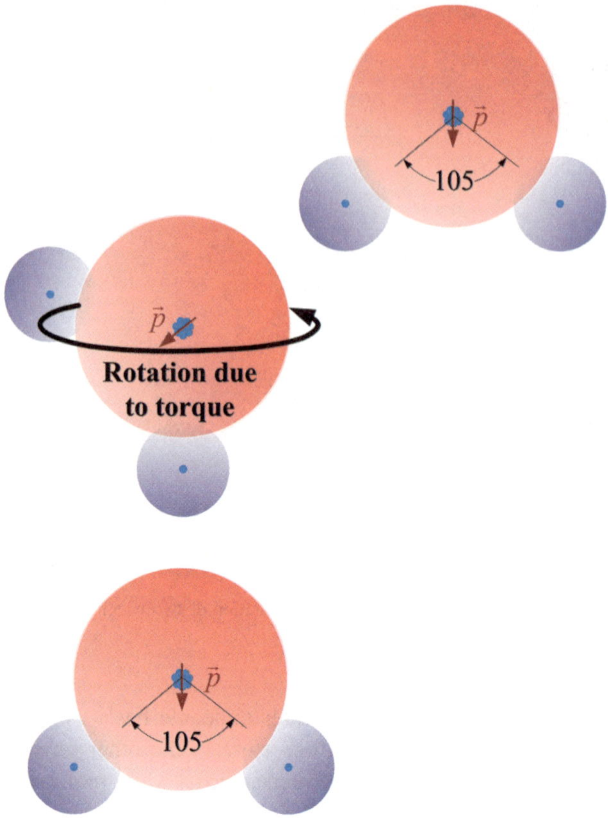

Figure 5.8 Relative orientation of three water molecules.

constant (discussed in the following section) associated with its possible rotations (called *librations*) and vibrations. Because there are likely to be many different environments in an electronic application (e.g., water absorbed in the propagating medium of a PCB), there are likely to be many different absorption peaks with different widths and resonant frequencies in a typical application. Even worse, the environment is likely to be dependent on the local temperature so the orientation will change with time as the input power of an incident electromagnetic wave changes.

Water Molecule in an External E&M Wave

A simple model of an isolated water molecule in an external electromagnetic field is found at a University of Colorado, physics department site, http://www.colorado.edu/physics/2000/applets/h2o.html. The applet permits the viewer to increase the applied power so that the electric dipole moment more closely responds to the applied electric field. In the applet, we can see that the water molecule lags the

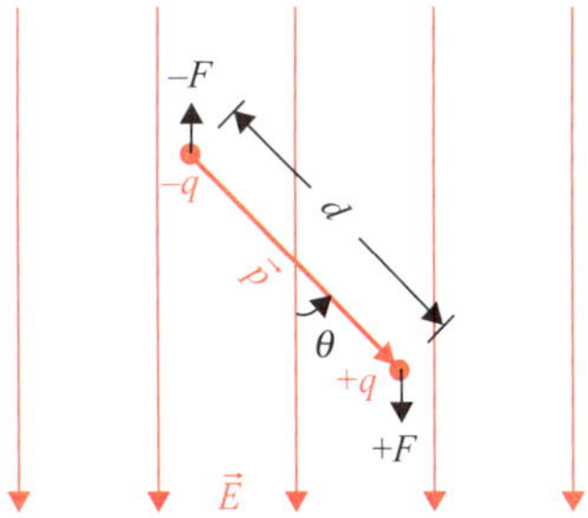

Figure 5.9 Torque on an electric dipole due to an external electric field intensity.

applied field and that the molecule rotation often overshoots the new equilibrium position because of its inertia. In this model, there is no other restoring torque than the external field and no damping torques.

In a more realistic model of Figure 5.8, there is only a very small restoring force for rotation, about one of the O-H axes, because there is a preferred orientation of the other proton relative to the other protons on the neighboring water molecule, but there is a large damping force brought about by neighboring atoms (especially in liquid water). The orientation polarization of water and other molecules was discussed by Debye in 1912 and published[3] in 1929 for molecules that possess a permanent electric dipole moment, $\vec{p}_0$, that can be oriented in an arbitrary direction. In the absence of external electric field intensity, thermal agitation will keep the molecules randomly oriented, so there will be no net electric dipole moment in a sample. When external electric field intensity is applied to the sample, there will be a tendency of the individual dipoles to align with the external field in a configuration that gives the lowest energy, and this will produce a net average electric dipole moment for the sample. The torque, $\vec{\tau} = qd|\vec{E}|\hat{a}_\perp = \vec{p}_0 \times \vec{E} = -|\vec{p}_0||\vec{E}|\sin\theta\hat{a}_\perp$, applied to an electric dipole moment, $\vec{p}_0$, in external electric field intensity, $\vec{E}$, is shown in Figure 5.9.

The potential energy of the electric dipole as shown aligned with the external electric field intensity to the angle θ relative to $\theta = 0$ is thus

$$U(\theta) = \int_0^\theta |\vec{p}_0||\vec{E}|\sin\theta d\theta = p_0 E(1-\cos\theta) \quad (5.2)$$

But the thermal energy associated with dipoles at temperature, T, will cause an assembly of such dipoles to fall into a Boltzmann distribution. Thus, the average electric dipole moment in the direction of the field will be given by

$$\langle|\vec{p}|\rangle = \frac{\int_0^{2\pi}\int_0^{\pi} e^{-\left(\frac{-p_0 E}{k_B T}\right)\cos\theta}(p_0\cos\theta)\sin\theta d\theta d\phi}{\int_0^{2\pi}\int_0^{\pi} e^{-\left(\frac{-p_0 E}{k_B T}\right)\cos\theta}\sin\theta d\theta d\phi} = p_0\left[\coth\left(\frac{p_0 E}{k_B T}\right) - \left(\frac{k_B T}{p_0 E}\right)\right], \quad (5.3)$$

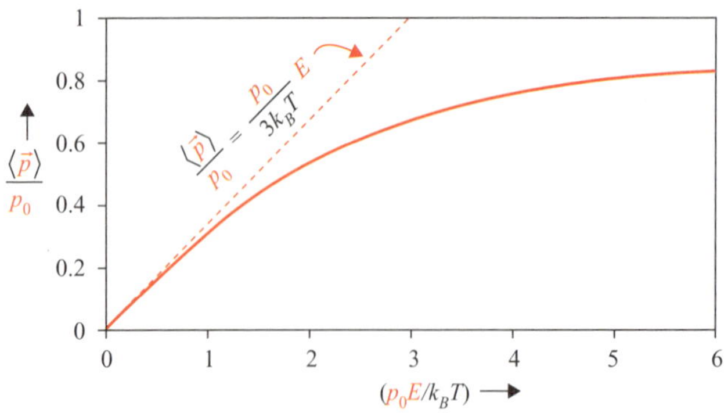

Figure 5.10 Fractional average electric dipole moment, $\langle \vec{P} \rangle / p_0$, in the direction of applied electric field intensity, E, as a function of the quantity (p_0E/k_BT).

where $p_0 = |\vec{p}_0|$, $E = |\vec{E}|$, and the function in the square brackets of Equation 5.3 is called the Langevin[4] function and is shown plotted in Figure 5.10. Note that the constant term p_0E in the potential energy Equation 5.2 plays no role in the statistical average Equation 5.3.

Note that Equation 5.3 may be found by introducing the constant $x = p_0E/k_BT$ and the variable $\mu = \cos\theta$ and carrying out the integral over the variable ϕ so that $\langle|\vec{p}|\rangle/p_0 = \int_{-1}^{1} e^{x\mu}\mu d\mu \Big/ \int_{-1}^{1} e^{x\mu} d\mu$. We observe that the numerator of this expression is the differential of the denominator with respect to x and that the value of the denominator is $\int_{-1}^{1} e^{x\mu} d\mu = (e^{x} - e^{-x})/2$.

At room temperature, p_0E will be small compared with k_BT, so we can write the polarization as[5]

$$\vec{P} = N\langle \vec{p}_{mol} \rangle \approx \frac{N}{3}\frac{p_0^{\,2}}{k_BT}\vec{E}, \tag{5.4}$$

where N is the number of electric dipole moments per cubic meter.

We can also write an equation of motion for the molecules that form the permanent electric dipole moments by equating the torque on the electric dipole with its moment of inertial times the angular acceleration as

$$\vec{\tau} = \vec{p} \times \vec{E} = -p_0E\sin\theta\,\hat{a}_\perp = I_\perp \frac{d^2\theta_\perp}{dt^2}\hat{a}_\perp, \tag{5.5}$$

where the $\perp$ symbol indicates rotation about an axis perpendicular to both $\vec{p}_0$ *and* $\vec{E}$ (out of the page) and $I_\perp$ is the moment of inertial relative to rotation about that axis.

For example, if free water molecules in a gas are isolated, as shown in Figure 5.11, then rotation can occur about any one of the three principal axes of rotation about the center of mass of the water molecule.

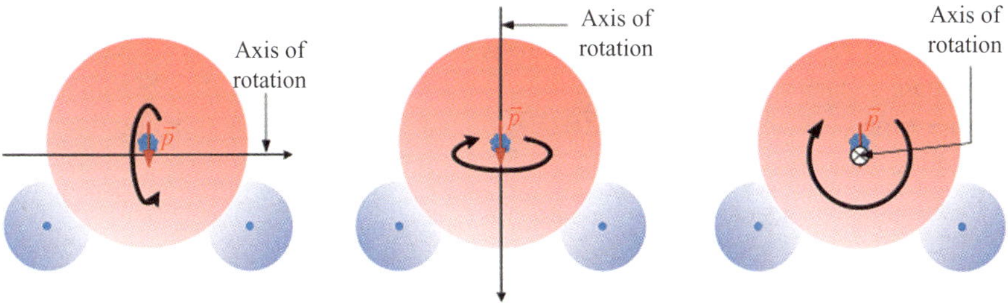

Figure 5.11 Principal moments of inertia for rotation of a free water molecule about the x-, y-, or z-axis.

It can be seen from Figure 5.11 that, if the electric field intensity is harmonic in the x-direction, the molecule can absorb energy by rotating about its lowest and highest moment of inertia axes but not its intermediate axis of rotation because that one always has its electric dipole moment down. Thus, if free water molecules existed in a microstrip, we would expect only two kinds of rotation when the frequency of the external electric field was resonant with those rotations. This topic will be considered again at the end of this chapter.

If the water molecules strongly interact with one another, as they certainly would in a liquid state, then drag forces or damping forces would exist and these often occur as proportional to the rotational velocity of the molecules. In that case, the equation of motion (Equation 5.5) would contain an additional term $-bd\theta/dt$ on the right-hand side, where b would be called the damping coefficient. If there were a restoring force, an additional term $-k\theta$ would exist on the right-hand side where k would be called the spring constant, Equation 5.5 could then be modified as the equation of motion:

$$\frac{d^2\theta}{dt^2} + \frac{b}{I}\frac{d\theta}{dt} + \frac{k\theta}{I} = -\frac{p_0}{I} E(t)\sin\theta \qquad (5.6)$$

Using the applied field $E(t) = \mathrm{Re}[E_0 e^{-i\omega t}]$ in Equation 5.6 would yield the frequency dependence.

In *Polar Molecules*, Debye presented a generalization of materials with orientational polarizability, in which he neglected the restoring force by assuming that the dielectric polarization is only *relaxing*; that is, the period of the electromagnetic wave is comparable with the alignment time of the molecule or atom. When the applied frequency was much greater than the reciprocal of the alignment time, Debye called the relative permittivity the infinite relative dielectric constant, $\varepsilon_{r\infty}$. As the frequency approached zero, Debye called the relative permittivity the static relative dielectric constant, ε_{rs}. In order to establish the frequency dependence of the relative permittivity, Debye found a relationship of the form $\varepsilon_r(\omega) = \varepsilon_{r\infty} + f(\omega)$, where $f(\omega)$

is a function of frequency for which $f(0) = \varepsilon_{rs} - \varepsilon_{r\infty}$. If a steady field is applied to align the molecules and then switched off, the polarization and the local field will diminish, so Debye *assumed* that the electric polarization, P, decayed exponentially with a time constant, τ, that is, $P(t) = P_0 e^{-t/\tau}$ and found that the function $f(\omega)$ was related to $P(t)$ by the Fourier transform:

$$f(\omega) = A\int_0^\infty P(t)e^{-i\omega t}dt = \frac{AP_0}{(1/\tau) - i\omega},$$

where A was a constant that ensured that $f(\omega)$ had the correct dimensions. Using the condition when $\omega = 0$ that $AP_0\tau = \varepsilon_{rs} - \varepsilon_{r\infty}$, he thus deduced that

$$\varepsilon_r(\omega) = \varepsilon_{r\infty} + \frac{(\varepsilon_{rs} - \varepsilon_{r\infty})}{1 - i\omega\tau_e}, \tag{5.7}$$

which is called the Debye *relaxation* equation.

Empirical Permittivity of Liquid Water Molecules

The Web site http://www.lsbu.ac.uk/water/microwav3.html summarizes work of others that parametrically shows that "the most appropriate description of the dielectric properties of liquid water is a bimodal relaxation time expression of the Debye form."

$$\varepsilon_r(\omega) = \varepsilon_\infty + \frac{\varepsilon_s - \varepsilon_2}{1 - i\omega\tau_D} + \frac{\varepsilon_2 - \varepsilon_\infty}{1 - i\omega\tau_2} \tag{5.8}$$

as long as the values of the fit parameters given in Figure 5.12 are used.

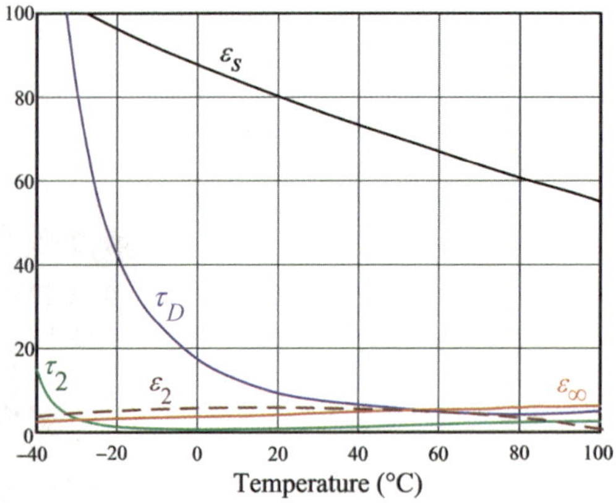

Figure 5.12 Values of the parameters to be used in Equation 5.8 for pure water over the range 20–40°C extrapolated (dashed lines) to indicate trends; relaxation times are in picoseconds.

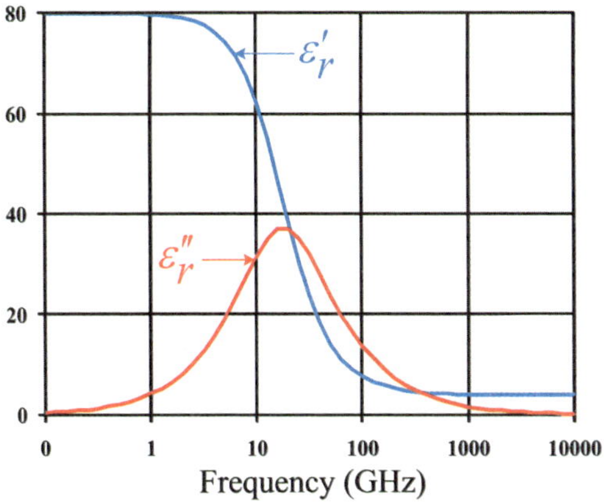

Figure 5.13 Real part, ε_r', and imaginary part, ε_r'', of the electric permittivity of water at 20°C.

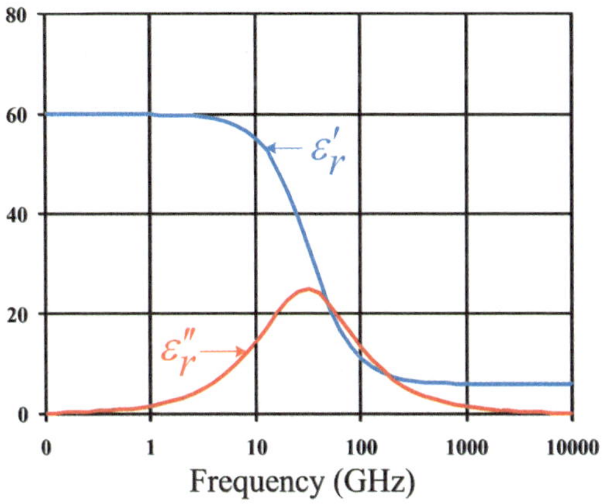

Figure 5.14 Real part, ε_r', and imaginary part, ε_r'', of the electric permittivity of water at 100°C.

Resolving the real and imaginary parts of Equation 5.8,

$$\varepsilon_r = \varepsilon_r' + i\varepsilon_r'' = \left[\varepsilon_\infty + \frac{\varepsilon_S - \varepsilon_2}{1+\omega^2\tau_D^2} + \frac{\varepsilon_2 - \varepsilon_\infty}{1+\omega^2\tau_2^2}\right] + i\left[\frac{(\varepsilon_S - \varepsilon_2)\omega\tau_D}{1+\omega^2\tau_D^2} + \frac{(\varepsilon_2 - \varepsilon_\infty)\omega\tau_2}{1+\omega^2\tau_2^2}\right] \tag{5.9}$$

which is shown plotted in Figure 5.13 for liquid water at a temperature of 20°C.

This analysis is adequate for water losses in the neighborhood of 2.45 GHz used for heating food in microwave ovens. We can see from the red curve in Figure 5.13 that the resonance of liquid water predicted by this fit would occur at about 20 GHz.

For a temperature of 100°C, the data of Equation 5.9 may be used to describe the permittivity of water, as shown in Figure 5.14. We can see from the blue curve that the classical value of relative permittivity decreases from its room temperature value of 80 to a value of 60 and, from the red curve in Figure 5.14, that the resonance of liquid water predicted by this fit would occur at about 32 GHz.

Good Questions

Is it possible to have absorbed water in PCB transmission lines? Does the relative humidity of air in Arizona differ from that in the Philippians or on a ship? Would the partial pressure of water in a microscopic void cause water to be in its liquid or vapor form? Is the resonance line width lower for gas than for liquid molecules? Would the absorbed humidity change with time? As the transmission line heats up after turning on a laptop, does the water absorbed in a PCB shift its resonance frequency? Would the water resonance change from location to location on the motherboard because of local heating? Would the absorbed water resonance in a PCB be the same as that of pure water or that of salt water? (See the web site http://www.lsbu.ac.uk/water/microwave.html for the change in Figure 5.12 constants with salt content.) Is it possible that some fraction of the water molecules interact with the walls of the PCB void in which it exists? Would the resonant frequency be the same for those molecules as for pure water? Would the effect change with the size and number of the voids? As a reference point, the phase diagram of water is shown in Figure 5.15.

Polar Molecules in Precipitation or Vapor State

The propagating medium in a PCB is designed to be relatively free of permanent dipole moments, but there are sure to be inclusions or voids that fill with liquids or atmosphere used in the manufacturing process. From planetary atmospheric radar studies (Ghosh), it is known that molecules such as H_2O, N_2O, NO_2, S_2, H_2S, and

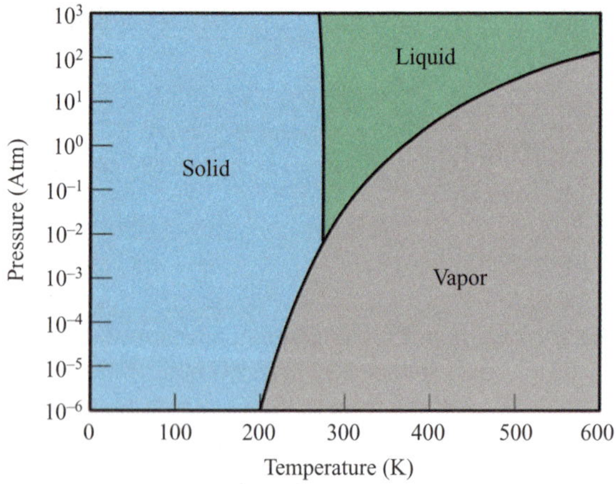

Figure 5.15 Phase diagram of water as a function of pressure and temperature.

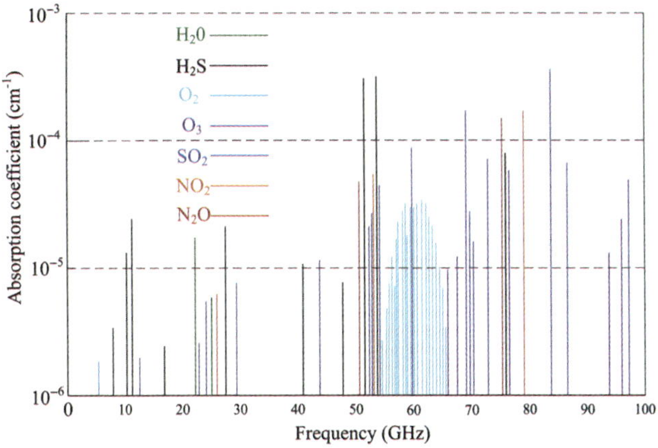

Figure 5.16 Absorption coefficients of atmospheric electromagnetic waves as a function of frequency for molecules with electric or magnetic dipole moments.

CO have rotational spectra that appear at microwave frequencies because of their electric dipole moments. The molecule O_2 does not have an electric dipole moment but possesses a magnetic dipole moment that gives rise to absorpti on lines between 55 and 65 GHz. N_2 does not have a dipole moment. Some known frequencies for molecular resonances are shown in Figure 5.16.

5.3 INDUCED DIPOLE MOMENTS

Materials used in practical media consist of atoms with unfilled electron shells. Incident electric field intensity will cause those electrons to oscillate relative to their nuclei, and that, in turn, will *induce* an electric dipole moment. The combined effects of N such atoms per unit volume will yield an additional polarization for the media, which will change its permittivity. Depending on how the electrons oscillate relative to the incident field, the additional polarization can be large (e.g., if the electrons have a resonant frequency). Depending on whether or not the driving electric field intensity stimulates oscillations below or above resonance, the additional polarization will be more in phase or out of phase with the incident field. This complicated behavior can cause a permittivity that is highly dependent upon frequency. Because the phase velocity, u_p, of a medium is inversely proportional to the square root of the permittivity, u_p will also depend on frequency. Thus, an electromagnetic pulse propagating in such a medium will have its high-frequency components traveling at a different phase velocity than its low-frequency components. The net result is that electromagnetic pulses will *disperse* in most media. In order to model electromagnetic wave propagation in media, we need to know the cause of such dispersion and how it will behave mathematically with frequency. Such a model is presented below, and restrictions on the real and imaginary parts of the analytic permittivity are shown through Kramers–Kronig relations that follow.

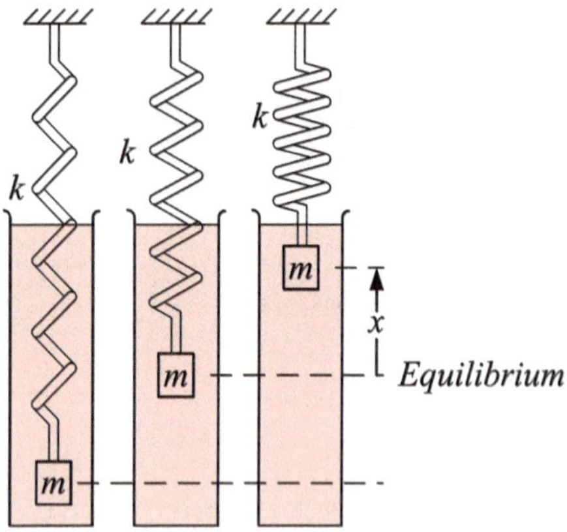

Figure 5.17 A mechanical oscillator of mass, m, and spring constant, k, immersed in a damping medium.

Causal Model

The analogy between electron charges in an unfilled shell and those of a mass on a damped mechanical spring yields the same set of mathematical equations, so their behavior is often described by means of a mechanical analog. For example, the charged mass, m, shown in Figure 5.17 is normally at its equilibrium position but, when displaced by an amount x through the application of an external force, experiences a restoring force, $\vec{F}_{spring} = -kx\hat{a}_x$, tending to bring it back to its equilibrium position. In this model, it is also assumed that there is a damping or drag force that provides a force in the direction opposite to and with a magnitude proportional to its velocity, $\vec{F}_{drag} = -b\dfrac{dx}{dt}$.

The equation of motion for the mass m is classically given by

$$m\frac{d^2x}{dt^2}\hat{a}_x = -k\,x\,\hat{a}_x - b\frac{dx}{dt}\hat{a}_x + \vec{F}_{external}, \tag{5.10}$$

which is in the form of a second-order, linear, inhomogeneous differential equation. If the external force is caused by a polarized electric field intensity, $\vec{F}_{external} = -eE(\vec{x}, t)\hat{a}_x$,

$$\frac{d^2x}{dt^2} + \frac{b}{m}\frac{dx}{dt} + \frac{k}{m}x = -\frac{e}{m}E(\vec{x}, t) \tag{5.11}$$

The solution to this equation is a linear combination of the general solution to the homogeneous differential equation (DE) and the particular solution to the inhomogeneous DE.

The Homogeneous DE

$$\frac{d^2x}{dt^2}+\frac{b}{m}\frac{dx}{dt}+\frac{k}{m}x=0 \tag{5.12}$$

has solutions

$$x_h(t)=e^{-(b/2m)t}\begin{pmatrix} e^{-i\omega' t} \\ e^{i\omega' t}\end{pmatrix} \quad \text{where } \omega'=\sqrt{k/m-b^2/4m^2} \tag{5.13}$$

PROBLEM

5.1 Verify that Equation 5.13 satisfies the homogeneous differential equation 5.12, discuss the solution for three cases, $\frac{k}{m}>\frac{b^2}{4m^2}$, $\frac{k}{m}=\frac{b^2}{4m^2}$, $\frac{k}{m}<\frac{b^2}{4m^2}$, sketch the displacement Re $[x_h(t)]$ as a function of time for small values of b, and show that, in the limit as $b\to 0$, the solution is that of an undamped oscillator.

The Inhomogeneous DE for the Special Case of a Harmonic External Field

For harmonic external electric fields, $E(\vec{x})e^{-i\omega t}$, so if we make the approximation that the amplitude of the electric field is the same for all displacements, x, that is, $E(\vec{x})=E_0$, we are assuming that displacements are much smaller than variations in the electric field (the wavelength, $\lambda=2\pi u_p/\omega$) and may use the method of undetermined coefficients to find a particular solution to the inhomogeneous differential Equation 5.11 to be

$$x_p(t)=\frac{-(e/m)E_0e^{-i\omega t}}{\left(k/m-\omega^2\right)-i(b/m)\omega} \tag{5.14}$$

PROBLEMS

5.2 Verify that Equation 5.14 satisfies the inhomogeneous DE 5.11.

5.3 Verify that the most general sum of Equations 5.13 and 5.14

$$x_h(t)+x_p(t)=Ae^{-\frac{b}{2m}t+i\sqrt{\frac{k}{m}-\frac{b^2}{4m^2}}\,t}+Be^{-\frac{b}{2m}t-i\sqrt{\frac{k}{m}-\frac{b^2}{4m^2}}\,t}-\frac{(e/m)E_0e^{-i\omega t}}{\left(k/m-\omega^2\right)-i(b/m)\omega} \tag{5.15}$$

with A and B as arbitrary constants chosen to satisfy the initial conditions, also satisfies Equation 5.11.

The displacement of an electron in a constituent atom or molecule by an amount $\vec{x}(t) = [x_h(t) + x_p(t)]\hat{a}_x$ will cause a time-dependent electric dipole moment, $\vec{p}(t) = -e\vec{x}(t)$, that can be computed through the use of Equation 5.15. For small values of damping coefficient b, we can see that the first two terms in Equation 5.15 are oscillating at a frequency $\omega' = \sqrt{k/m - b^2/4m^2}$ and are exponentially damped in time. For values of ω much lower than ω', these oscillations will time-average to a null value of $\vec{p}_{avg}(t)$ For values of ω close to ω', the last term in Equation 5.15 yields a resonance which in turn produces a large value

$$\vec{p}(t) = \frac{e^2/m}{(k/m - \omega^2) - i(b/m)\omega} E(\vec{x}, t)\hat{a}_x \tag{5.16}$$

If there are Z polarizable electrons in each atom with different spring constants, k_i, and different damping coefficients, b_i, and there are N_i atoms per unit volume with individual polarizability[1] of α_i then the volume average polarization in the medium (the macroscopic polarization, $\vec{P}(t)$) can be written as

$$\vec{P}(t) = \sum_{i=1}^{Z} \frac{N_i \alpha_i e^2/m}{(k_i/m - \omega^2) - i(b_i/m)\omega} E(\vec{x}, t) a_x \tag{5.17}$$

The permittivity of the medium can thus be written[2] as

$$\varepsilon_r'(\omega) + i\varepsilon_r''(\omega) = 1 + \frac{\vec{P}(t)}{\varepsilon_0 \vec{E}(\vec{x}, t)} = 1 + \sum_{i=1}^{Z} \frac{N_i \alpha_i e^2/\varepsilon_0 m}{(k_i/m - \omega^2) - i(b_i/m)\omega} \tag{5.18}$$

The real and imaginary parts of Equation 12.18 are thus found to be

$$\varepsilon_r'(\omega) = 1 + \sum_{i=1}^{Z} \frac{N_i \alpha_i e^2/\varepsilon_0 m (k_i/m - \omega^2)}{(k_i/m - \omega^2)^2 + ((b_i/m)\omega)^2} \tag{5.19}$$

$$\varepsilon_r''(\omega) = \sum_{i=1}^{Z} \frac{N_i \alpha_i e^2/\varepsilon_0 m ((b_i/m)\omega)}{(k_i/m - \omega^2)^2 + ((b_i/m)\omega)^2} \tag{5.20}$$

The complex permittivity is sometimes written as

$$\varepsilon_r(\omega) = \frac{\varepsilon(\omega)}{\varepsilon_0} = \varepsilon_r'(\omega) + i\varepsilon_r''(\omega) = \varepsilon_r'(\omega)[1 + i\tan\delta(\omega)], \tag{5.21}$$

[1] In quantum theory, this term corresponds to the **oscillator strength** of an electric dipole transition between atomic states. In classical theory, this term corresponds to the probability that an electron in a molecule or atom will effectively take part in the polarization.

[2] The effective mass of polarizable electrons in an atom might vary or there might be a contribution of a charged ion with respect to other ions in the material. These effects may be accounted for by changing the value of the ith resonant frequency to $\omega_i = \sqrt{k_i/m_i}$.

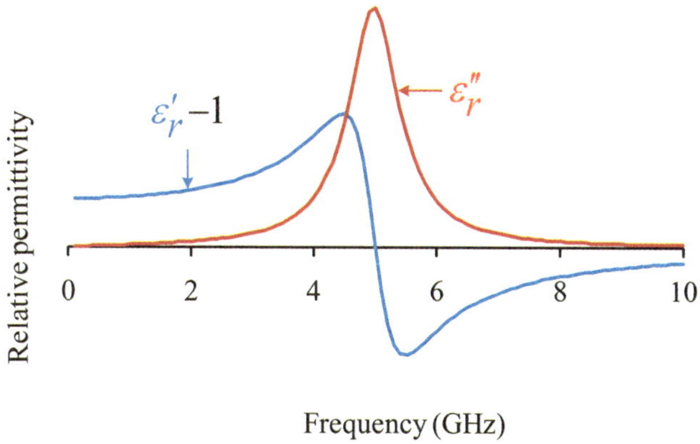

Figure 5.18 Relative real part of the permittivity, $\varepsilon_r'(\omega)-1$, and imaginary part, $\varepsilon_r''(\omega)$, for values of $\sqrt{k_1/m}=5\,GHz$ and $b_1/m = 1$ GHz.

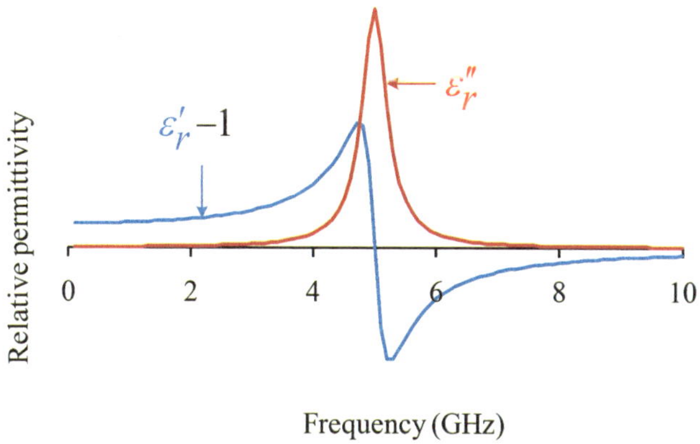

Figure 5.19 Relative real part of the permittivity, $\varepsilon_r'(\omega)-1$, and imaginary part, $\varepsilon_r''(\omega)$, for values of $\sqrt{k_1/m}=5$ GHz and $b_1/m = 0.5$ GHz.

where

$$\tan\delta(\omega)=\varepsilon_r''(\omega)/\varepsilon_r'(\omega) \tag{5.22}$$

Let us make several observations about these answers:

- $\varepsilon_r''(\omega)$ is normally much smaller than $\varepsilon_r'(\omega)$ because b_i is normally much smaller than k_i so $\tan\delta(\omega)\approx\delta(\omega)$.
- $\varepsilon_r''(\omega)$ is the sum of a set of Lorentzian functions, each of which are always a positive quantity as a function of ω.
- $\varepsilon_r'(\omega)$ is larger than 1 for $\omega<\sqrt{k_i/m}$ and is smaller than 1 for $\omega>\sqrt{k_i/m}$.

The quantities $\varepsilon_r'(\omega)$ and $\varepsilon_r''(\omega)$ are shown plotted as a function of ω for several values of spring constant and drag coefficient in Figures 5.18, 5.19, and 5.20.

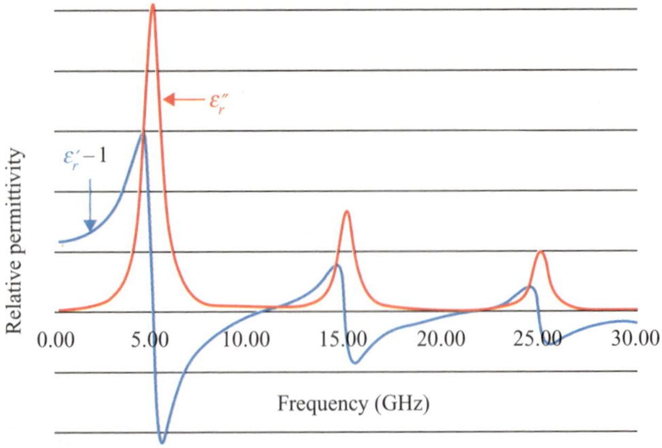

Figure 5.20 Relative real part of the permittivity, $\varepsilon_r'(\omega)-1$, and imaginary part, $\varepsilon_r''(\omega)$, for $b_i/m = 1.0$ GHz, $\sqrt{k_1/m} = 5$ GHz, $\sqrt{k_2/m} = 15$ GHz, and $\sqrt{k_3/m} = 25$ GHz.

In Figure 5.18, higher-frequency resonances, if any, have been ignored to understand the character of the individual resonance at the lowest frequency. It is seen that the choice of constant $\sqrt{k_1/m} = 5$ GHz approximately corresponds to the first moment of the $\varepsilon''_r(\omega)$ spectrum.

Figure 5.19 shows the same plot with $b_1/m = 0.5$ GHz in order to demonstrate that this value corresponds to the second moment of the frequency distribution of $\varepsilon''_r(\omega)$.

Figure 5.20 shows a plot of relative permittivity with $\sqrt{k_1/m} = 5$ GHz, $\sqrt{k_2/m} = 15$ GHz, and $\sqrt{k_3/m} = 25$ GHz with $b_1/m = b_2/m = b_3/m = 1.0$ GHz to show the additive effect of three different resonances. Note that the low-frequency value of $\varepsilon'_r(\omega) - 1$ is higher than the value above each resonance so that, if there are even higher-frequency resonances, they will each contribute to the real part of the permittivity at low frequencies.

Response Function

A consequence of the frequency dependence of permittivity is that

$$\vec{D}(\vec{x},\omega) = \varepsilon(\omega)\vec{E}(\vec{x},\omega) \tag{5.23}$$

By using the Fourier integral representations of $\vec{D}$ and $\vec{E}$,

$$\vec{D}(\vec{x},t) = (1/2\pi)\int_{-\infty}^{\infty} \vec{D}(\vec{x},\omega)e^{-i\omega t}d\omega \tag{5.24}$$

$$\vec{E}(\vec{x},t) = (1/2\pi)\int_{-\infty}^{\infty} \vec{E}(\vec{x},\omega)e^{-i\omega t}d\omega \tag{5.25}$$

and their inverse transformations

$$\vec{D}(\vec{x},\omega)=\int_{-\infty}^{\infty}\vec{D}(\vec{x},t)e^{i\omega t}dt \tag{5.26}$$

$$\vec{E}(\vec{x},\omega)=\int_{-\infty}^{\infty}\vec{E}(\vec{x},t)e^{i\omega t}dt, \tag{5.27}$$

it is found that

$$\vec{D}(\vec{x},t)=(1/2\pi)\int_{-\infty}^{\infty}\varepsilon(\omega)\vec{E}(\vec{x},\omega)e^{-i\omega t}d\omega \tag{5.28}$$

and, by using the inverse Fourier transform of $\vec{E}(\vec{x}, \omega)$ from Equation 5.27 in Equation 5.23,

$$\vec{D}(\vec{x},t)=(1/2\pi)\int_{-\infty}^{\infty}\varepsilon(\omega)\int_{-\infty}^{\infty}\vec{E}(\vec{x},t')e^{i\omega t'}dt'e^{-i\omega t}d\omega \tag{5.29}$$

and, exchanging the order of integration,

$$\vec{D}(\vec{x},t)=\int_{-\infty}^{\infty}\vec{E}(\vec{x},t')\left[(1/2\pi)\int_{-\infty}^{\infty}\varepsilon(\omega)e^{-i\omega(t-t')}d\omega\right]dt' \tag{5.30}$$

Now, if the value of $\varepsilon(\omega)$ as given by Equation 5.18 is used in Equation 5.30,

$$\vec{D}(\vec{x},t)=\int_{-\infty}^{\infty}\vec{E}(\vec{x},t')\left\{(1/2\pi)\int_{-\infty}^{\infty}\left[\varepsilon_0+\sum_{i=1}^{Z}\frac{N_i\alpha_i e^2/m}{(k_i/m-\omega^2)-i(b_i/m)\omega}\right]e^{-i\omega(t-t')}d\omega\right\}dt' \tag{5.31}$$

The first of the integrals (with ε_0) is a delta function, so

$$\vec{D}(\vec{x},t)=\varepsilon_0\vec{E}(\vec{x},t)+\int_{-\infty}^{\infty}\vec{E}(\vec{x},t')\left\{(1/2\pi)\int_{-\infty}^{\infty}\left[\sum_{i=1}^{Z}\frac{N_i\alpha_i e^2/m}{(k_i/m-\omega^2)-i(b_i/m)\omega}\right]e^{-i\omega(t-t')}d\omega\right\}dt'$$

and, changing the variable in the time integral to $\tau=(t-t')$,

$$\vec{D}(\vec{x},t)=\varepsilon_0\vec{E}(\vec{x},t)-\int_{-\infty}^{\infty}\vec{E}(\vec{x},t-\tau)\left\{(1/2\pi)\int_{\infty}^{-\infty}\left[\sum_{i=1}^{Z}\frac{N_i\alpha_i e^2/m}{(k_i/m-\omega^2)-i(b_i/m)\omega}\right]e^{-i\omega\tau}d\omega\right\}d\tau$$

or

$$\vec{D}(\vec{x},t)=\varepsilon_0\vec{E}(\vec{x},t)+\int_{-\infty}^{\infty}\varepsilon_0\vec{E}(\vec{x},t-\tau)G(\tau)d\tau \tag{5.32}$$

$$G(\tau)=\frac{e^2}{2\pi\varepsilon_0 m}\sum_{i=1}^{Z}N_i\alpha_i\oint_{C+C_R}\frac{e^{-i\omega\tau}}{(\omega-\omega_-)(\omega-\omega_+)}d\omega \quad \text{for } \tau>0$$
$$=0 \quad \text{for } \tau<0 \tag{5.40}$$

Because there are two poles of order 1 within the closed path of Figure 5.22, we may use the Cauchy integral theorem for its evaluation, so

$$G(\tau)=-i\frac{e^2}{\varepsilon_0 m}\sum_{i=1}^{Z}N_i\alpha_i\left[\left.\frac{e^{-i\omega\tau}}{(\omega-\omega_+)}\right|_{\omega_-}+\left.\frac{e^{-i\omega\tau}}{(\omega-\omega_-)}\right|_{\omega_+}\right] \text{for } \tau>0 \tag{5.41}$$

$$G(\tau)=i\frac{e^2}{\varepsilon_0 m}\sum_{i=1}^{Z}N_i\alpha_i\left[\frac{e^{-(b_i/2m)\tau}e^{i\sqrt{k_i/m-(b_i^2/4m^2)}\,\tau}}{2\sqrt{k_i/m-b_i^2/4m^2}}-\frac{e^{-(b_i/2m)\tau}e^{-i\sqrt{k_i/m-(b_i^2/4m^2)}\,\tau}}{2\sqrt{k_i/m-b_i^2/4m^2}}\right] \text{for } \tau>0$$

$$G(\tau)=\frac{e^2}{\varepsilon_0 m}\sum_{i=1}^{Z}N_i\alpha_i e^{-(b_i/2m)\tau}\frac{\sin\sqrt{k_i/m-(b_i^2/4m^2)}\,\tau}{\sqrt{k_i/m-b_i^2/4m^2}} \text{ for } \tau>0 \tag{5.42}$$

Equation 5.42 tells us that the response function is exponentially damped with time constant $2m/b_i$ and oscillates in time with a frequency $\omega'=\sqrt{k_i/m-b_i^2/4m^2}$ of the different oscillators. The response connection $\vec{D}(\vec{x}, t) = \varepsilon_0\vec{E}(\vec{x}, t) + \int_0^\infty \varepsilon_0\vec{E}(\vec{x}, t-\tau)G(\tau)d\tau$ thus says that the electric flux density is different from $\varepsilon_0\vec{E}(\vec{x}, t)$ by an amount computed from the convolution of the response function for times of order $2m/b_i$. However, we have seen from Figures 5.18 and 5.19 that b_i/m corresponds to the second moment (the width) in frequency of the different spectral lines.

PROBLEMS

5.4 Evaluate and plot $G(\tau)$ vs τ for the two cases given in Figures 5.18 and 5.19.

5.5 Verify that $G(\tau)$ as given in Equation 5.42 when used in Equation 5.34, yields Equation 5.18.

5.5 FREQUENCY CHARACTER OF THE PERMITTIVITY

From Equation 5.36, it can be deduced that

$$[\varepsilon^*(\omega)/\varepsilon_0-1]=\int_0^\infty G(\tau)e^{-i\omega\tau}d\tau \tag{5.43}$$

for a real response function $G(\tau)$ such as the one found in Equation 5.42 so that

$$\varepsilon^*(\omega)=\varepsilon(-\omega) \tag{5.44}$$

from which the real part of $\varepsilon(\omega)$ is even in ω, while the imaginary part of $\varepsilon(\omega)$ is odd:

																	He $1s^2$ **0.223**
Li^+ 2s **0.032**	Be^{2+} $2s^2$ **0.01**											B^{3+} $2s^22p$ **0.003**	C^{4+} $2s^22p^2$ **0.001**		O^{3+} $2s^22p^4$ **2.67**	F^- $2s^22p^5$ **0.953**	Ne $2s^22p^6$ **0.433**
Na^+ 3s **0.322**	Mg^{2+} $3s^2$ **0.104**											Al^{3+} $3s^23p$ **0.058**	Si^{4+} $3s^23p^2$ **0.018**		S^{2-} $3s^23p^4$ **6.11**	Cl^- $3s^23p^5$ **3.27**	Ar $3s^23p^6$ **1.80**
K^+ 4s **1.259**	Ca^{2+} $4s^2$ **1.22**	Sc^{3+} $4s^23d$ **3.18**	Ti^{4+} $4s^23d^2$ **1.00**												Se^{3+} $4s^24p^4$ **7.80**	Br^- $4s^24p^5$ **4.55**	Kr $4s^24p^6$ **2.73**
Rb^+ 5s **1.87**	Sr^{2+} $5s^2$ **1.78**	Y^{3+} $5s^24d$ **0.61**	Zr^{4+} $5s^24d^2$ **0.41**												Te^{2-} $5s^25p^4$ **10.0**	I^- $5s^25p^5$ **6.80**	Xe $5s^25p^6$ **4.43**
Cs^+ 6s **3.05**	Ba^{2+} $6s^2$ **2.78**	La^{3+} $6s^25d$ **1.16**															
			Ce^{4+} $6s^24f^2$ **0.81**														

Figure 5.23 Measured electronic polarizability, α_i, in 10^{-40} m^3/atom.

$$\varepsilon_r'(-\omega) = \varepsilon_r'(\omega)$$
$$\varepsilon_r''(-\omega) = -\varepsilon_r''(\omega) \tag{5.45}$$

NOTE Equations 5.19 and 5.20 are consistent with Equation 5.45.
NOTE If Equation 5.36 represents $\varepsilon(\omega)$ in the complex ω-plane, then $\varepsilon(\omega)$ is analytic in the entire ω-plane (including the upper half) as long as $G(\tau)$ is finite for all τ. This requires only that $\lim_{\tau\to\infty} G(\tau) = 0$, which for lossy dielectrics, is true, as given by Equation 5.42.

Measured Values of Polarizability, α

We have considered only relative values of permittivity in the above analysis because values of the quantities, Z, N_i, and α_i are difficult to evaluate from *ab initio* calculations. Mossotti (in 1850), Clausius (in 1879), and Debye (in 1929) tried to evaluate the collective effects of these quantities through measured values of the permittivity. Pauling (in 1927), Tressman, Kahn, and Shockley (in 1953), and Jaswal and Sharma (in 1973) have measured the electronic polarizability of various ions (Figure 5.23) at optical frequencies. The schemes are not entirely self-consistent because the polarizability of an ion depends on the environment and on the sample in which it is placed. We have always selected the latest measurement.

Measured values depend upon the sample and Lorentz cavity polarization field, P, that opposes the applied field, E_0, in a sample, and that depends on its geometry (see Chapter 3) according to the expression[6]

$$E = E_0 - NP/\varepsilon_0, \tag{5.46}$$

where N is the depolarization factor: for a sphere $N = 1/3$; for a thin slab normal to its axis $N = 1$; for a thin slab in the plane of its axis $N = 0$; for a long, circular cylinder in the longitudinal axis $N = 0$; and for a long, circular cylinder transverse to its axis $N = 1/2$. Lorentz used the local electric field experienced by an atom at the center of a sphere as

$$E_{local} = E_{Macroscopic} + P/3\varepsilon_0 \tag{5.47}$$

because of the arrangement of its neighbors will be the geometric field plus the fields caused by all other dipoles in the material.[7] For ions in a cubic symmetry, the contributions of atoms inside the sphere vanish. However, in a noncubic crystal, the dielectric response must be described by the components of the dielectric constant tensor, and they will, in general, lead to variations in measurements that depend upon sample orientation relative to the lattice structure.

As found above, the polarizability, α_i, of an atom is defined in terms of the local electric field intensity at an atom by $p_i = \alpha_i E_{local}(i)$ and the polarization of a crystal may be expressed approximately as $P = \sum_i N_i \alpha_i E_{local}(i)$, where N_i is the concentration of atoms of type i per unit volume and α_i is the polarizability of atoms of type i. If the local field is given by Equation 5.47,

$$P = \left(\sum_i N_i \alpha_i\right)\left(E_{Macroscopic} + P/3\varepsilon_0\right) \tag{5.48}$$

from which we can evaluate $\varepsilon_r - 1 = \chi_e = P/E_{Macroscopic}$ to find

$$\frac{\varepsilon_r - 1}{\varepsilon_r + 2} = \frac{1}{3\varepsilon_0}\sum_i N_i \alpha_i \tag{5.49}$$

the **Clausius–Mossotti** relation. This equation gives us the sum of the number of atoms of type i per unit volume times their electronic polarizability by measuring the permittivity of the material for crystal structures for which Equation 5.47 holds.

At optical frequencies, the terms that contribute to the permittivity, ε, are due primarily to the electronic polarizability as shown for case (b) above, and we may use the index of refraction relation, $n^2 = \varepsilon_r$, to evaluate α_i of different ions with known concentration.

In a complex medium like FR-4, Figure 5.23 gives us some trends about the character of different constituents, but it would be unlikely that the conditions of measurements would be the same in this medium because of the different σ and π bonds involved in the organic molecules. Nevertheless, we see that constituent atoms (O, S, Se, Te, F, Cl, Br, and I) with nearly complete electronic shells give relatively large contributions to the electronic polarization. We would not expect to find atoms of the rare gasses (He, Ne, Ar, Kr, or Xe) in a propagating medium used for electronic devices, but it is surprising that these filled shell atoms (particularly the heavier ones) can contribute to the electronic polarization in other applications. Lighter elements in the periodic table contribute relatively small amounts to the electronic polarization.

We can reason that the relative electric permittivity for a single atomic species, as shown in Figure 5.23 with i = 1, can be written as

$$\varepsilon_r(\omega) = 1 + \frac{N_1 \alpha_1 e^2 / \varepsilon_0 m}{(k_1/m - \omega^2) - i(b_1/m)\omega} \tag{5.50}$$

and that the last term in Equation 5.50 goes to zero as $\omega \to \infty$ so the relative electronic permittivity at $\omega = 0$ can be written as

$$(\varepsilon_{rs} - \varepsilon_{r\infty}) = (N_1 \alpha_1 e^2 / \varepsilon_0 m) / (k_1/m), \tag{5.51}$$

where $\varepsilon_{r\infty}$ is the **measured** value of $\varepsilon(\omega)$ at very high frequency (e.g., optical frequencies) and ε_{rs} is the **measured** "static" value of ε_r. Thus, setting $\omega_i = \sqrt{k_i/m}$ and $\tau_e = (b_1/m)/(k_1/m) = (b_1/k_1)$,

$$\varepsilon_r(\omega) = \varepsilon_{r\infty} + \frac{(\varepsilon_{rs} - \varepsilon_{r\infty})}{1 - (\omega/\omega_1)^2 - i\tau_e\omega} \tag{5.52}$$

The real and imaginary parts of this solution are

$$\varepsilon_r'(\omega) = \varepsilon_{r\infty} + \frac{(\varepsilon_{rs} - \varepsilon_{r\infty})\left[1 - (\omega/\omega_1)^2\right]}{\left[1 - (\omega/\omega_1)^2\right]^2 + \tau_e^2 \omega^2} \tag{5.53}$$

$$\varepsilon_r''(\omega) = \frac{(\varepsilon_{rs} - \varepsilon_{r\infty})\tau_e \omega}{\left[1 - (\omega/\omega_1)^2\right]^2 + \tau_e^2 \omega^2} \tag{5.54}$$

DeBye Equation

In his work, *Polar Molecule* (1929), Debye examined materials with orientation polarization in the region where the dielectric polarization is *relaxing*. Debye *assumed* that the polarization decays only exponentially with a time constant τ, the characteristic relaxation time of the dipole moment of the molecule (see Problem 5.1 for critically-damped or overdamped motion):

$$P(t) = P_0 e^{-t/\tau_e}, \tag{5.55}$$

which yields a Fourier transform of the form

$$\varepsilon_r(\omega) = \varepsilon_{r\infty} + \frac{(\varepsilon_{rs} - \varepsilon_{r\infty})}{1 - i\tau_e\omega} \tag{5.56}$$

now known as the Debye equation. We note that this equation is the same as Equation 5.52 if we ignore the term $(\omega/\omega_1)^2$ in the denominator. This approximation would be valid if this quantity were small in comparison with 1, which is true for resonant frequencies ω_l in the optical range of 10^{15} Hz if we are measuring the permittivity in the range below 10^{11} Hz. The values of polarizability found in Figure 5.23 were all measured by optical means, and, thus, the resonances for the induced electric dipole moments for these atoms are all at or above 10^{15} Hz. Authors who use the Debye equation to fit permittivity measurements in the sub-terahertz range are thus justified in its application for *these* atoms even though the dipoles induced in these atoms experience strong restoring forces.

5.6 KRAMERS–KRONIG RELATIONS FOR INDUCED MOMENTS

Kramers (1927) and Kronig (1926) independently derived the relationship between the real and imaginary parts of the permittivity (and other complex functions) by recognizing that the analytic function $[\varepsilon(\omega)/\varepsilon_0 - 1]$ must satisfy Cauchy's integral theorem:

$$[\varepsilon(z)/\varepsilon_0 - 1] = \frac{1}{2\pi i}\oint_C \frac{[\varepsilon(\omega')/\varepsilon_0 - 1]}{\omega' - z}\,d\omega', \tag{5.57}$$

where the integrand has a pole of order 1 in the complex ω' plane at $\omega' = z$.

Kramers and Kronig chose the path C to be the real ω' axis and completed the closed path in the integral by a large semicircle in the upper half of the complex ω' plane, as shown in Figure 5.24.

Equation 5.57 gives the value of the permittivity at any complex point in the upper half of the ω' plane in terms of its integral along the path C (the real ω' axis) because the integral over the path C_R goes to zero. We can thus rename the point z as ω and write Equation 5.57 as

$$[\varepsilon(\omega)/\varepsilon_0 - 1] = \frac{1}{2\pi i} P\int_{-\infty}^{\infty} \frac{[\varepsilon(\omega')/\varepsilon_0 - 1]}{\omega' - \omega}\,d\omega', \tag{5.58}$$

where $\omega' = \omega$ is any point in the upper half ω' plane (including points z on the real ω' axis). Kramers and Kronig pointed out that, if the pole of order 1 is on the path

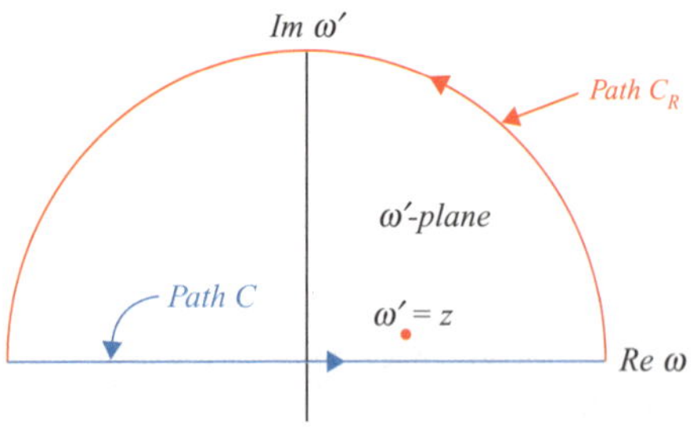

Figure 5.24 Path of integration for the evaluation of the Kramers–Kronig integration in Equation 5.57.

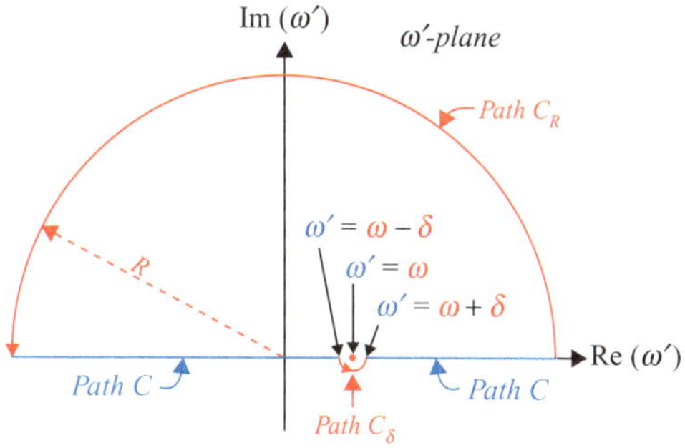

Figure 5.25 Principle value of the integral in Equation 5.59 including the remainder of the path on C_R and the small semicircle around the point $\omega' = \omega$.

of integration, we must use the *Principle value*[4] to describe the answer because the integral is not defined for a path that passes through a singularity. As shown in the footnote, P means the integral over all of the path C *except* the point $\omega' = \omega$. Because we are considering real values for ω, we may take the integral path C to be that shown in Figure 5.25.

Because the integral over the path C_R yields zero, the *Principle value* will be given by Cauchy's integral theorem *minus* the integral over the small semicircle. This amounts to an answer that is half as large as the integral over a whole circle and gives

$$[\varepsilon(\omega)/\varepsilon_0 - 1] = \frac{1}{\pi i} P\int_{-\infty}^{\infty} \frac{[\varepsilon(\omega')/\varepsilon_0 - 1]}{\omega' - \omega} d\omega' \tag{5.59}$$

The real and imaginary parts of this equation are

$$\mathrm{Re}[\varepsilon(\omega)/\varepsilon_0 - 1] = \frac{1}{\pi} P\int_{-\infty}^{\infty} \frac{\mathrm{Im}\,\varepsilon(\omega')/\varepsilon_0}{\omega' - \omega} d\omega' \tag{5.60}$$

$$\mathrm{Im}[\varepsilon(\omega)/\varepsilon_0 - 1] = -\frac{1}{\pi} P\int_{-\infty}^{\infty} \frac{\mathrm{Re}[\varepsilon(\omega')/\varepsilon_0 - 1]}{\omega' - \omega} d\omega' \tag{5.61}$$

Integrals 5.60 and 5.61 can thus be transformed to span only positive values of ω by using the symmetry properties given in Equation 5.45 in which case they can be written as

$$\varepsilon_r'(\omega) = 1 + \frac{2}{\pi} P\int_0^{\infty} \frac{\omega' \varepsilon_r''(\omega')}{\omega'^2 - \omega^2} d\omega' \tag{5.62}$$

[4] The Principle value is defined as

$$P\int_{-\infty}^{\infty} \frac{[\varepsilon(\omega')/\varepsilon_0 - 1]}{\omega' - \omega} d\omega' \equiv \lim_{\delta\to 0}\int_{-\infty}^{\omega-\delta} \frac{[\varepsilon(\omega')/\varepsilon_0 - 1]}{\omega' - \omega} d\omega' + \lim_{\delta\to 0}\int_{\omega+\delta}^{\infty} \frac{[\varepsilon(\omega')/\varepsilon_0 - 1]}{\omega' - \omega} d\omega'$$

$$\varepsilon_r''(\omega) = \frac{2\omega}{\pi} P\int_0^{\infty} \frac{1-\varepsilon_r'(\omega')}{\omega'^2 - \omega^2} d\omega' \tag{5.63}$$

Equations 5.62 and 5.63 are known as the Kramers–Kronig relations. These equations have very general applicability and follow from the *causal connection* between the real and the imaginary parts of the permittivity. They are useful in practical applications because it is sometimes possible to measure the loss component of the permittivity, $\varepsilon_r''(\omega)$, from absorption studies of the medium of interest. With this information and Equation 5.62, it is then possible to construct the value of the real part of the permittivity, $\varepsilon_r'(\omega)$, that is compliant with causal connections. The inverse statement is also true that, if the real part of an analytic function is known, then the imaginary part may be found from the other Kramers–Kronig relation.

5.7 ARBITRARY TIME STIMULUS

We previously solved Equation 5.11, $\frac{d^2x}{dt^2} + \frac{b_i}{m}\frac{dx}{dt} + \frac{k_i}{m}x = -\frac{e}{m}E(\vec{x}, t)$, by assuming external harmonic electric field intensity driving fields of the form, $E(\vec{x},t) = E(\vec{x})\, e^{-i\omega t}$. In the case of an *arbitrary* time-dependent electric field intensity, we can consider $E(\vec{x},t) = E(\vec{x}) f(t)$, where $f(t)$ could be a delta function, a Heavyside function, a square pulse, a Gaussian, a trapezoidal function, a harmonic function, or any other arbitrary time dependence. George Green solved such problems by solving a related equation, now called Green's differential equation:

$$\frac{d^2G(t;t')}{dt^2} + \frac{b_i}{m}\frac{dG(t;t')}{dt} + \frac{k_i}{m}G(t;t') = \delta(t-t') \tag{5.64}$$

Green then pointed out that, once we find $G(t;t')$ that satisfies Equation 5.64, we can find the particular solution to

$$\frac{d^2x}{dt^2} + \frac{b_i}{m}\frac{dx}{dt} + \frac{k_i}{m}x = -\frac{e}{m}E_0(\vec{x})f(t) \tag{5.65}$$

to be

$$x_p(t) = \int_{-\infty}^{\infty} -\frac{e}{m}E_0(\vec{x})f(t')G(t;t')\,dt' \tag{5.66}$$

PROBLEMS

5.6 Show that substitution of $x_p(t)$ as given in Equation 5.66 satisfies differential equation 5.65 as long as $G(t;t')$ satisfies Equation 5.64.

We can solve Equation 5.64 by writing $G(t;t')$ in terms of its Fourier transform and the delta function by its integral representation:

$$G(t;t') = \int_{-\infty}^{\infty} g(\omega) e^{-i\omega(t-t')} d\omega \tag{5.67}$$

$$\delta(t-t') = (1/2\pi)\int_{-\infty}^{\infty} e^{-i\omega(t-t')} d\omega \tag{5.68}$$

Putting these into Equation 5.64, we get

$$g(\omega) = \frac{1}{2\pi} \frac{1}{\left(k_i/m - i\omega(b_i/m) - \omega^2\right)} = \frac{-1}{2\pi} \frac{1}{(\omega-\omega_-)(\omega-\omega_+)}, \tag{5.69}$$

where

$$\omega_\pm = -i(b_i/2m) \pm \sqrt{k_i/m - (b_i/2m)^2} \tag{5.70}$$

and, putting this back into the Fourier transform 5.67,

$$\begin{aligned} G(t;t') &= \int_{-\infty}^{\infty} -\frac{1}{2\pi} \frac{1}{(\omega-\omega_-)(\omega-\omega_+)} e^{-i\omega(t-t')} d\omega \\ &= -\frac{1}{2\pi} \int_{\text{Path}\,C} \frac{e^{-i\omega(t-t')}}{(\omega-\omega_-)(\omega-\omega_+)} d\omega, \end{aligned} \tag{5.71}$$

where path C and the two singularities are the same as those shown in Figure 5.22.

Using the two contours shown in Figure 5.22, we conclude that

$$\begin{aligned} G(t;t') &= e^{-\frac{b_i}{2m}(t-t')} \frac{\sin\sqrt{k_i/m - (b_i/2m)^2}\,(t-t')}{\sqrt{k_i/m - (b_i/2m)^2}} && \text{for } t > t' \\ &= 0 && \text{for } t < t' \end{aligned} \tag{5.72}$$

5.7 Show that substitution of $G(t;t')$ as given by Equation 5.72 into the integral 5.66 with $f(t') = \varepsilon^{-j\omega t'}$ yields the same answer we previously deduced for $x_p(t)$ in Equation 5.14.

For the *arbitrary* time dependence, $f(t)$, we can conclude that

$$\begin{aligned} x_p(t) &= -\frac{e}{m} E(\vec{x}) \int_{-\infty}^{\infty} f(t') e^{-(b_i/2m)(t-t')} \frac{\sin\sqrt{k_i/m - (b_i/2m)^2}\,(t-t')}{\sqrt{k_i/m - (b_i/2m)^2}} dt' && \text{for } t > t' \\ &= 0 && \text{for } t < t' \end{aligned} \tag{5.73}$$

and, by substituting the variable $\tau = (t - t')$,

$$x_p(t) = \frac{-e}{m} E(\vec{x}) \int_0^\infty f(t-\tau) e^{-(b_i/2m)\tau} \frac{\sin\sqrt{k_i/m-(b_i/2m)^2}\,\tau}{\sqrt{k_i/m-(b_i/2m)^2}} d\tau \tag{5.74}$$

The integral is called the convolution of $f(t)$ and $G(t;t')$ and is written $f(t) * G(t)$.

EXAMPLES

5.1 Suppose an electric field intensity is produced uniformly in the x direction at $z = 0$ and undergoes a step function in time, $f(t)$, as shown in Figure 5.26.

As the electric field intensity begins, we can assume it propagates in the z-direction at the propagation velocity, $u_p = c/\sqrt{\varepsilon_r}$. Thus, an electric dipole oscillator at a point z in the medium will experience an electric field intensity beginning at time $t_1 = z\sqrt{\varepsilon_r}/c$. For the dipole at point z, we can compute the displacement, $x(t - t_1)$, from Equation 5.74 to be

$$x(t-t_1) = \frac{-e}{m} E_p \int_0^\infty f(t-\tau) e^{-\frac{b_i}{2m}\tau} \frac{\sin\sqrt{k_i/m-(b_i/2m)^2}\,\tau}{\sqrt{k_i/m-(b_i/2m)^2}} d\tau, \tag{5.75}$$

where $f(t)$ is given by Figure 5.27.

The displacement of the bound charge at point z will be the convolution of $f(t)$ and the Green's function shown in Figure 5.28.

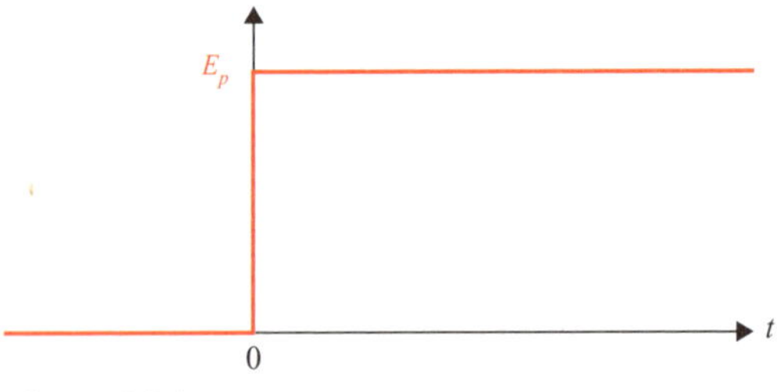

Figure 5.26 Example 1.1 electric field intensity in the x-direction at $z = 0$ as a function of time (E_p specifies the magnitude of the pulse).

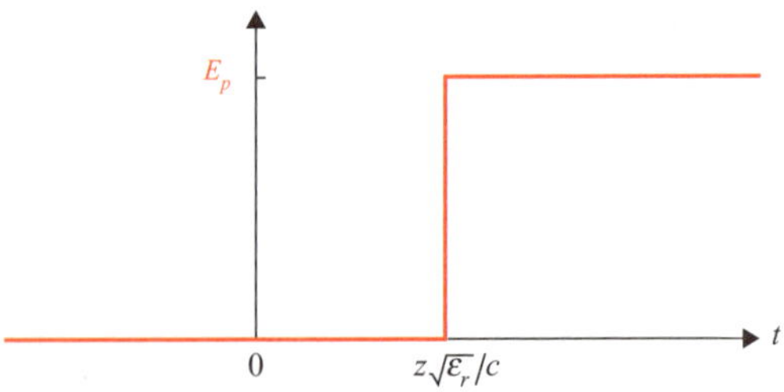

Figure 5.27 Time dependence of the electric field intensity for a bound charge at point z.

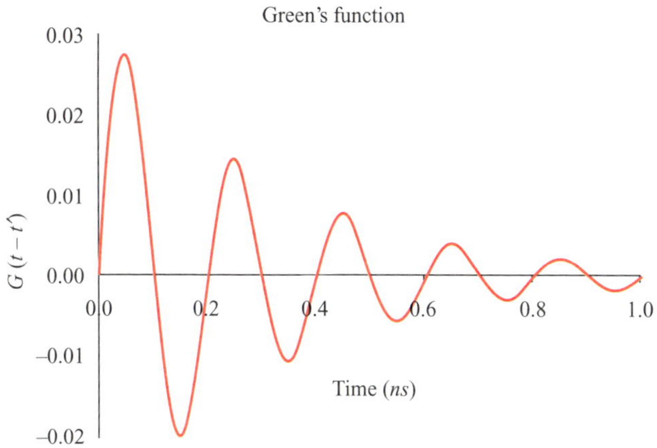

Figure 5.28 Green's function for the charge with spring constant $\sqrt{k_1/m} = 5$ GHz and damping constant $b_1/m = 1.0$ GHz.

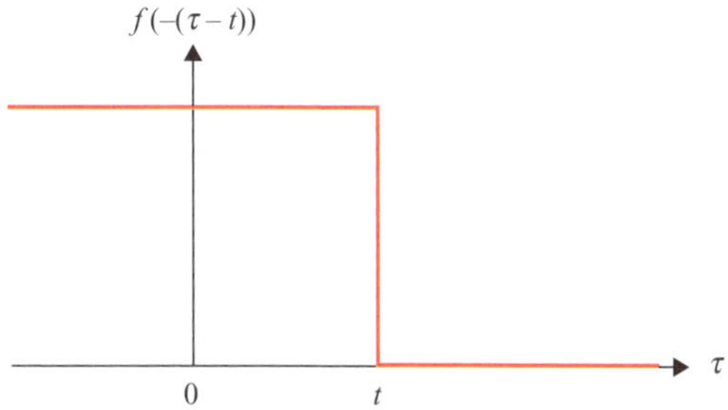

Figure 5.29 Time reversal and subsequent time shift of $f(t - \tau)$.

We may find the convolution by forming the equivalent integral:

$$f(t) * G(t) = \int_0^\infty f(t-\tau)G(\tau)\,d\tau = \int_0^\infty f(-(\tau - t))G(\tau)\,d\tau, \qquad (5.76)$$

where $f(-(\tau - t))$ is shown in Figure 5.29.

If we multiply Equation 5.29 by Equation 5.28 we form the integrand of the integral. Doing this for every time t and integrating the product we get the form shown in Figure 5.30.

EXAMPLE #2

5.2 Suppose an electric field intensity is produced uniformly in the x direction at $z = 0$ and undergoes a linear ramp in time, $f(t)$, as shown in Figure 5.31.

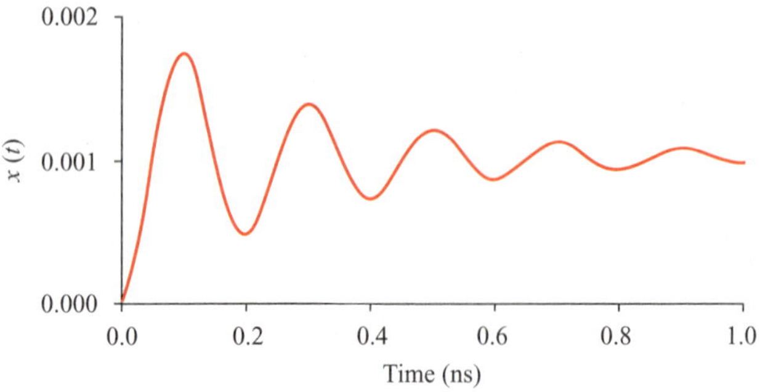

Figure 5.30 Convolution of Figures 5.26 and 5.28.

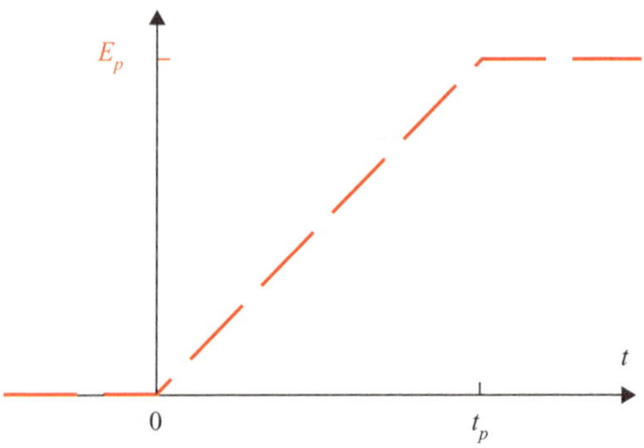

Figure 5.31 Sample ramp electric field intensity in the x-direction at $z = 0$ as a function of time.

As the electric field intensity begins to grow, it will propagate in the z-direction at the propagation velocity, $u_p = c/\sqrt{\varepsilon_r}$, and it will travel a distance $d_p = ct_p/\sqrt{\varepsilon_r}$ before it reaches its maximum E_p. Thus, an electric dipole oscillator at a point z in the medium will experience a growth in electric field intensity beginning at zero at time $t_1 = z\sqrt{\varepsilon_r}/c$ and rising to $E_p = E_p\hat{a}_x$ by time $t_2 = z\sqrt{\varepsilon_r}/c + t_p$; i.e., the maximum will be delayed by an amount t_p. This will in turn cause a delay in the response electric flux density in the same amount of time. The convolution integral is carried out in a manner similar to that for the step function and the results are shown in Figure 5.32.

In Figure 5.32 we see that the main influence of the more realistic ramp function is to delay the response in time but the absolute displacement is the same for both functions. This result implies that the electric dipoles formed by a ramp will occur later in time that those of a step function and thus the total field formed by the incident electric field intensity and that caused by the induced electric dipoles will occur later in time.

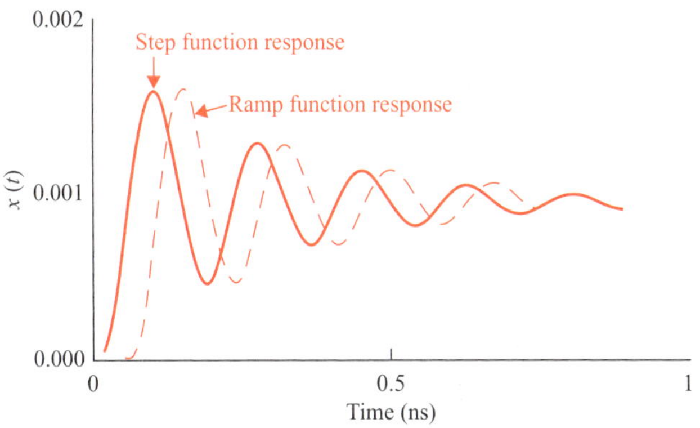

Figure 5.32 Convolution of the Green's function in Figure 5.28 with the ramp function in Figure 5.31 shown on the same graph with the previous result for the step function.

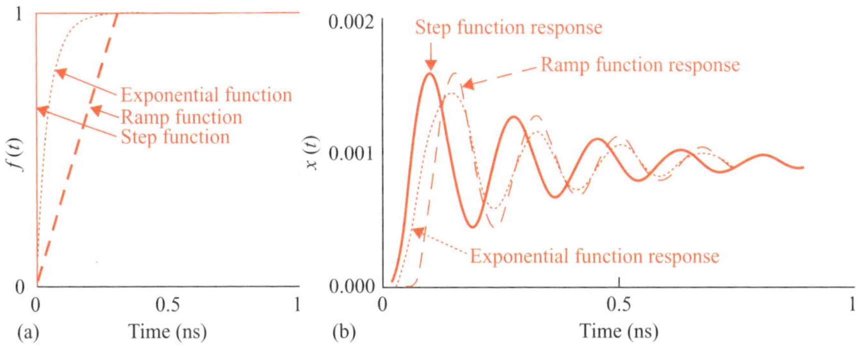

Figure 5.33 (a) Sample exponential electric field intensity in the x-direction at $z = 0$ as a function of time for a signal that has a ramp and an exponential plateau; (b) convolution of the Green's function in Figure 5.26, with the exponential plateau function in Figure 5.31 shown on the same graph with the previous results for the step and the ramp functions.

EXAMPLE #3

5.3 Suppose an electric field intensity is produced uniformly in the x direction at $z = 0$ and undergoes a linear ramp in time with an exponential plateau, $f(t)$, as shown in Figurc 5.33a.

The convolution integral for the ramp and the exponential plateau growth are shown in Figure 5.33b.

From Figure 5.33b we see that the effect of an even more realistic signal waveform (exponential plateau on a ramp function) is to produce the same absolute displacement of the induced electric dipoles, delay the response in time and to reduce the overshoot.

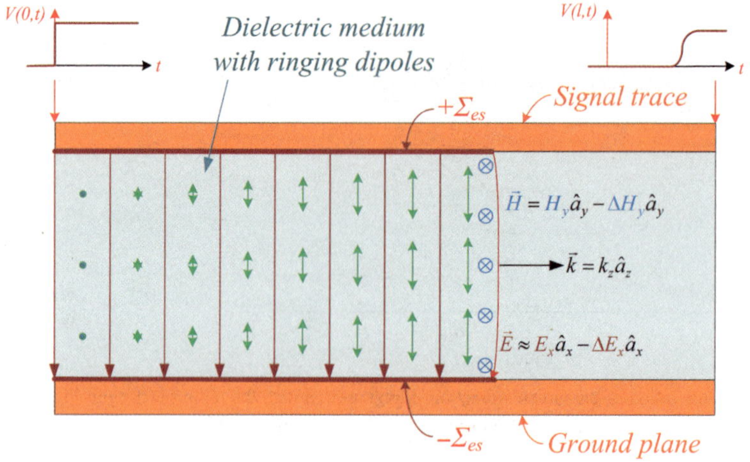

Figure 5.34 Ringing dipoles: snapshot of an initial propagating electric field intensity step function in time as it stimulates exponentially ringing and decaying electric dipoles in its "wake."

Conclusions

Electric dipoles in a electric field intensity (oriented in the x-direction but propagating in the z-direction), experience a transverse displacement relative to the direction of propagation. The size of the displacement depends on the number of electric dipoles per unit volume in the propagating medium. Electric dipoles of a given species oscillate at their resonant frequency and their amplitude decays according to a damping constant characteristic of that species and its environment; both characterize the constants in a response function. The signal pulse thus produces a trail of dipoles in its wake that oscillate and decay with time. The oscillating electric dipoles produce a subsequent electric field intensity that contributes to the incident wave front and leads to dispersion of the pulse. The dispersion is a property of the medium dipole species composition and their volume density. We can see the effects of dispersion by examining the dynamics of propagating electric field intensity as it stimulates dipoles in a microstrip or stripline as shown in Figure 5.34.

As electric field intensity passes by polarizable atoms or molecules it induces electric dipole moments which subsequently oscillate and decay as shown in Figure 5.33, depending on the leading edge nature of the waveform; in a sense it "rings the bell" of each induced dipole. With a small differential pickup probe in the propagating medium, and with repetitive positive and negative electric field intensity wavefronts phase locked to the front, could we sense the fluorescent fields produced by those dipoles in the "wake" of the front? By measuring their lifetimes, could we distinguish between over and under-damped oscillators? If so, would the decay constants yield loss factors from those induced dipoles? Would it be possible to distinguish between species of oscillators with different natural frequencies?

5.8 CONDUCTION ELECTRON PERMITTIVITY

For lossy insulators some fraction of the conduction electrons[8] are "free" to move when a DC [$\omega = 0$] electric field or a low frequency field is imposed. The free electrons produced a current density according to Ohm's law:

$$\vec{J} = \sigma_e \vec{E} \quad (\mathrm{A/m^2}) \tag{5.10}$$

and when we put this into Ampere's equation for a time harmonic field (with time dependence $e^{-i\omega t}$)

$$\begin{aligned}\vec{\nabla} \times \vec{H}(\vec{x}, t) &= \vec{J}(\vec{x}, t) + \partial \vec{D}(\vec{x}, t)/\partial t = \sigma_e \vec{E}(\vec{x}, t) - i\omega[\varepsilon' + i\varepsilon''] \vec{E}(\vec{x}, t) \\ &= -i\omega(\varepsilon' + i\varepsilon'' + i\sigma_e/\omega) \vec{E}(\vec{x}, t) \\ &= -i\omega\varepsilon_0 (\varepsilon_r' + i\varepsilon_r'' + i\sigma_e/\varepsilon_0\omega) \vec{E}(\vec{x}, t)\end{aligned} \tag{5.77}$$

A common practice is to attribute all of the properties of the medium to the dielectric constant by regarding the term in large parenthesis in Equation 5.77 as a complex relative permittivity in which the "normal" permittivity has been modified by the term $i\sigma_e/\varepsilon_0\omega$.

It is further commonly argued that for ***weakly conducting*** dielectrics (insulators) $\sigma_e \rightarrow$ small so that $\sigma_e/\varepsilon_0\omega << 1$. However, as we note below, even for dielectrics this interpretation causes a singularity in ε_r'' at $\omega = 0$. This formulation is also complicated by the fact that σ_e is a complex quantity as was recognized as early as 1900 by the Paul Drude model for electrical conductivity which yields the frequency dependence as

$$\begin{aligned}\sigma_e(\omega) &= \frac{N_f e^2}{m_e} \tau_s \left(\frac{1 + i\omega\tau_s}{1 + \omega^2\tau_s^2} \right) = \sigma_e(0) \left(\frac{1 + i\omega\tau_s}{1 + \omega^2\tau_s^2} \right) \\ &= \sigma_{real}(\omega) + i\sigma_{imaginary}(\omega)\end{aligned} \tag{5.78}$$

In the classical Drude model, the electron scattering time, τ_s, is related to the electron scattering time with defects, $\tau_{defects}$, and the scattering time with phonons, $\tau_{phonons}$, as $1/\tau_s = (1/\tau_{defects} + 1/\tau_{phonons})$ and accurately describes the residual resistivity of a metal as well as its temperature dependence as shown in Figure 5.35. At low temperatures the resistivity is mostly determined by the defects and is proportional to T^5. At higher temperatures the resistivity is mostly determined by the phonon collisions and is proportional to T.

Experimental measurements give values τ_s of order 2.5×10^{-14}s for copper and τ_s of order 1.5×10^{-12}s for silicon. The frequency dependence of the electrical conductivity of Equation 5.78 is shown plotted in Figure 5.36.

Surprisingly, the Drude model gives the same form as the 1990 quantum mechanical Sommerfeld model of conductivity. Thus, for frequencies up to 10^{11} Hz, the conductivity is thought to be essentially real, current is in phase with the field, and permittivity modification is inversely dependent on frequency. At higher fre-

$$\vec{D}(\vec{x},\omega)=\varepsilon_{dipoles}(\omega)\vec{E}(\vec{x},\omega)+\left[\sigma_e(0)\tau_s+\frac{j\sigma_e(0)}{\omega(1-j\omega\tau_s)}\right]\vec{E}(\vec{x},\omega) \tag{5.82}$$

As in the case of induced dipoles, the electric flux density because of the conductivity can be written in the time domain as a convolution, $\vec{D}(\vec{x},t) = \varepsilon_0\vec{E}(\vec{x},t) + \int_0^\infty \varepsilon_0\vec{E}(\vec{x},t - \tau)\, G(\tau)\, d(\tau)$ but here the integral has an additional term $\vec{D}(\vec{x},t) = \varepsilon_0\vec{E}(\vec{x},t) + \vec{D}_{dipoles}(\vec{x},t) + \vec{D}_e(\vec{x},t)$ where

$$\vec{D}_e(\vec{x},t)=\varepsilon_0\int_{-\infty}^{\infty}\vec{E}(\vec{x},t-\tau)\left\{\frac{1}{2\pi}\int_{-\infty}^{\infty}\left[\frac{\sigma_e(0)}{\varepsilon_0(1/\tau_s)}+\frac{i\sigma_e(0)}{\varepsilon_0\omega(1-i\omega\tau_s)}\right]e^{-i\omega\tau}d\omega\right\}d\tau \tag{5.83}$$

The term in square brackets ([]) is the conductivity response function and can be written as $G_e(\tau) = G_e^{'}(\tau) + jG_e^{''}(\tau)$. The convolution integral Equation 5.83 includes $\tau = (t - t') < 0$, which implies that the electric flux density is influenced by the interaction of the electric field intensity with the material properties *at a future time*. However, causality requires that $G_e(\tau) = 0$ for $\tau = (t - t') < 0$; there cannot be a response to a source that has not yet occurred. This condition will permit the evaluation of the integral in parenthesis as part of a contour integral in the ω-plane.

In Equation 5.83, the first term leads to a small shift in ε_0:

$$\vec{D}(\vec{x},t)=\varepsilon_0\left[1+\frac{\sigma_e(0)}{\varepsilon_0 2\pi(1/\tau_s)}\right]\vec{E}(\vec{x},t) + \varepsilon_0\int_{-\infty}^{\infty}\vec{E}(\vec{x},t-\tau)\left[\frac{1}{2\pi}\int_{-\infty}^{\infty}\frac{i\sigma_e(0)}{\varepsilon_0\omega(1-i\omega\tau_s)}e^{-i\omega\tau}d\omega\right]d\tau,$$

while the second term in the response function to be evaluated can be written as

$$G_e^{''}(\tau)=\frac{-\sigma_e(0)}{\varepsilon_0 2\pi\tau_s}\int_{-\infty}^{\infty}\frac{e^{-i\omega\tau}}{(\omega-0)(\omega+i/\tau_s)}d\omega, \tag{5.84}$$

where the integrand has a pole of order 1 at $\omega = 0$ and $\omega = -i/\tau_s$. The integrand is otherwise an analytic function so that it may be evaluated as part of a contour integral that is closed by an infinite semicircle at $\omega = \lim_{R\to\infty} Re^{i\theta}$ where $0 \le \theta \le \pi$ if $\tau = (t - t') < 0$ shown in Figure 5.38.

In Figure 5.38, the small semicircle around $\omega = 0$ is assumed to have an infinitesimal radius. An alternate way to picture this path is to say the pole at $\omega = 0$ has a vanishingly small negative imaginary component that causes it to lie just below the path of integration along the real ω-axis that needs to be interpreted in Equation 5.84. For $\tau = (t - t') \ge 0$, the integrand may be evaluated as part of a contour integral that is closed by an infinite semicircle at $\omega = \lim_{R\to\infty} Re^{i\theta}$ where $\pi \le \theta \le 2\pi$ if as shown in Figure 5.38. Line integrals that pass through a pole are not defined, so care must be used when evaluating Equation 5.84 at $\omega = 0$.

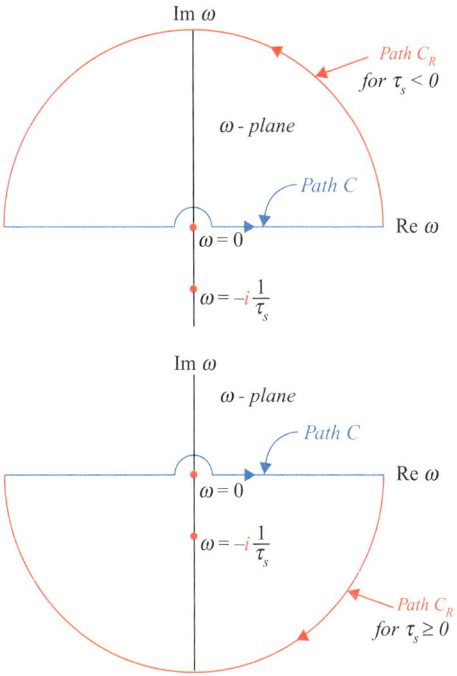

Figure 5.38 The top contour is used to evaluate the conductivity response function for $\tau = (t - t') < 0$. The bottom contour is used to evaluate the conductivity response function for $\tau = (t - t') \geq 0$.

Integration along the real ω-axis *cannot* be interpreted as its principal value because this interpretation gives a finite value for $G_e(\tau)$ for $t < t'$ that would violate causality. We can, however, evaluate the integral along *path C*, as shown in Figure 5.38, because this yields

$$G_e(\tau) = \begin{bmatrix} 0 \\ \dfrac{-\sigma_e(0)}{2\pi\tau_s}\displaystyle\int_{path\,C}\frac{e^{-i\omega\tau}\,d\omega}{(\omega-0)(\omega+i/\tau_s)} \end{bmatrix} \quad \text{for} \quad \begin{matrix} \tau < 0 \\ \tau \geq 0 \end{matrix} \tag{5.85}$$

Thus, there is an additional contribution to the permittivity caused by *conducting* dielectric materials:

$$\begin{aligned} &\vec{D}_e(\vec{x},t) = \int_0^\infty \vec{E}(\vec{x},t-\tau)G_e(\tau)d\tau \textit{ where} \\ &G_e(\tau) = \sigma_e(0)\left[1-e^{-\tau/\tau_s}\right] \end{aligned} \tag{5.86}$$

This result is a consequence of the simple pole of order 1 at $\omega = 0$ that we created in Equation 5.77 when we interpreted $\varepsilon_r''(\omega) = j\text{Re}\sigma_e/\varepsilon_0\omega$. This response function is also real so that we can conclude the same symmetry properties for the "conduction" permittivity as we found in Equation 5.45. However, we see that $\lim_{\tau \to \infty} G_e(\tau) \neq 0$ as was the case for induced dipoles, so we cannot assure analytic continuation into the complex ω–plane for this quantity.

Conduction Electron Conclusions and Frequency Symmetry

Electric fields that propagate in dielectrics induce electric dipole moments in the material that disperses the fields according to a convolution integral with a dipole response function in time, $G_{dipole}(\tau)$. If the propagating medium has conductivity, an additional response term, $G_e(\tau)$, gives an additional contribution to dispersion from the conduction electrons. Both response functions are real, so symmetry properties for the "dipole" and "conduction" permittivity are

$$\begin{aligned} \varepsilon_r'(-\omega) &= \varepsilon_r'(\omega) \\ \varepsilon_r''(-\omega) &= -\varepsilon_r''(\omega) \end{aligned} \tag{5.87}$$

5.10 PERMITTIVITY OF PLASMA OSCILLATIONS

Materials used in conducting media can be thought of as a lattice of positive ions balanced by a uniform distribution of conduction electrons that behave like a plasma cloud. An exaggerated graphic depiction of this configuration is shown in Figure 5.39, in which the red region represents the conduction electrons as a uniform negative charge cloud, and the blue spheres represent the ion cores in a lattice, each with a positive charge.

Longitudinal Plasmons

Upon the application of an external electric field intensity, $\vec{E}(z,t) = E_0 e^{j(k_z z - \omega t)} \hat{a}_z$, the electron cloud density at point z will experience a force to the left (as indicated by

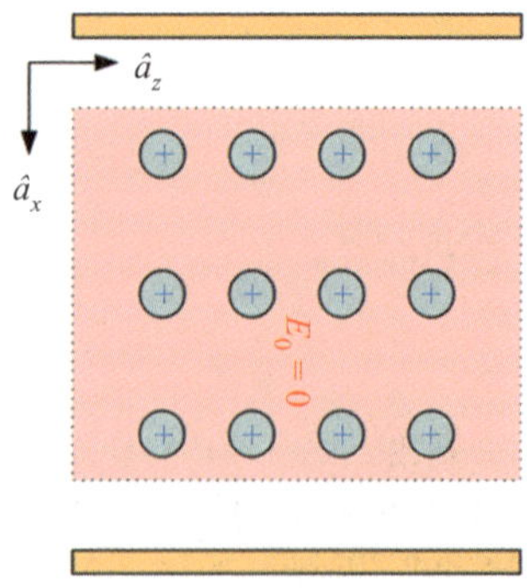

Figure 5.39 Model of a propagation material in a neutral plasma state; red represents a uniform electron charge density and blue represents the ion cores.

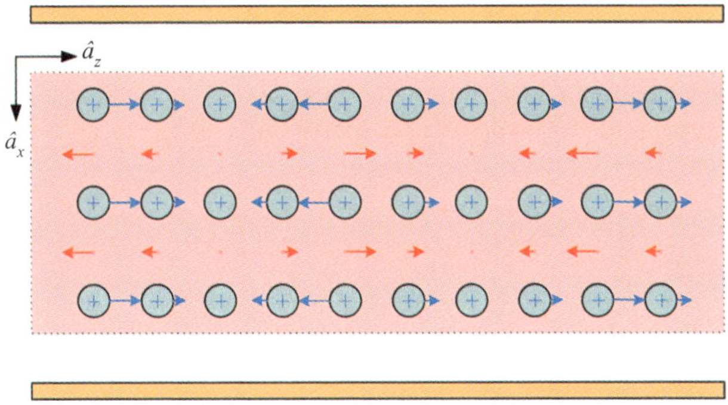

Figure 5.40 Forces on electrons and ion cores as they experience electric field intensity in the z-direction that varies harmonically with time. Here, the electric field intensity wave is assumed to be propagating in the z-direction.

the red arrows) in proportion to the magnitude of the electric field at the point z, as shown in Figure 5.40. The positive ion cores at point z will similarly experience a force to the right (as indicated by the blue arrows). The wave is propagating in the z-direction, so the electron cloud and the ions will be oppositely accelerated in accordance with the force and inversely to their mass. The electron cloud density will dilate and contract in a manner called a *longitudinal plasmon*. These displacements will alter the local electric neutrality and will cause Coulomb restoring forces on the respective masses, but, because the mass of the ions is so much larger than the mass of the electrons, the displacement of the electrons from their neutral position will be proportionately larger. We can calculate the electronic displacement by considering the force on a free electron at point z as

$$\vec{F}_e = m_e\vec{a} = m_e\left(d^2x/dt^2\right)\hat{a}_z = -eE(z)e^{-i\omega t}\hat{a}_z \tag{5.88}$$

Equation 5.88 is the same differential equation as 5.11 for the harmonic oscillator with $b_i = 0$, and $k_i = 0$ (*there is no drag or restoring force*) so the particular solution, dipole moment, Polarization, and permittivity for the plasma will be

$$x_p(t) = (e/m)E(z)e^{-i\omega t}/\omega^2, \qquad \vec{p}(t) = \left[-\left(e^2/m\right)/\omega^2\right]E(z)e^{-i\omega t}\,\hat{a}_x,$$

$$\vec{P}(t) = -\left[\left(Ne^2/m_e\right)/\omega^2\right]E(\vec{x},t)\,\hat{a}_x,$$

and

$$\varepsilon_r'(\omega) + i\varepsilon_r''(\omega) = 1 - \omega_p^2/\omega^2 \quad \text{with } \omega_p = \sqrt{Ne^2/\varepsilon_0 m_e} \tag{5.89}$$

We can also recall that for source-free space, $\vec{E}(\vec{x},t)$ satisfies the wave equation, $\nabla^2\vec{E}(\vec{x} = \mu\varepsilon\partial^2\vec{E}(\vec{x},t)/\partial t^2$, so that, if $\vec{E}(\vec{x},t)$ is polarized in the z-direction and propagates in the z-direction, $\vec{E}(z,t) = E_0e^{i(k_zz-\omega t)}\hat{a}_z$ with the condition that $k_z^2 = \omega^2\varepsilon_r/c^2$, or

$c^2k_z^2 = \omega^2 - \omega_p^2$. We conclude that the electric field intensity propagates in the z-direction with no attenuation for $\omega > \omega_p$ and that it is purely attenuated for $\omega < \omega_p$ (i.e., ω_p is a cutoff frequency). We can also calculate the phase velocity, u_p, and group velocity, u_g, of the wave to see how it will disperse with ω:

$$u_p = \omega/k_z = c\Big/\sqrt{1-\omega_p^2/\omega^2} \quad \text{and}$$
$$u_g = d\omega/dk_z = c^2\,k_z\Big/\sqrt{c^2\,k_z^2+\omega_p^2} = c\sqrt{1-\omega_p^2/\omega^2} \tag{5.90}$$

Conclusion

Longitudinal plasma oscillations cause incident electromagnetic waves with $\omega < \omega_p$ to reflect when they impinge on a material with conduction electrons and to transmit when $\omega > \omega_p$.

Measurements of the reflectance of photons is shown as a function of energy for indium antimonide with $N = 4 \times 10^{24}$ electrons/m^3 in Figure 5.41.

PROBLEM

5.8 Compare the value of $\omega_p = \sqrt{Ne^2/\varepsilon_0 m_e}$ calculated for indium antimonide to the measured value from Figure 5.41.
Answer: $\omega_p \approx 2.2 \times 10^{13}\,Hz$ (0.09eV)

Transverse Plasmons

Upon the application of an external electric field intensity, $\vec{E}(z,t) = E_0 e^{j(k_z z-\omega t)}\hat{a}_x$, the electron cloud density at point z will experience a force in the negative x-direction

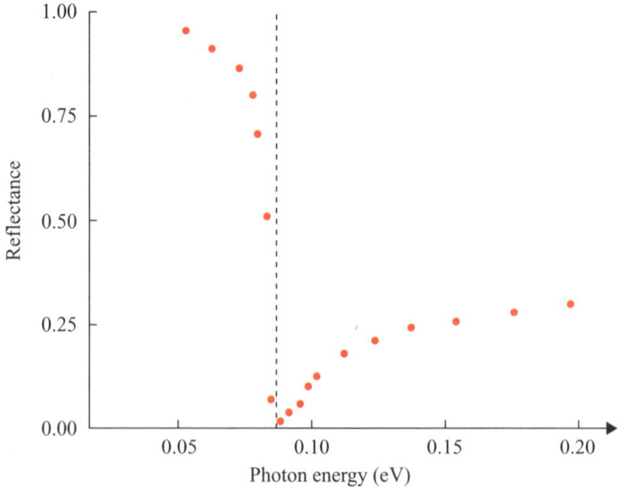

Figure 5.41 Reflectance of indium antimonide as a function of incident wave energy (frequency).

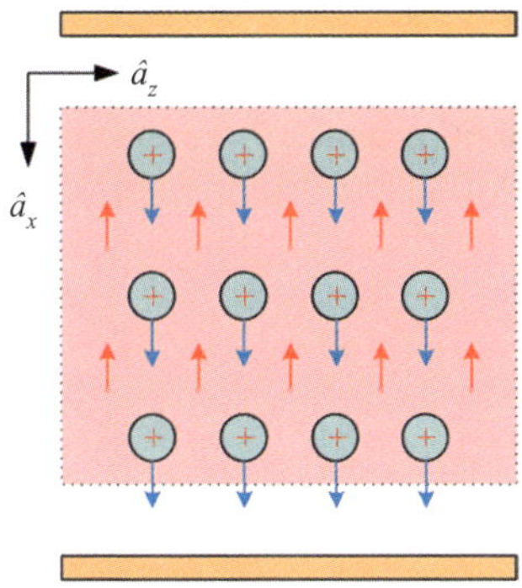

Figure 5.42 Transverse forces on electrons and ion cores.

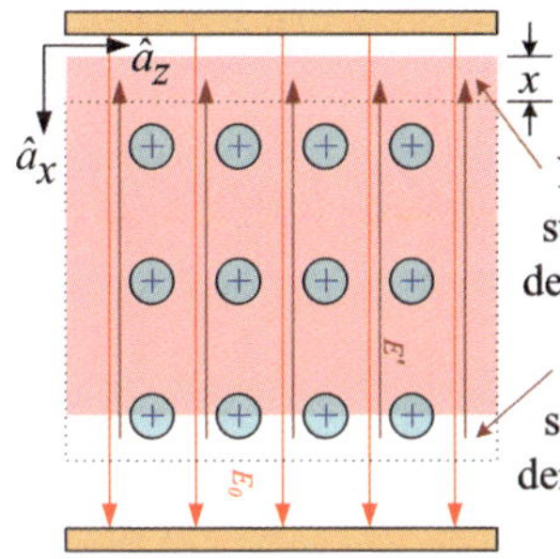

Figure 5.43 Displacement of the electron cloud and positive ions as a result of an applied electric field intensity in the x-direction.

(as indicated by the red arrows) in proportion to the magnitude of the electric field at the point z, as shown in Figure 5.42.

In Figure 5.42, the ions and electrons experience electric field intensity in the x-direction that varies harmonically with time. The electric field intensity wave is assumed to be propagating in the z-direction.

The positive ion cores at point z will similarly experience a force in the positive x-direction (as indicated by the blue arrows). The wave is propagating in the z-direction, so the electron cloud and the ions will be oppositely accelerated in accordance with the force and inversely to their mass. The electron cloud density will move in a manner called a *transverse plasmon*. In Figure 5.42, we have shown the relative forces on ions and electrons for electric field intensity wavelength in the z-direction, $\lambda_z = 2\pi/k_z = c/f$, that is large compared with the separation of the plates that create the external electric field intensity (i.e., the forces are all about the same magnitude for the volume segment shown). These forces will cause the ions to accelerate in the x-direction and the electron cloud to accelerate in the negative x-direction so that they produce an electronic displacement, as shown in Figure 5.43.

Here, we have shown the displacement of the electrons to be much larger than the displacement of the ion cores because their mass is so much smaller. Because the negative conduction electron cloud is displaced by an amount, x, it (and the uncovered positive ion cores) creates an electric field intensity, $\vec{E}' = -(Nex/\varepsilon_0)\hat{a}_x$, that opposes the applied field and acts as a restoring force on the electron cloud (and the ion cores). The classic equation of motion for a unit volume of the electrons is thus

$$\vec{F} = m\vec{a} = m_e N (d^2x/dt^2)\hat{a}_x = -NeE_0 e^{-i\omega t}\hat{a}_x - Ne\left(Nex/\varepsilon_0\,\hat{a}_x\right) \qquad (5.91)$$

or

$$\frac{d^2x}{dt^2} + \frac{Ne^2}{\varepsilon_0 m_e}x = -\frac{e}{m_e}E_0 e^{-i\omega t} \qquad (5.92)$$

Equation 5.92 is of the same form as Equation 5.11, so its solution is the same as that previously obtained in Equation 5.16 with $b = 0$ (*no drag force*) and $\sqrt{k/m} = \sqrt{Ne^2/\varepsilon_0 m_e}$ so

$$\varepsilon_r'(\omega) + i\varepsilon_r''(\omega) = 1 + \frac{\vec{P}(t)}{\varepsilon_0 \vec{E}(\vec{x}, t)} = 1 + \frac{N e^2/\varepsilon_0 m_e}{\omega_p^2 - \omega^2} = 1 + \left[1 - \omega^2/\omega_p^2\right]^{-1}, \qquad (5.93)$$

where $\omega_p = \sqrt{Ne^2/\varepsilon_0 m_e}$, the transverse plasmon frequency. Note that ω_p for the transverse plasmon is the same as it was for the longitudinal plasmon.

The positive ions of mass M are also be free to move, but their mass is large compared with m_e, so that the resulting resonance is at a lower frequency (typically a factor of 100), and the magnitude of their oscillations is much smaller (typically a factor of 10,000). We can see from Equation 5.93 that $\varepsilon_r(\omega)$ for a transverse plasmon is real, singular at $\omega = \omega_p$, positive for $\omega < \omega_p$, and is negative for $\omega > \sqrt{2}\,\omega_p$. We conclude that the electric field intensity propagates in the z-direction with *no attenuation* for $\omega < \omega_p$ and that it is purely attenuated for $\omega > \sqrt{2}\,\omega_p$ (i.e., $\sqrt{2}\,\omega_p$ is a cutoff frequency). We can also calculate the phase velocity, u_p, and group velocity, u_g, of the wave to see how it will disperse with ω:

$$u_p = \frac{\omega}{k_z} = c\frac{(1 - \omega^2/\omega_p^2)}{(2 - \omega^2/\omega_p^2)} \quad \text{and}$$

$$u_g = \frac{d\omega}{dk_z} = c\sqrt{1 + \frac{\omega_p^2}{(\omega_p^2 - \omega^2)}\left\{(1 - \omega^2/\omega_p^2)^2 \Big/ \left[2 - 2\omega^2/\omega_p^2 + (\omega^2/\omega_p^2)^2\right]\right\}} \qquad (5.94)$$

Conclusion

For either transverse or longitudinal waves, the plasmon frequency varies with the concentration of conduction electrons in the propagating medium. In metals, semimetals, and semiconductors, the concentration of conduction electrons is highly variable, as shown in Figure 5.44.

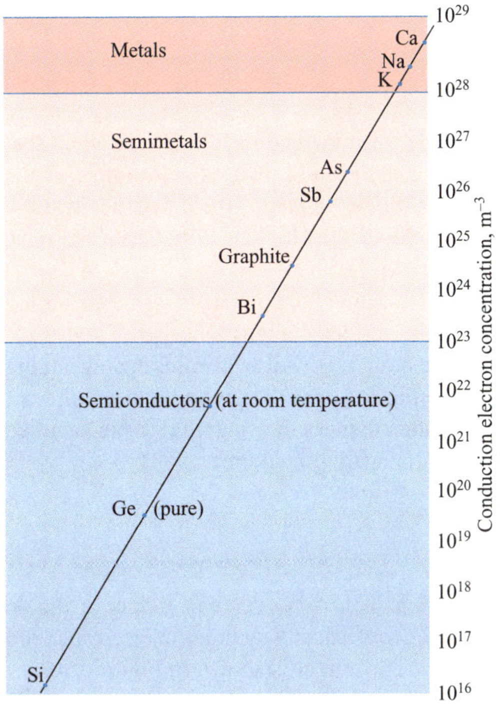

Figure 5.44 Concentration of conduction electrons at the Fermi level in metals, semimetals, and semiconductors.

As projected from this figure, insulators will have an even lower concentration of conduction electrons at the Fermi level, but we will include the effect of plasmons because there is a possibility that conducting impurities contribute to the permittivity.

PROBLEMS

5.9 Develop an ω_p-scale for the abscissa in Figure 5.44.

5.10 Compute the value of $\omega_p' = \sqrt{Ne^2/\varepsilon_0 M}$ for the ion cores of the materials in Figure 5.44.

5.11 Electrons in a plasmon are normally assumed to experience no drag.

a) How would the inclusion of a drag term effect the permittivity in Equation 5.94?

b) Can we justify the drag free assumption given the Sommerfeld model for electron scattering?

c) Electrons that undergo Coulomb collisions emit energy in the form of Bremsstrahlung. Discuss the magnitude of this emission and its relation to a drag force for plasmons at frequency ω_p.

5.11 PERMITTIVITY SUMMARY

Permanent Dipoles

Properties of the propagating medium have an influence on the electric field that shows up in the space between a trace and the ground plane. For example, if the material has molecules with a permanent dipole moment, they will be partially aligned by the application of the external electric field intensity between the trace and the ground plane, as shown in Figure 5.45.

In practical applications, the external electric field intensity is never strong enough to saturate the dipoles (make them perfectly align) because of the finite temperature of the lattice in which they are bound. However, the thermal averaging leaves a net positive charge density of the dipoles near the ground plane and a net negative charge density near the trace, and this reduces the magnitude of the applied electric field intensity at any point in the propagating medium.

Induced Dipoles

In addition, the presence of applied electric field intensity can induce a dipole moment in neutral atoms by displacing the geometrically symmetric electron cloud relative to the positive nucleus of the atom. This effect is shown in Figure 5.46.

A lattice of neutral atoms in an applied field would then have the cumulative displacements shown in Figure 5.47. In many cases, the induced dipole species will occur not as the pure lattice shown but as dispersed impurities, and, in many cases, there may be several species of impurities with different effective spring constants and drag coefficients.

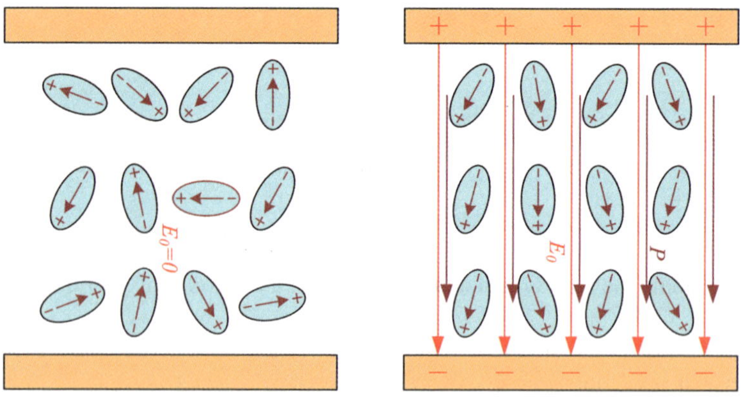

Figure 5.45 (a) Random orientation of permanent dipoles in the absence of an external electric field intensity; (b) the partial alignment of the permanent dipoles in the presence of an applied external field E_0 (in thermal equilibrium with their environment).

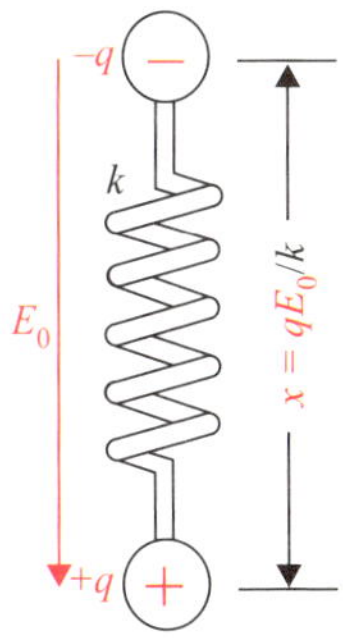

Figure 5.46 Electric dipoles induced in otherwise neutral atoms by an electric field intensity, E_0.

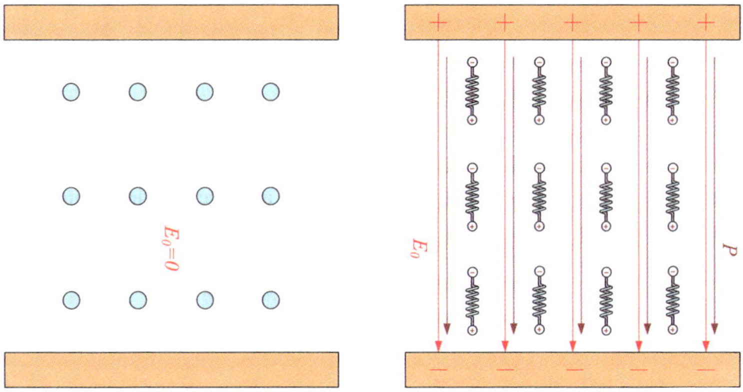

Figure 5.47 (a) Lattice of neutral atoms in the absence of electric field intensity, $E_0 = 0$; (b) Lattice of induced atomic electric dipoles in an externally applied electric field, E_0, and their induced polarization, P.

Conduction Electrons in the Propagating Medium

Some propagating materials will have concentrations of electrical conductivity that permit "free" electrons at the Fermi level to move relative to their ion cores in response to applied electric field intensity. The conduction electrons will experience a resistive loss because of collisions with impurities, phonons, voids, grain boundariesm or lattice irregularities[9] but will not experience a restoring force. A graphic to illustrate their drift velocity is shown in Figure 5.48.

QUESTIONS What is the amplitude of oscillation of conduction electrons in an externally applied harmonic field relative to their mean free path? If the amplitude is a function of frequency, will the resistive loss change with frequency? If the amplitude is much less than the mean free path, will electrons experience collisions? Will that make the electrical conductivity, σ, depend on frequency?

and drag coefficients, and the strength of the attenuation near resonance will depend on the concentration of each molecule, N_i, in the medium. At these frequencies atoms have a polarizability, α_i, that permits an external electric field intensity to differentially displace electrons from their positive cores with large atomic restoring forces and modest relaxation or drag forces. There may be many types of atoms in a medium, each with their own concentration, N_i, and resonance frequency (often in the optical range), so the net effect on permittivity is a sum over the atomic types.

- **Plasmons** At frequencies above 10^{16} Hz, propagating transverse electromagnetic (TEM^z) waves can drive transverse plasmons with small restoring forces and little or no drag with a resonance at $\omega = \omega_p$ and with pure attenuation for $\omega \geq \sqrt{2}\omega_p = \sqrt{2Ne^2/\varepsilon_0 m_e}$. By contrast, transverse magnetic TM^z waves can drive longitudinal plasmons that are purely attenuated for frequencies $\omega < \omega_p = \sqrt{Ne^2/\varepsilon_0 m_e}$. For both types, N is the number of free electrons per cubic meter in the plasma.

For a real-world manufactured propagating medium, the only way to determine how the permittivity will change is to make careful measurements of the real or imaginary part over the entire frequency spectrum and then use the analytic or graphical form of the Kramers–Kronig relations to find the other part. The use of the graphical form is shown in section 5.13 that applies theory to empirical measurements.

5.12 EMPIRICAL PERMITTIVITY

Many industrial researchers have measured the electric permittivity of PCB materials to determine their frequency dependence over many orders of magnitude in frequency. These measurements sometimes include the fiberglass weave in the material, some intentionally include proprietary impurities to reduce the loss (tan δ), some expose the materials to humid conditions, and some measure at a variety of temperatures. Because of this, most measurements among different groups are hard to compare because not all measurements specify the whole suite of environmental conditions under which they were measured. One set of measurements made[10] over a large range of frequencies is shown in Figure 5.50.

Another set of measurements made[11] to complement the first set at higher frequencies is shown in Figure 5.51. These measurements were made on a sample of FR-4 with an imbedded 1080 fiberglass weave of a standard circuit board stack-up with the copper etched off, leaving only the bare dielectric. Because of the affinity of polymer-based PCB materials to absorb water, the samples were conditioned to equilibrium and measured at 21°C and less than 5% relative humidity.

In Figure 5.51, for frequencies between 1 and 10 GHz, a four split-post dielectric resonator operating in the TE011 mode was used. For frequencies between 15

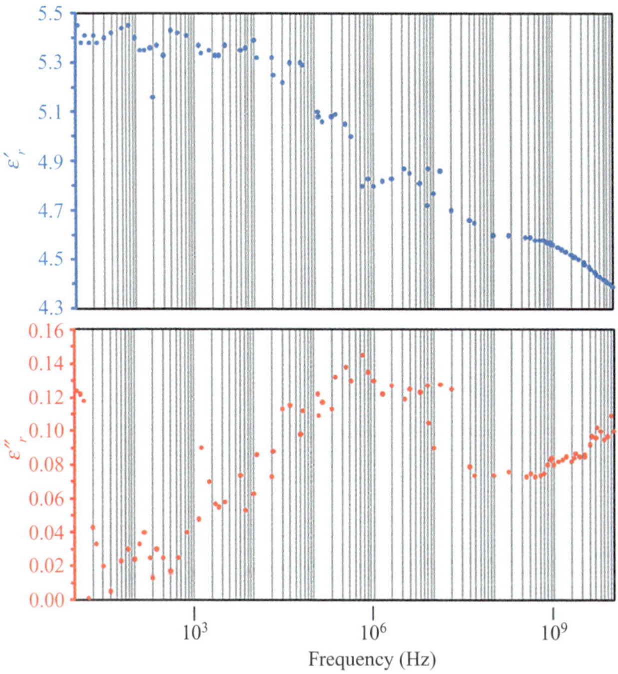

Figure 5.50 Real and imaginary parts of the electric permittivity of FR-4 as a function of frequency between 10 Hz and 10 GHz.

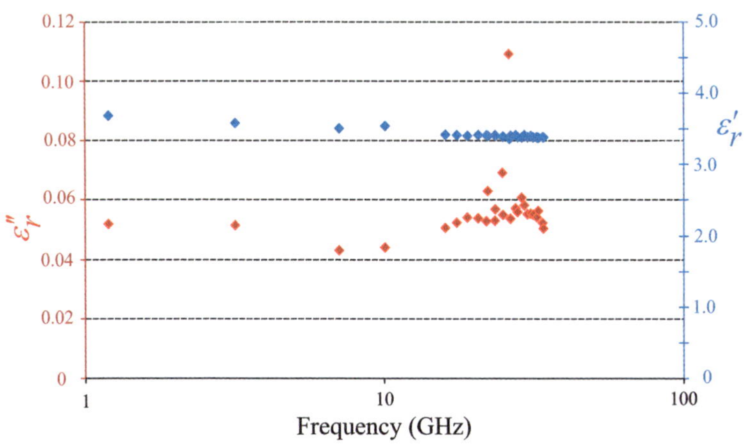

Figure 5.51 Real and imaginary parts of the electric permittivity of FR-4 with imbedded fiberglass in a 1080 weave as a function of frequency between 1 and 40 GHz.

Figure 5.52 Fabry–Pérot resonance cavity used for the measurement of electrical permittivity of a sample placed in its geometric center.

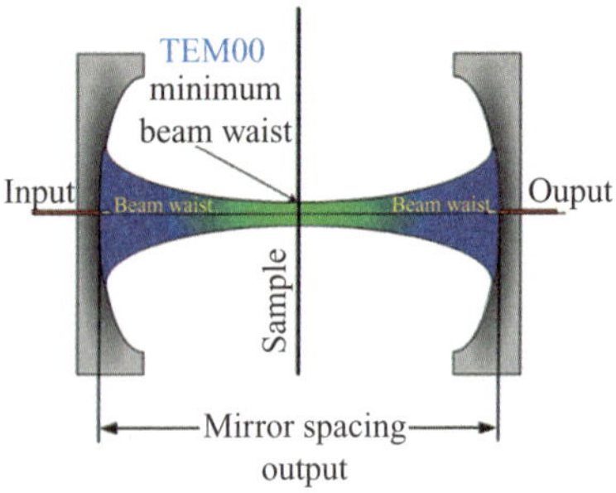

Figure 5.53 Schematic of the displacement of spherical mirrors that support a TEM00 mode of resonance with a sample in their geometric center.

and 35 GHz, a Fabry–Pérot resonator operating in the TEM00 mode, as shown in Figure 5.52, was used.

Values of permittivity are extracted in this instrument by tuning an empty cavity to its resonant frequency and measuring the change in Q-value in the presence of a sample loaded in the cavity. The largest absolute error in these measurements comes about because of the uncertainty in the sample thickness (0.400 ± 0.025 mm). A schematic of the cavity operating in the TEM00 mode is shown in Figure 5.53.

The ε_r'' data of Figure 5.51 suggests that a very narrow resonance might occur for the FR-4 fiberglass stack-up of material at a frequency of about 25 GHz, but this anomaly can be explained as the possible interference of another (non-TEM) mode of resonance that is inadvertently introduced by the interferometer. However, this possibility is especially intriguing because vector network analyzer (VNA) measurements have also shown material resonances in this same frequency range, as is shown in Figure 5.54.

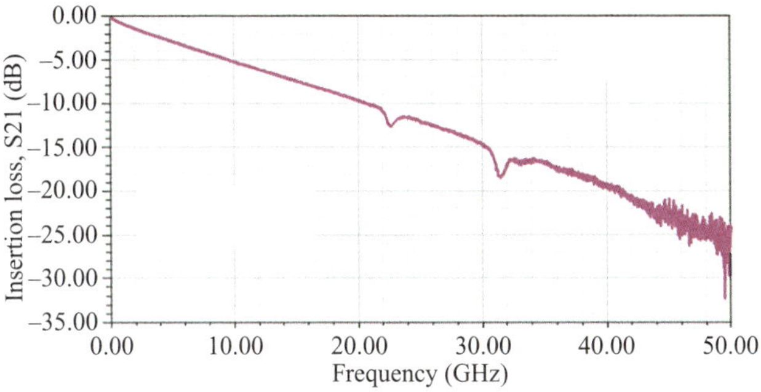

Figure 5.54 S21 measurements of a 5-in-long microstrip with a low-profile surface roughness on an Isola propagating medium with a 1080 fiber weave.

Resonances can be seen to occur at 22.5 and 31.5 GHz for this sample. Here, the absorption coefficient (or insertion power loss) is related to signal attenuation, which depends upon the imaginary component of the electric permittivity. The reason for these resonances is unknown but are possibly caused by included dielectric contaminants such as the atmospheric molecules presented in Figure 5.16. While Figure 5.16 shows a strong water absorption line at 22.235 GHz, there are no molecular resonances in the range 30–40 GHz in that figure. If the 22.5 GHz absorption in Figure 5.54 is ascribed to liquid water at 20°C (room temperature), as shown in Figure 5.13, one would need to explain why the line width is so much larger in that figure than the line width found in Figure 5.54. The resonance at 31.5 GHz in Figure 5.54 might be ascribed to water vapor because the resonant frequency corresponds to that for water at 100°C, but, if that is the case, why is the resonance line width so narrow by comparison to the much larger 100°C water resonance shown in Figure 5.14? Some chemists conclude that the drag coefficient of water in the vapor state is much smaller than that in the liquid state and exhibit smaller resonant line widths.

Some of the characteristics of the resonances shown in Figure 5.54 are that they have been measured in different materials (FR-4, Rogers or Isola) but are not always observed (suggesting manufacturing details are responsible). The resonances do not shift with the length of the line measured, as shown in Figure 5.55.

However, the resonant frequencies for a 5-inch length line show a slight shift because of their propagating material and a larger shift because of their presence in a stripline versus that of a microstrip with or without a solder mask, as shown in Figure 5.56.

Permittivity of Mixtures

Because empirical measurements involve real specimens, they usually include two or more materials in a mixture. For example, a multilayer PCB is formed as a stack-up of alternating layers of laminate and copper sheets. The laminate sheets are often

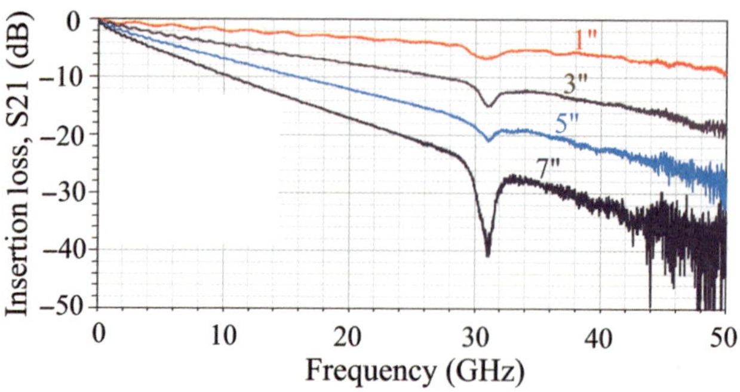

Figure 5.55 Insertion losses for four different line lengths of a microstrip with the same surface roughness on FR-4 material as a function of frequency.

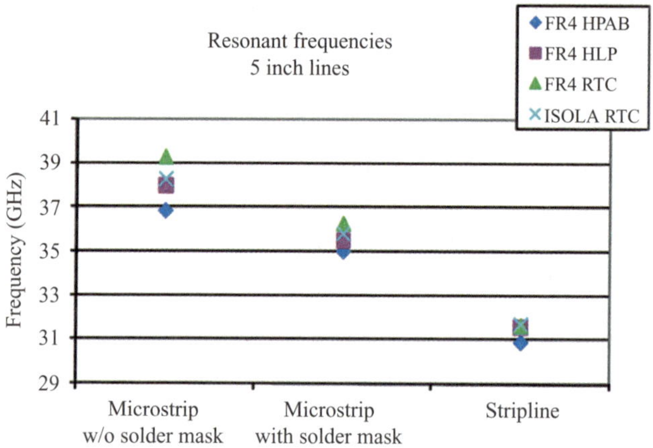

Figure 5.56 Measured resonant frequency as a function of environment for traces with different roughness and material type.

composed of a weave of glass fiber yarns that are preimpregnated with a resin that is intentionally allowed to only partially cure (hence the name "*prepreg*" laminate). The sheets of laminate or copper come in various sizes and yarn styles classified by the number and diameter of glass threads, the weave pattern, and the percentage of resin impregnation. Typical resin content of the mats is in the 45–65% range by volume. Copper foil attached to one or both sides of a fully cured prepreg sheet forms a "*core*." Fabricators choose various sizes and thicknesses of cores and prepreg sheets from which to construct a PCB under heat and pressure. It is common for layers to flow and bond to the cores. While FR-4 epoxy resin is a generic specification for some of the cores, fabricators select the laminate stack-up to meet specific electrical and mechanical or thermal characteristics. For example, the FR-4 epoxy can be blended with multifunctional resins to improve the materials coefficient of expansion, the glass transition temperature, and the rate of moisture absorption. For high-frequency applications, the fabricator may choose alternate resins to intentionally change the dielectric constant or its loss tangent (tan δ) of particular layers in the stack-up. The resins are often composed of proprietary mixtures of polyphenylene oxide, bismaleimide triazine, polyamide or cyanate esters. One series of laminates from the Rogers Corporation is a reinforced hydrocarbon ceramic material. Together with materials from Nelco Park, Isola–USA, and Matsushita, a fabricator can customize a PCB with a relative real permittivity, ε_r', at 1 GHz that ranges between 3.4 and 4.3 and a relative imaginary permittivity, ε_r'', as low as 0.001. Corresponding values at 1 MHz have an ε_r' between 3.8 and 4.6 and an ε_r'' as low as 0.002. High glass content can also be added to further improve loss tangent but it usually has a higher permittivity constant ($\varepsilon_r' = 6.2$ for E glass and $\varepsilon_r' = 5.2$ for S glass). Some glass fibers consist of silicon oxide, aluminum oxide and magnesium oxide in place of calcium oxide or very short Kevlar® fibers (BFG Industries, Inc., Greensboro, NC). Some stack-ups consist of hybrid construction of various options above.

Conclusion

Electromagnetic waves in a PCB do not propagate in a homogeneous medium (even for striplines) but move in an aggregate formed of a mixture of materials and with various density of resins. In order to compare measurements from one PCB fabricator to the next, we must know what materials are specifically used in each PCB and how the permittivity of a mixture behaves.

Effective Permittivity of a Mixture

Textbooks[12] often give the zeroth order approximation of the effective complex permittivity of a mix of constituent materials to be the volume average of the individual complex permittivity values of the materials in a mixture. The electric flux density that takes into account the material polarization is then given in terms of the external applied electric field intensity as

$$\langle \vec{D} \rangle = \varepsilon_{mix} \langle \vec{E} \rangle = \langle \varepsilon \rangle \langle \vec{E} \rangle = \left(\sum_n v_n \varepsilon_n \right) \langle \vec{E} \rangle, \tag{5.95}$$

where the coefficients, v_n, are the volumetric filling factors (assumed to sum to unity) of the various materials involved so that

$$\varepsilon_{r,mix} = \langle \varepsilon \rangle = \sum_n v_n \varepsilon_{r,n} = \left(\frac{1}{V} \int_V \varepsilon_r \, dV \right) \tag{5.96}$$

Landau and Lifshitz pointed out[13] that Equations 5.95 and 5.96 hold only for mixtures in which the individual permittivity values are close to one another because it does not take into account the local field differences inside the medium constituents. In the event that the mixture is isotropic and the differences in their permittivity values are small in comparison with individual permittivity, then it is possible to calculate ε_{mix} in a general form that is correct to terms of the second order in these differences by taking the local field as $\vec{E} = \langle\vec{E}\rangle + \delta\vec{E}$ and calculating the electric flux density as

$$\langle\vec{D}\rangle = \langle(\langle\varepsilon\rangle + \delta\varepsilon)(\langle\vec{E}\rangle + \delta\vec{E})\rangle = \left(\frac{1}{V}\int_V \varepsilon\, dV\right)\langle\vec{E}\rangle + \langle\delta\varepsilon\delta\vec{E}\rangle \tag{5.97}$$

because the mean values of $\delta\varepsilon$ and $\delta\vec{E}$ are zero by definition. Thus in the zeroth order approximation, Equation 5.95 is correct, but the first nonzero correction term will be of the second order in $\delta\varepsilon$. The averaging of the product $\delta\varepsilon\delta\vec{E}$ was carried out in two stages:

a. For a given $\delta\varepsilon$ they first averaged over the volume of particles of a given kind, then

b. the value of $\delta\vec{E}$ was found from the nonaveraged divergence equation, $\vec{\nabla}\cdot\vec{D} = 0$ using small terms of the first order:

$$\vec{\nabla}\cdot[(\langle\varepsilon\rangle + \delta\varepsilon)(\langle\vec{E}\rangle + \delta\vec{E})] = \langle\varepsilon\rangle\vec{\nabla}\cdot\delta\vec{E} + \langle\vec{E}\rangle\vec{\nabla}\delta\varepsilon = 0 \tag{5.98}$$

So, because of the isotropy of the mixture as a whole,

$$\partial\langle\delta E_x\rangle/\partial x = \partial\langle\delta E_y\rangle/\partial y = \partial\langle\delta E_z\rangle/\partial z = \vec{\nabla}\cdot\langle\delta\vec{E}\rangle/3 \tag{5.99}$$

Now, if $\langle\vec{E}\rangle$ is in the x-direction, we have from Equation 5.98: $3\langle\varepsilon\rangle\partial\langle\delta E_x\rangle/\partial x = -\langle E_x\rangle\partial\delta\varepsilon/\partial x$ or $\langle\delta E_x\rangle = -(\langle E_x\rangle/3\langle\varepsilon\rangle)\delta\varepsilon$. Multiplying by $\delta\varepsilon$ we obtain $\langle\delta\varepsilon\delta\vec{E}\rangle = -(1/3\langle\varepsilon\rangle)\vec{E}\langle\delta\varepsilon\rangle^2$ and putting this back into Equation 5.97 and comparing it to 5.96 the result is

$$\varepsilon_{mix} = \langle\varepsilon\rangle - \langle\delta\varepsilon\rangle^2/3\langle\varepsilon\rangle \tag{5.100}$$

Landau and Lifshitz also found the permittivity of an emulsion having an arbitrarily large difference between the permittivities ε_l of the medium and ε_2 of the disperse phase but only a small concentration of the latter, whose particles were assumed spherical. Using the Lorentz local field,[14] they obtained the proportionality constant between $\vec{D}$ and $\vec{E}$ to be

$$\varepsilon_{r,mix} = \varepsilon_{r,1} + 3v\,\varepsilon_{r,1}(\varepsilon_{r,2} - \varepsilon_{r,1})/(\varepsilon_{r,2} + 2\varepsilon_{r,1}), \tag{5.101}$$

which they found to be correct to terms of the first order in v, the volume concentration of the emulsion.

The history of dielectric mixing goes back[15] at least to the mid-1800s. As authors like Clausius, Maxwell, L. V. Lorenz, H. A. Lorentz, Rayleigh, and Garnett began to apply their own analysis, they gave specialized results for various optical properties or for dielectric materials immersed in a number of liquids. Shape effects were considered especially important in the case of liquids partly because electric fields in spherical or ellipsoidal inclusions could be solved analytically. Some of the work appears to contradict others for the same mixtures, but, as often as not, it is easy to fail to notice special conditions that were imposed in a given analysis. A review[16] of the mixing rules shows a few of those results below.

The **Clausius–Mossotti formula** for the polarizability, α_i, of N_i molecules per unit volume was developed from the dipole moment, $\vec{P}_i$, created in a local (Lorentz) spherical cavity[14] in a macroscopic external electric field, $\vec{P}_i = \alpha_i \vec{E}_P$, within a homogeneous medium of relative permeability, ε_r, as 3.145 of *Maxwell's Equations*[17]:

$$\frac{\varepsilon_r - 1}{\varepsilon_r + 2} = \frac{N_i \alpha_i}{3\varepsilon_0}$$

If those molecules are immersed in a cavity filled with an inclusion dielectric, then the macroscopic average is found in terms of the emulsion dielectric, ε_e, from Equation 5.101 as

$$\frac{\varepsilon_{r,mix} - \varepsilon_{r,e}}{\varepsilon_{r,mix} + 2\varepsilon_{r,e}} = \frac{N_i \alpha_i}{3\varepsilon_{r,e}\varepsilon_0} \tag{5.102}$$

In applications with $v_i = N_i V$, the volume fraction of the inclusions in the mixture, **Rayleigh** obtained the mixing formula

$$\frac{\varepsilon_{r,mix} - \varepsilon_{r,e}}{\varepsilon_{r,mix} + 2\varepsilon_{r,e}} = v_i \frac{\varepsilon_{r,i} - \varepsilon_{r,e}}{\varepsilon_{r,i} + 2\varepsilon_{r,e}} \tag{5.103}$$

Note that only the volume fraction and the permittivities are included in Equation 5.103 so that the spheres need not be of equal size. However, there is a restriction that the sizes must all be small compared with the wavelength.

The most common mixing formula is that developed by **Maxwell and Garnett,** which comes from the Rayleigh formula explicitly written for the effective permittivity:

$$\varepsilon_{r,mix} = \varepsilon_{r,e} + 3v_i \varepsilon_{r,e} \left(\varepsilon_{r,i} - \varepsilon_{r,e}\right) / \left[\varepsilon_{r,i} + 2\varepsilon_{r,e} - v_i \left(\varepsilon_{r,i} - \varepsilon_{r,e}\right)\right] \tag{5.104}$$

The Maxwell–Garnett formula is used in many applications because it satisfies the limiting processes for vanishing inclusion and emulsion phases:

$$\begin{aligned} \varepsilon_{r,mix} &\to \varepsilon_{r,e} \quad \text{for } v_i \to 0 \\ \varepsilon_{r,mix} &\to \varepsilon_{r,i} \quad \text{for } v_i \to 1 \end{aligned} \tag{5.105}$$

and for small values of v_i the Maxwell–Garnett rule gives the permittivity as

$$\varepsilon_{r,mix} \approx \varepsilon_{r,e} + 3v_i\varepsilon_{r,e}(\varepsilon_{r,i} - \varepsilon_{r,e})/[\varepsilon_{r,i} + 2\varepsilon_{r,e}] \tag{5.106}$$

The Maxwell–Garnet formula can also be used for n different kinds of spherical inclusions as

$$\frac{\varepsilon_{r,mix} - \varepsilon_{r,e}}{\varepsilon_{r,mix} + 2\varepsilon_{r,e}} = \sum_{i=1}^{n} v_i \frac{\varepsilon_{r,i} - \varepsilon_{r,e}}{\varepsilon_{r,i} + 2\varepsilon_{r,e}} \tag{5.107}$$

with an expression for the average mixture permittivity:

$$\varepsilon_{r,mix} \approx \varepsilon_{r,e} + 3\varepsilon_{r,e}\left[\sum_{i=1}^{n} v_i(\varepsilon_{r,i} - \varepsilon_{r,e})\right] \Bigg/ \left[1 - \sum_{i=1}^{n} v_i \frac{(\varepsilon_{r,i} - \varepsilon_{r,e})}{(\varepsilon_{r,i} + 2\varepsilon_{r,e})}\right] \tag{5.108}$$

Sihvola gives extended versions of the above expressions for shapes that are not spherical such as ellipsoids of revolution, for aligned mixtures, and for random mixtures as well as for some anisotropic materials and for an inclusion substructure; a treatment is also provided for losses of conducting inclusions in terms of their skin depths, for moist and high-loss materials. Dispersion results are written in terms of a convolution operator like that shown in Equation 5.34.

5.13 THEORY APPLIED TO EMPIRICAL PERMITTIVITY

It has been shown above that four basic types of physical phenomena contribute to electric permittivity. We have shown that because permittivity due to a plasmon is applicable only to very high frequencies (10^{13} Hz or above) where VNA measurements cannot be made; it should contribute at most a constant real term to the permittivity for measurements at VNA frequencies. We have also shown that electrical conductivity that is mathematically described as permittivity is applicable only to very low frequencies (100 Hz or below) and that it contributes only an inverse frequency-dependent term to the imaginary part of the permittivity. Thus, it would appear to the novice that, once the real part of the background permittivity is accounted for and a fit has been made to the imaginary part at low frequencies, measured values of permittivity should reveal their physical basis from their character in a frequency domain. Even better, the Kramers–Kronig relationship has shown that measurement of the real part or of the imaginary part when integrated as a Hilbert transform yields the other.

Unfortunately, we have also shown that real-world propagation media are very complex structures that contain many components, compounds, atomic species, impurities, bubbles, inclusions, and intentional mixtures like fiberglass, ceramic, or semiconductor components so that the medium is certainly not homogeneous. We have also seen that variability in resin hardness, material density fluctuations, or geometric effects of the boundaries (especially the air or solder mask boundary) for a microstrip also contributes to the effective permittivity of a real transmission line.

In addition, there are many formulas (some contradictory) used to treat the mixtures to correct the measurements for multiple phases. Finally, the process of obtaining good VNA data includes difficult and frequent calibrations to remove or de-imbed geometric materials such as the contact pad or vias that the data always contain degrees of statistical error that can disguise itself as a real resonance. Thus, many kinds of frequency-dependent fits have been made to empirical data such as that found in Figures 5.50 and 5.51 that come to different conclusions. One such fit for a finite number of resonances is described below along with one extension to an infinite number of resonances.

Finite Number of Debye Resonances

Observation of empirical data can suggest frequency resonances, especially in the imaginary part, but one must use some judgment about which variations are physical and which are statistical. One such fit[18] was made by using only Debye resonances. This treatment explained how different configurations of parallel-plate capacitors, cavity resonators, half-wave-length, open-ended microstrip lines, ring resonators were used to find insertion and reflection losses. This group assumed that dielectric losses dominate at microwave frequencies so that an accurate characterization of the conductor losses was not crucial. The topic of conductor surface roughness will be covered in Chapters 6 and 7, but we note here that this assumption is far from conclusive. Nevertheless, the data in Figure 5.50 were fit to eight terms in a Debye relaxation formula:

$$\varepsilon_r(\omega) = \varepsilon'_{r,\infty} + \sum_{i=1}^{8} \frac{\Delta\varepsilon'_{r,i}}{1 - i\,\omega/\omega_i} + i\frac{\sigma}{\omega\varepsilon_0} \tag{5.109}$$

using $\varepsilon'_{r,\infty} = 4.2$ and $\sigma = 80$ pS/m. Parameter values for the eight resonances are given in Table 5.1.

The "fit" of Equation 5.109 using the parameters from Table 5.1 to the data in Figure 5.50 is shown in Figure 5.57.

The quality of this "fit" to the empirical data within the stipulations stated in the three paragraphs above is left to the judgment of the reader. However, it should be noted that the choice of the Debye equation for each of the resonances makes the choice that only a relaxation term enters the material properties of FR-4.

Table 5.1 Parameters for the Debye sum 5.109 that "fit" the data in Figure 5.50

i	1	2	3	4	5	6	7	8
f_i GHz	2×10^{-5}	2×10^{-4}	2×10^{-3}	2×10^{-2}	2×10^{-1}	2×10^{0}	2×10^{1}	2×10^{2}
$\Delta\varepsilon'_i$	0.12	0.14	0.22	0.18	0.12	0.10	0.10	0.24

graphical equivalent of the Kramers–Kronig relationship in which the imaginary part is chosen to fit the ε_r'' data and the real part is chosen to fit those parameters through Equation 5.110. Note also that both Equations 5.109 and 5.110 are analytic functions so they have real and imaginary parts that satisfy the Kramers–Kronig relationship. Thus, they both yield a causal relationship in the time domain.

In Figure 5.58, we chose to make a best fit to the imaginary part of the measured permittivity, and, in Table 5.2, we found that the values of $b_i/2m$ (GHz) were greater than $\sqrt{k_i/m}$ (GHz) for all resonances below 1 GHz. Thus, the term $\sqrt{k_i/m-(b_i/2m)^2}$ in the response function, Equation 5.42, is imaginary. This implies that the terms in the response function below 1 GHz are overdamped or purely relaxing as was assumed in the Debye fit to the same data in Figure 5.57. Thus, we should not be surprised that the fit of the two functions to the real part of the permittivity looks substantially similar. However, it can be noted that the statistical variation of the data in the neighborhood of a resonance is hardly random in its variance with the functional fit of either Equation 5.109 or 5.110.

Underdamped Dipoles

If underdamped permittivity terms are included in the response functions, Equation 5.13 shows that the resonant frequency of the terms shift to a lower value

$$\omega_i' = \sqrt{k_i/m - b_i^2/4m^2} = \sqrt{1-(\tau_i^2\omega_i^2/4)}\,\omega_i \tag{5.111}$$

and the permittivity changes from the Debye mathematical form

$$\frac{[\varepsilon_r'(\omega)-\varepsilon_{r\infty}]}{\Delta\varepsilon} = \frac{1}{1+\omega^2\tau_e^2} \qquad \frac{\varepsilon_r''(\omega)}{\Delta\varepsilon} = \frac{\omega\tau_e}{1+\omega^2\tau_e^2} \tag{5.112}$$

to the Lorentzian function form as shown in Equation 5.113:

$$\frac{[\varepsilon_r'(\omega)-\varepsilon_{r\infty}]}{\Delta\varepsilon} = \frac{\left[1-(\omega/\omega_1)^2\right]}{\left[1-(\omega/\omega_1)^2\right]^2+\omega^2\tau_e^2} \qquad \frac{\varepsilon_r''(\omega)}{\Delta\varepsilon} = \frac{\omega\tau_e}{\left[1-(\omega/\omega_1)^2\right]^2+\omega^2\tau_e^2} \tag{5.113}$$

The effect of these changes on the real and imaginary part of the permittivity is shown in Figure 5.59.

Finite Number of Underdamped Lorentz Resonances

In Figure 5.59, the resonance frequency has been chosen to create an underdamped resonance that oscillates and relaxes in time to an equilibrium final value. We see that the effect relative to a Debye pure relaxation is to shift and narrow the imaginary part of the permittivity and to cause the real part to exhibit "overshoot" before and

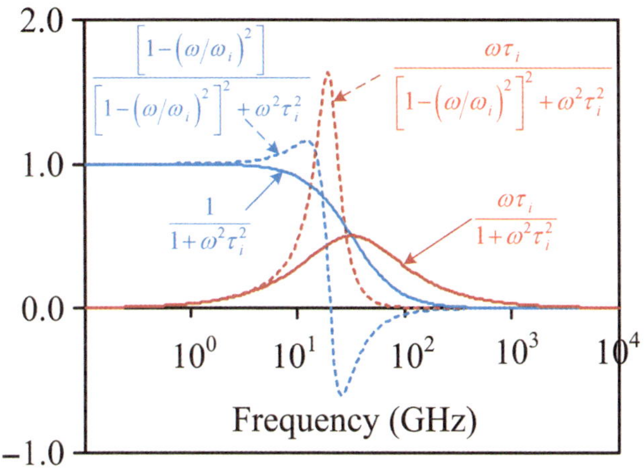

Figure 5.59 Solid curves are the real and imaginary part of the Debye permittivity function 5.112 for τ_i = 5 ps. Dotted are the real and imaginary part of the Lorentzian permittivity function 5.113 for τ_i = 5 ps and ω_i = 20 GHz.

Table 5.3 Parameters for the Lorentzian sum in Equation 5.110 that "fit" the real data in Figure 5.50 to underdamped dipoles. The value chosen for σ = 40 pS/m

i	0	1	2	3	4	5	6	7	8
$\sqrt{k_i/m}$ (GHz)	1.0E-7	1.3E-6	2.1E-5	1.1E-4	5.5E-4	9.0E-3	2.7E-4	3.1	6.9
$\Delta\varepsilon'_{r,i}$	0.030	0.045	0.065	0.155	0.250	0.120	0.160	0.110	0.110
$b_i/2m$ (GHz)	1.2E-8	2.0E-7	2.5E-6	4.0E-5	2.5E-4	4.3E-3	1.0E-6	2.9	4.45

after the resonance frequency. Both Lorentzian dotted curves are also asymmetric compared with the Debye solid curves.

A fit of Equation 5.110 that makes a best fit to the real part of the measured permittivity is shown in Figure 5.60 by using the fit parameters given in Table 5.3.

In Table 5.3, values of $b_i/2m$ are smaller than $\sqrt{k_i/m}$ for all resonances. Thus, the term $\sqrt{k_i/m-(b_i/2m)^2}$ in the response function, Equation 5.42, is real. This implies all terms in the response function are underdamped and therefore oscillate while relaxing unlike the behavior assumed in the Debye fit to the same data in Figure 5.57 or the overdamped Lorentzian fit to the same data in Figure 5.58.

In Figure 5.60, we can see that the resonances are relatively sharp but that they approximate the measured real values of permittivity in better detail. Because the damping term $b_i/2m$ is less than (but not much less than) the resonant frequency $\sqrt{k_i/m}$, the response function is said to be weakly damped. The real part of the permittivity is associated with the index of refraction and with the energy density of the propagating medium caused by electron oscillations in the constituent atoms

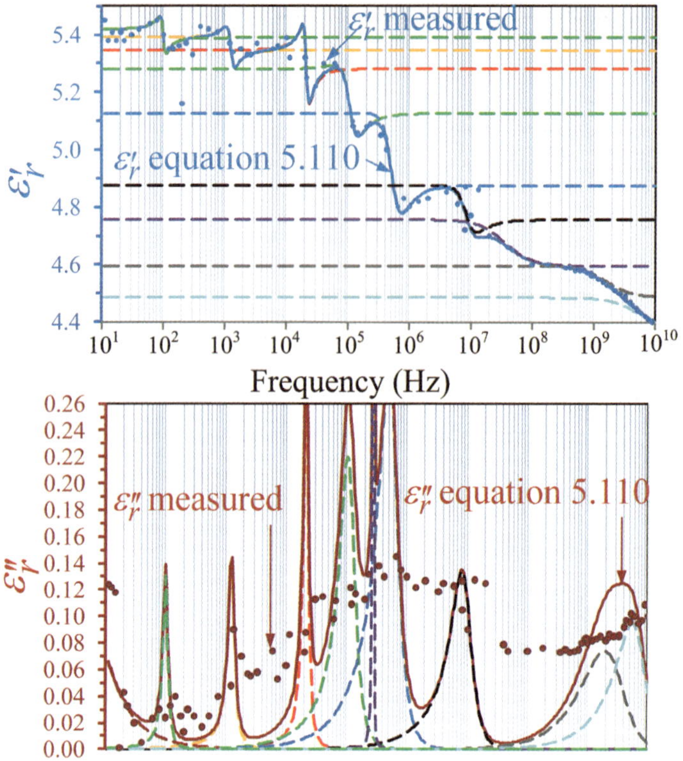

Figure 5.60 Real and imaginary parts of the permittivity of FR-4 compared with the nine-term underdamped Lorentzian Equation 5.110 using parameters of Table 5.3.

and molecules. Because the electrons radiate or loose energy through local friction (damping), the imaginary part of the permittivity is associated with dissipative (energy loss) processes. The electromagnetic radiation of the electron oscillations causes fields that are coherent (but delayed) from the incident fields that induced the dipole moments. The fields radiated by the electrons (except for an anomalous effect described below) retard the phase of the incident fields. In general, they produce a phase velocity that is less than the speed of light, c, and have an index of refraction greater than 1.

5.14 DISPERSION OF A SIGNAL PROPAGATING THROUGH A MEDIUM WITH COMPLEX PERMITTIVITY

We have shown[19] that the electric field intensity of an arbitrary time signal that propagates through a dispersive medium with Lorentzian resonances in the z-direction with velocity $u_p = c/\sqrt{\varepsilon_r}$ yields an electric flux density beginning at time $t_1 = z\sqrt{\varepsilon_r}/c$.

Table 5.4 Values of the constants used in the underdamped Lorentzian response function 5.42 in (radians/s)^2 or (radians/s) as determined by their empirical fit values from Table 5.3

i	0	1	2	3	4	5	6	7	8
$\frac{N_i \alpha_i e^2}{\varepsilon_0 m}$	1.18E4	2.8E6	1.1E9	7.4E10	3.0E12	3.8E14	4.6E11	4.2E19	2.0E20
$b_i/2m$	1.45E2	2.51E3	3.14E4	5.03E5	3.14E6	5.34E7	1.26E4	3.64E10	5.59E10
$\sqrt{\frac{k_i}{m} - \frac{b_i^2}{4m^2}}$	6.1E2	7.4E3	1.3E5	4.7E5	1.4E6	1.9E6	1.7E6	i3.1E10	i3.5E10

$$\vec{D}(\dot{x}, t) = \varepsilon_0 \vec{E}(\vec{x}, t) + \int_0^{\infty} \varepsilon_0 \vec{E}(\vec{x}, t - \tau) G(\tau)\, d\tau$$

with

$$G(\tau) = \frac{e^2}{\varepsilon_0 m} \sum_{i=1}^{Z} N_i \alpha_i e^{-(b_i/2m)\tau} \frac{\sin\sqrt{k_i/m - (b_i^2/4m^2)}\,\tau}{\sqrt{k_i/m - b_i^2/4m^2}} \quad \text{for } \tau > 0$$

We can determine the values of the individual coefficients, $N_i\alpha_i e^2/\varepsilon_0 m$, for each of the individual resonances by multiplying the $\Delta\varepsilon_i'$ parameters in the second row of Table 5.3 by the corresponding value of $\omega_i^2 = k_i/m = (2\pi f_i)^2$, as given in the first row. Because $\sqrt{k_i/m - (b_i/2m)^2}$ is real for all of the resonance lines, the constants in Equation 5.42 are determined. The values of the constants determined by this fit are given in Table 5.4.

Thus, in Equation 5.42, the exponential decay constants, $b_i/2m$, are increasing with frequency; that is, higher frequency terms are more attenuated than lower order terms. Propagating at a phase velocity of $u_p(\omega) = c/\sqrt{\varepsilon_r(\omega)}$, the mean distance for decay of e^{-1}, δ, is frequency dependent.

PROBLEM

5.12 Calculate the coefficient of absorption for each of the frequencies listed in Tables 5.2 and 5.3. Explain what happens to the two resonances above 1 GHz where the argument of the sine function is complex. Would the FR-4 that provided the data for Figure 5.50 be a good candidate for transmission of signals above 1 GHz? Would the FR-4 used for Figure 5.51 be a better candidate?

Phase and Group Velocity of an Electromagnetic Pulse

In section 1.5, phase velocity, $u_p(\omega) = c/n(\omega) = c/\sqrt{\varepsilon_r(\omega)}$, and group velocity, $u_g = c/[n(\omega) + \omega dn(\omega)/d\omega]$, for a pulse of light were considered in a medium with an index of refraction is the square root of the real part of the medium permittivity; the quantity in square brackets was called the "group refractive index." If $n(\omega)$ varied linearly with frequency, the effect of the modified interference for a pulse (having several Fourier components) was to shift the peak of the pulse in time, but with the pulse shape staying the same. We have now observed in this section that the real part of the relative permittivity of a medium, Equation 5.109 or 5.110, can be more complicated than a linear dependence on frequency. If a pulse is described by an incident time-dependent wave train, $E_i(0, t)$, that has a well-defined beginning at $z = 0$ (called Port 1), how would the signal reconstruct itself at some later time at a remote point, $z = l$ (called Port 2), if the real part of its relative permittivity is not linear with frequency?

Some authors[20] assume electric field intensity of the wave is orthogonally incident from air into a medium at Port 1 and include the transmission coefficient 2.44b for normal incidence ($\cos\theta_i = 1$) to write the amplitude of the electric field of the wave for $z > 0$ in terms of its Fourier transform as

$$E_x(z,t) = \int_{-\infty}^{\infty} \left[\frac{2}{1+n(\omega)}\right] A(\omega) e^{i k_z(\omega) z - i\omega t} d\omega, \tag{5.114}$$

where

$$A(\omega) = \frac{1}{2\pi}\int_{-\infty}^{\infty} E_i(0,t) e^{i\omega t} dt \tag{5.115}$$

is the Fourier transform of the real incident electric field intensity evaluated just outside the propagating medium, and the term in square brackets is the transmission coefficient. Here, the wave number, $k_z(\omega) = \omega\sqrt{\varepsilon_r(\omega)}/c$, is generally complex, with the positive imaginary part corresponding to absorption of energy during propagation. Using frequency as the variable of integration in Equation 5.114 permits us to utilize Equation 5.110 in the integrand exponent. Landau and Lifschitz[21] solve the real and imaginary part of the complex permittivity function to express the final integral. Jackson uses a one resonance model to prove that within our multiresonance analysis for permittivity, no signal can propagate with a velocity greater than c, whatever the medium.

Gauthier and Boyd[22] have shown how the group velocity of a pulse of light propagating through a dispersive material can exceed the speed of light in vacuum. The author recommends students read this article for a sense of enjoyment and understanding that, in very special circumstances, such as that which occurs when the derivative of the index of refraction becomes negative in the neighborhood of one of the resonances in Equation 5.110, group velocity, $u_g = c/[n(\omega) + \omega dn(\omega)/d\omega]$, can exceed the speed of light. These authors claim that the possibility of "fast light" has been known for nearly a century and show that "fast light behavior is completely consistent with Maxwell's equations that describe pulse propagation through a dispersive material and, hence, *does not violate* Einstein's special theory of relativity, which is based on Maxwell's equations."

ENDNOTES

1. Paul G. Huray, *Maxwell's Equations* (Hoboken, NJ: John Wiley & Sons, 2009), Chapter 3.
2. S. J. Mumby, and D. A. Schwarzkopf, "Dielectric Properties and High-Speed Electrical Performance Issues," in M. L. Menges, *Electronic Materials Handbook*, Vol. 1: *Packaging* (Materials Park, OH: ASM International, 1989).
3. P. Debye, *Polar Molecules* (New York: Dover Publications, 1929).
4. P. Langevin, *Journal de Physique et Le Radium* 4 (1905): 678–93; *Annales des Chimie et des Physique* 5 (1905): 70.
5. J. D. Jackson, *Classical Electrodynamics*, 3rd ed. (Hoboken, NJ: John Wiley & Sons, 1999), 164.
6. Charles Kittel, *Introduction to Solid State Physics*, 7th ed. (Hoboken, NJ: John Wiley & Sons, 1996), 385.
7. Huray, *Maxwell's Equations*, Equation 3.115.
8. Ibid., Chapter 5.
9. Ibid.
10. A. Djordjevic, R. Biljic, V. Likar-Smiljanic, and T. Sarkar, "Wideband Frequency-Domain Characterization of FR-4 and Time Domain Causality," *IEEE Transactions on Electromagnetic Compatibility* 3, no. 4 (2001): 662–67.
11. S. Pytel, G. Barnes, D. Hua, A. Moonshiram, G. Brist, R. Mellitz, S. Hall, and P. G. Huray, "Dielectric Modeling, Characterization, and Validation up to 40 GHZ," in *10th IEEE Workshop on Signal Propagation on Interconnects*, Berlin, Germany, 2006.
12. H. Johnson and M. Graham, *High-Speed Signal Propagation: Advanced Black Magic* (Upper Saddle River, NJ: Prentice-Hall, 2003), 99.
13. L. D. Landau and E. M. Lifshitz, *Electrodynamics of Continuous Media*, translated from the Russian by J. B. Sykes, J. S. Bell, and M. J. Kearsley, 2nd ed. revised and enlarged by E. M. Lifshitz and L. P. Pitaevskii (Oxford, UK: Elsevier, 2006), 43.
14. Huray, *Maxwell's Equations*, Figure 3.39.
15. O. F. Mossotti, "Discussione analitica sull'influenza che l'azione di un mezzo dielectrico ha sulta distribuzione dell'elettricità alla superficie di piu corpi elettrici disseminati in esso," Memoire di matematica e di fisica della Societa Italiana delle scienze, residente in Modena, vol. 24, part 2, pp. 49–74, 1850.
16. A. Sihvola, "Mixing Rules with Complex Dielectric Coefficients," *Subsurface Sensing Technologies and Applications* 1, no. 4 (2000): 393–415.
17. Huray, *Maxwell's Equations*.
18. A. R. Djordjevic, R. M. Biljic, V. D. Likar-Smiljanic, and T. K. Sakar, "Wideband Frequency-Domain Characterization of FR-4 and Time-Domain Causality," *IEEE Transactions on Electromagnetic Compatibility* 43, no. 4 (Nov. 2001): 662–7.
19. Huray, *Maxwell's Equations*, Equation 5.34.
20. J. D. Jackson, *Classical Electrodynamics*, 3rd ed. (Hoboken, NJ: John Wiley & Sons, 1999), 336.
21. Landau and Lifshitz, *Electrodynamics of Continuous Media*, 285.
22. D. J. Gautheir and R. W. Boyd, "Fast Light, Slow Light, and Optical Precursors: What Does It All Mean?" *Photonics Spectra Magazine* (January 2007): 82.

Chapter 6

Surface Roughness

LEARNING OBJECTIVES

- Model real-world physical properties of a typical printed circuit board PCB high- or low-profile microstrip or stripline transmission line
- Develop a three-dimensional model for surface roughness that is a sufficient match to the observed conductor roughness and estimate the error made in the approximations
- Find induced electric dipole moments and magnetic dipole moments for perfect electrically conducting (PEC) and good conducting spheres in a propagating E&M field intensity
- Construct magnetic vector potential, electric field intensity, and magnetic field intensity created from a Green's function multipole moment formulation
- Express incident, scattered, and absorbed electromagnetic fields in terms of vector spherical harmonics and give the dominant terms in a multipole expansion of the power scattered and absorbed by uniform conducting spheres
- Evaluate the fraction of power lost in a propagating wave because of surface roughness

INTRODUCTION

Scanning electron microscope (SEM) photographs of typical copper conductors prepared[1] by the printed circuit board (PCB) industry exhibit a three-dimensional (3-D) "snowball" structure of copper surface distortions as shown in Chapter 4. The "snowballs" are approximately made up of a distribution of different size spheres whose surfaces have a different composition[2] from their interior. The stacking[3] of the spheres creates[4] many voids of 25–100 nm in size. In this chapter, we will develop an analytical[5] basis for the electromagnetic scattering by individual copper "snowballs" and then construct a field of pyramid like structures to represent the

The Foundations of Signal Integrity, by Paul G. Huray

rough surface of a microstrip or stripline. A random phase approximation will permit a prediction of scattered and absorbed power losses by the surface that are added to those presented by the propagating medium, as discussed in Chapter 5. In this chapter, the fundamental concepts of scattering by conducting spheres are first described. Then the analysis turns to multiple scattering centers and a vector harmonics approach used in the high-energy community. Finally, a power absorption[6] argument is developed[7] by using 3-D scattering analysis.

As shown in Figures 4.1, 4.2, 4.3, and 4.4, an electromagnetic pulse caused by a voltage $V(0, t)$ is assumed to be applied to one end (Port 1 in the language of Chapter 8) of a copper microstrip. And, as is concluded in Chapter 5, the pulse components can be followed as they propagate down the transmission line (in the z-direction) at a phase velocity $c_2 = c/\sqrt{\varepsilon_r(\omega)} = c/\sqrt{\varepsilon_r'(\omega) + i\varepsilon_r''(\omega)}$ and so it is assumed that the field lines will fringe and disperse as they propagate in the medium. Contrary to the approach of Chapter 5, the influence of the electric dipoles that are induced in the propagating medium[8] is ignored in this chapter so that the influence of the surface roughness or scattering irregularities is the main focus and so that only one variable at a time is considered. By permitting the scattering/absorbing spheres to have their own conductivity or dielectric properties, additional spherical inclusions, impurities, or spherical clumps of any kind of scattering or absorbing event in the propagating material can be considered. Even if the scattering center is a bubble, it will be described as "not medium" for the purpose of accounting for voids in an otherwise homogeneous medium.

In the case of conducting materials shown in Figures 4.19 and 4.20, particular attention is paid to the skin depth of the electromagnetic field intensity inside the conductor, as shown in Figure 6.1, which is an SEM photograph of the "snowballs" on the underside of the microstrip in comparison with the skin depth of pure copper at several frequencies.

Figure 6.1 shows that even the largest copper snowball may be considered small compared with the copper skin depth at 1 GHz but that most of the snowballs would be considered large compared with the copper skin depth at 100 GHz. Thus, it is said that the size of the copper snowball depends on the frequency of the incident electromagnetic wave.

6.1 SNOWBALL MODEL FOR SURFACE ROUGHNESS

Idcal Model

For the real-world image shown in Figure 6.1 to be described in terms that can be used in a scattering/absorbing analysis, an inverted stack-up of spheres is considered in cross section, as shown in Figure 6.2.

In Figure 6.2, a random distribution of spherical "snowballs" is shown in two pyramidlike stack cross sections. Each of the snowballs is identified by its location, x_i, below the flat plane of a conductor (so that we can use our ideal model in Chapter 4 as a starting point). The field at a random point, P, below the stacks will be

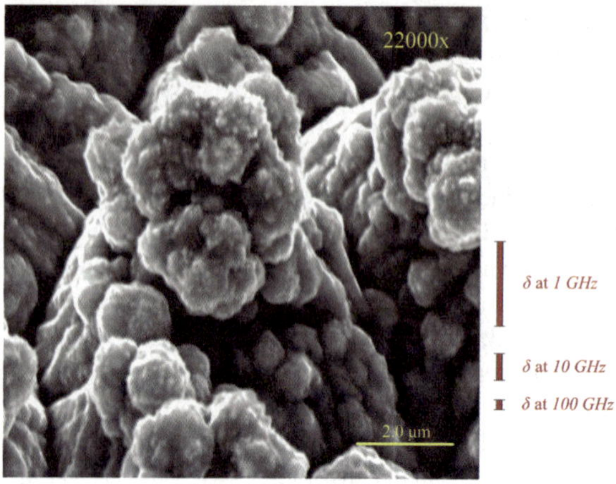

Figure 6.1 Scanning electron microscope photograph of surface distortions for a rough copper surface: copper skin depths, $\delta = \sqrt{2/\mu\sigma\omega}$, for 1, 10, and 100 GHz are shown for relative scale.

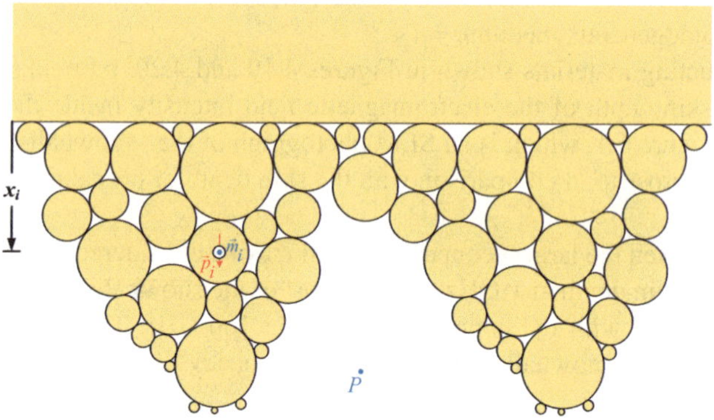

Figure 6.2 Ideal model cross section of a distribution of spheres for analyzing power losses because of surface irregularities.

calculated. Each of the snowballs is subject to applied external electric field intensity and magnetic field intensity so it will have an electric dipole moment, $\vec{p}_i$, and a magnetic dipole moment, $\vec{m}_i$ induced on it.

Periodic electric field intensity and magnetic field intensity will behave in a manner similar to that shown in Figure 6.3, where a single, exaggerated copper sphere (a "snowball") of radius $\boldsymbol{a}_i$ has been added in the path of a propagating elec-

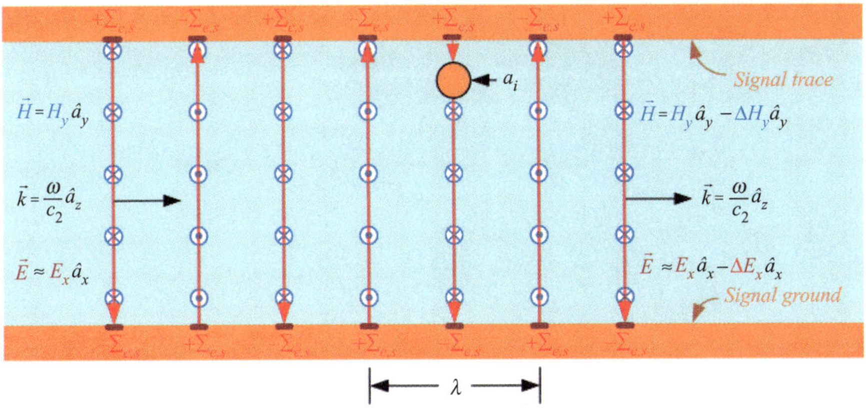

Figure 6.3 Snapshot in time of the cross section of periodic electromagnetic waves as they impinge upon an isolated "snowball" below a copper trace (side view).

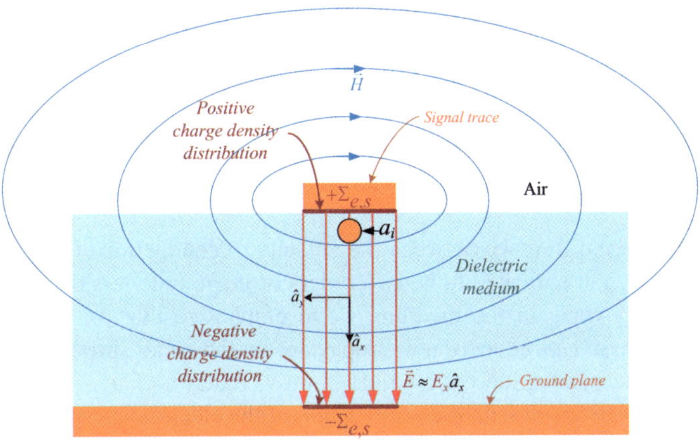

Figure 6.4 Cross section of a periodic electromagnetic wave as it impinges upon an isolated "snowball" below a copper trace (rear view) before the field lines begin to fringe.

tromagnetic wave for an ideal trace model. Note that $\lambda >> \boldsymbol{a}_i$ in this sketch as it is for frequencies below 1 THz.

The surface charge density, $+\Sigma_{e,s}\cos(2\pi z/\lambda)$, required by Gauss's law is shown in a side perspective for a snapshot in time above and below the electric field lines in Figure 6.3. For clarity of field directions, Figure 6.4 shows the same cross section from the rear perspective.

An exaggerated side perspective view (recall from Chapter 4 that the typical displacement of the electron cloud is less than one *nuclear* diameter) of the charge density required to support the periodic fields is shown in Figure 6.5.

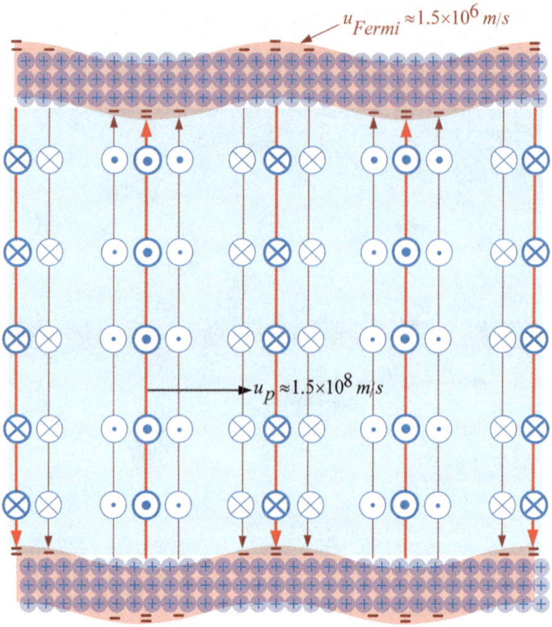

Figure 6.5 Transverse displacement of conduction electrons relative to copper ion cores as they produce a surface charge density to support the propagating electric field intensity near the medium interface. Magnetic field intensity lines are shown as arrows into or out of the page.

In Figure 6.5, the large difference between the velocity of conduction electrons at the Fermi surface, u_{Fermi}, is indicated to be two orders of magnitude less than the speed of the propagating wave in the medium, c_2, to emphasize that the charge density must be a *transverse charge wave* that propagates on the metal surface. The wave displacement of the conduction electron cloud thus creates a charge density on the surface that moves in synch with the propagation of the electromagnetic field intensity. However, because no electrons actually move at this speed, we need *not* take into account relativistic effects.

Figure 6.6 shows electric field intensity incident on an isolated perfect electrically conducting (PEC) sphere inducing an electric dipole moment, $\vec{p}_i$, and a magnetic dipole moment, $\vec{m}_i$, to produce scattering of the incident electromagnetic field outside the sphere. This approach to scattering is called the Born approximation (after Max Born, who in 1954 won the Nobel Prize in Physics). It consists of taking the incident field in place of the total field as the driving field at each point on a scattering object. The approximation is most valid when the wavelength is long compared with the size of the scattering object and when the scattered field intensity is a small fraction of the incident field intensity. In quantum mechanics, the approach uses a perturbation method when applied to scattering by an extended body in which the 0th order fields (fields in the absence of a scattering object) are used to find scattered fields and then that combined 1st order field is put back into the analysis

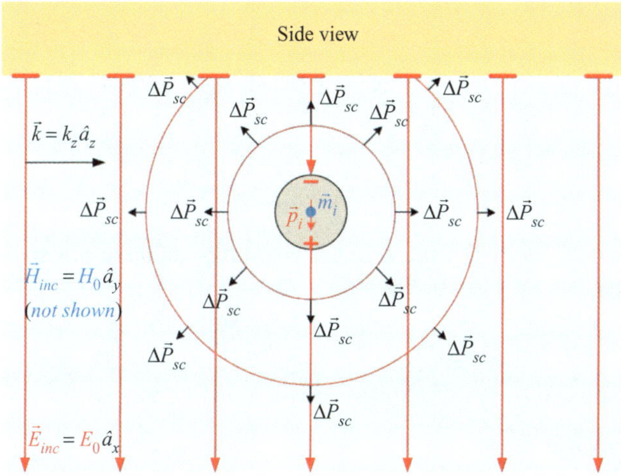

Figure 6.6 Electric dipole moment and magnetic dipole moment induced by an incident electromagnetic wave as it propagates past a perfect electrically conducting "snowball." Magnetic field intensity lines are not shown, and it is assumed that $\lambda >> a_i$.

to find 2nd order terms, and so forth. For small scattering fractions the approximation series usually converges to an answer, but it is not guaranteed that it will converge to the correct answer. Later in this chapter, we will solve Maxwell's equations to find an exact solution to the scattered fields. It is comforting that, in the limit in which the Born approximation is valid, the two techniques give the same answer.

In a Born approximation, we take the incident electric field intensity to be of the form

$$\begin{aligned} \vec{E}_{inc}(\vec{x}, t) &= E_0 e^{-j(\vec{k}\cdot\vec{x}-\omega t)}\hat{a}_x \\ \vec{E}_{inc}(\vec{x}, t) &= E_0 e^{i(\vec{k}\cdot\vec{x}-\omega t)}\hat{a}_x \end{aligned} \quad \text{with } \vec{k} = k_2\hat{a}_z. \tag{6.1}$$

The second form of Equation 6.1 uses the notation of the physics community ($-j \rightarrow i$), and because this chapter deals with scattering that was developed by that community, that form shall be adopted here. In this form, the incident magnetic field intensity is described by

$$\vec{H}_{inc} = H_0 e^{i(\vec{k}\cdot\vec{x}-\omega t)}\hat{a}_y, \tag{6.2}$$

which are not shown in Figure 6.6 to reduce the number of symbols. There are several ways to express the amount of scattered power in any given direction, one of which is to treat the scattering sphere as having sufficient induced surface charge density so as to cause the electric field intensity inside to be zero. From Chapter 1, the incident power density (flux) at any location and any instant in time was given by the Poynting vector in Equation 1.74,

$$\vec{\mathcal{P}}_{inc} = \vec{E} \times \vec{H} \tag{6.3}$$

and the time average of the incident power density over an integer number of cycles 1.85:

$$\vec{\mathcal{P}}_{Avg} = (1/2)\mathrm{Re}\left[\vec{E} \times \vec{H}^*\right] \tag{6.4}$$

Thus, for electric field intensity and magnetic field intensity propagating in a *lossless medium* with permeability, μ, and permittivity, ε,

$$\vec{\mathcal{P}}_{Avg} = (1/2)E_0 H_0 \tag{6.5}$$

Real Propagating Medium

Chapter 5 showed that permittivity, ε_r, is complex and is a function of frequency, ω, due to the fact that losses occur in a real propagating medium. Thus, the electric and magnetic field intensities, $\vec{E}$ and $\vec{H}$, are related by a complex propagation constant, $\tilde{k} = \beta - j\alpha$, as $(\beta - j\alpha)\hat{a}_k \times \vec{E}_0 = \omega\mu\vec{H}_0$ or

$$\vec{H}_0 = [(\beta - j\alpha)/\omega\mu]\hat{a}_k \times \vec{E}_0 = (1/\tilde{\eta})\hat{a}_k \times \vec{E}_0 \tag{1.79}$$

$$\text{where} \quad \tilde{\eta}(\omega) \equiv \omega\mu/(\beta - j\alpha) = |\tilde{\eta}(\omega)| e^{j\theta_\eta(\omega)} \tag{1.80}$$

Generally, the propagating medium is, at best, a *weakly conducting* material so

$$\tilde{k} = \beta - j\alpha \approx \omega\sqrt{\mu\varepsilon}\,[1 - j(\sigma_{medium}/2\omega\varepsilon)] \tag{1.30b}$$

In the case of a lossy propagating medium, we can use Equation 1.84 to revise Equation 6.5:

$$\vec{\mathcal{P}}_{Avg}(\omega) = (1/2)|\tilde{\eta}_{medium}(\omega)|\, H_o^2 e^{-2\alpha_{medium} z} \cos\theta_\eta(\omega)\,\hat{a}_z \left(W/m^2\right) \tag{6.6}$$

We thus see that there is a power density loss per unit area because of the term $e^{-2\alpha_{medium} z}$ as the incident wave propagates in a weakly conducting propagating medium and if $|\sigma_{medium}/\omega\varepsilon| << 1$ then $\cos\theta_\eta \approx 1$.

Conclusion

Power will decay exponentially because of the absorption coefficient, α_{medium}, as $e^{-2\alpha_{medium} z}$ (trace width, w, trace-ground spacing, t) as the incident wave propagates in the z-direction. Here, we make the assumption that the plane wave shows little fringing, as shown in Figure 4.2. As the wave propagates, it will of course, fringe, with width on the ground plane being larger than w, but the total integrated power should not change because of the fringing so the approximation is approximately valid. The fraction of incident power lost per differential length, dz, is thus

$$\Delta P_{medium}/P_{inc} = \frac{[|\tilde{\eta}_{medium}(\omega)|\alpha_{medium} H_o^2(w \cdot dz)]}{[(1/2)|\tilde{\eta}_{medium}| H_0^2(w \cdot t)]} = (2\alpha_{medium})(dz/t) \tag{6.7}$$

Field Penetration into the *Good* Conductor Trace and Ground Plane

It has been shown[9] that magnetic field intensity tangent to a flat *good* conductor (i.e., in the $\hat{a}_y$ direction) penetrate into the conductor as

$$\vec{H}_c(\vec{x},t) = H_{medium}(0)e^{-\xi/\delta}e^{-j\xi/\delta}e^{j\omega t}\hat{a}_y,$$

where ξ is the distance into the conductor. In addition, there is induced electric field intensity in the flat conductor

$$\vec{E}_c(\vec{x},t) = [(1+j)/\sigma_1\delta]H_{medium}(0)e^{-\xi/\delta}e^{-j\xi/\delta}e^{j\omega t}\hat{a}_z,$$

where the variable ξ is the distance into the flat conductor (in this case, the $-x$-*direction*). Because of the continuity of the induced tangential electric field intensity inside the good conductor, the presence of the flat conductor produces additional electric field intensity in the medium, parallel, or antiparallel to the direction of propagation. We call the ratio of the tangential component of electric field intensity to magnetic field intensity the surface impedance[10] Z_s, which is given by

$$Z_s = E_c(0)/H_c(0) = [(1+j)/\sigma_c\delta]$$

The electric field intensity induced in the conductor trace and ground plane is *caused* by the propagating magnetic field intensity and moves into the conductor at phase velocity $u_p = \left(c/\sqrt{\varepsilon_c}\right)\sqrt{2}\sqrt{\omega\varepsilon_0/\sigma_c}$ (slow by comparison to the medium propagation phase velocity) according to the complex wave vector as given by:

$$\vec{\tilde{k}}_c = (\beta - j\alpha)\hat{a}_\xi \approx \omega\sqrt{\mu\varepsilon_c}\left[\sqrt{(\sigma_c/\omega\varepsilon_0)}(1-j)/\sqrt{2}\right]\hat{a}_\xi \tag{1.30c}$$

Recall from Chapter 1 that this wave vector also describes electric field intensity that leads magnetic field intensity in a conductor with a phase angle of 45°. This means that the tangential component of the electric field intensity induced at the surface in the propagating medium also leads the magnetic field intensity by 45°. If the trace and ground planes are thick enough to absorb all of the incident power, then the lost power is the same as the power at $\xi = 0$; $\vec{\mathcal{P}}_{Avg,c} = (1/2)\mathrm{Re}[\vec{E}_c(0) \times \vec{H}_c(0)^*]$. The power density (flux) propagating into each conductor is thus power that is lost from the incident wave:

$$\vec{\mathcal{P}}_{Avg,c} = (1/2)\mathrm{Re}[\vec{E}_c \times \vec{H}_c^*] = (1/2)\mathrm{Re}[(1+j)/\sigma_c\delta]|\vec{H}_c(0)|^2 = (1/2\sigma_c\delta)|\vec{H}_{medium}(0)|^2$$

The incident power lost, ΔP_c, to the *two* conductors for a trace width, w, and a differential length dz is thus

$$\Delta P_c = (1/\sigma_c\delta)\,H_0^2(w \cdot dz) \tag{6.8}$$

so the fraction of incident power lost per differential length, dz, for a perfectly flat trace and ground plane is

$$\Delta P_c/P_{inc} = [(1/\sigma_c\delta)\,H_0^2(w \cdot dz)]/[(1/2)|\tilde{\eta}_{medium}|\,H_0^2(w \cdot t)] = (2/\sigma_c\delta)(dz/t) \tag{6.9}$$

where t is the separation distance between the trace and the ground plane.

Scattering by a PEC

Physicists describe the power radiated in a given direction relative to the incident power flux as the differential scattering cross section:

$$\Delta P_{sc}(\theta, \phi)/\mathcal{P}_{inc} = d\sigma_{sc}(\theta, \phi)/d\Omega \tag{6.10}$$

and they define the scattering cross section as

$$\sigma_{sc} = \int_0^{2\pi}\int_0^{\pi}(d\sigma_{sc}/d\Omega)\sin\theta d\theta d\phi \tag{6.11}$$

We can see that this expression will produce a loss of power because of scattering of incident electromagnetic flux due to a PEC sphere that will add to the losses we obtained above for losses due to the propagating medium and losses as a result of penetration of fields into the flat conducting trace and ground plane. However, *good* conducting spheres will also permit fields to penetrate, and there will be both *absorption* and *scattering* losses by *good* conducting spheres. To study each part separately, we will first consider PEC-conducting spheres because they permit no penetration and thus have no absorption losses. This will also give us a limiting expression when conductivity approaches infinity.

6.2 PERFECT ELECTRIC CONDUCTORS IN STATIC FIELDS

Figure 6.7 shows the surface charge density required to produce zero electric field intensity inside a PEC and still maintain the boundary condition that there is no tangential electric field intensity on the surface and that the normal component of the electric field intensity does not penetrate the surface. The electric dipole moment, $\vec{p}_i$, that results from this charge density distribution is

$$\vec{p}_i = 4\pi a_i^3 \varepsilon_2 E_0 \hat{a}_x \tag{6.12}$$

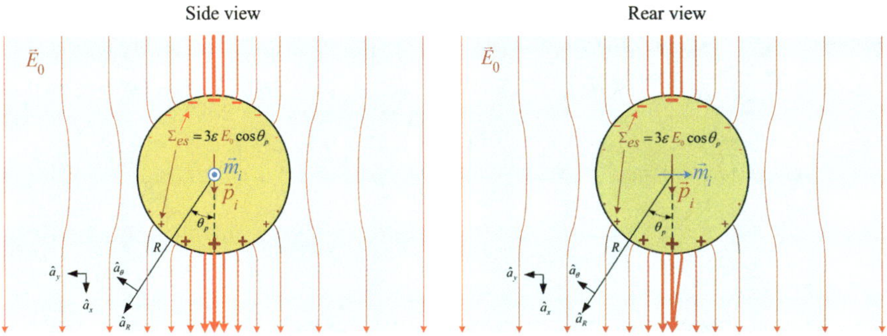

Figure 6.7 Side and rear views of electric field intensity as it induces an electric dipole moment, $\vec{p}_i = 4\pi\varepsilon a_i^3 E_0 \hat{a}_x$, in a perfect electrically conducting sphere; note the angle θ_p is measured relative to the direction of the electric dipole moment, $\hat{a}_x$.

This moment is thus induced in the PEC sphere of radius, a_i, by incident electric field intensity at an instant in space and time, as specified in Equation 6.1. As the incident electric field intensity oscillates, the periodic oscillation of the electric dipole moment causes power to be radiated, as indicated in Figure 6.6.

PROBLEMS

6.1 Show that charge distribution, $\Sigma_{e,s} = 3\varepsilon E_0 \cos\theta_p$, in Figure 6.7 is required to maintain the correct boundary conditions (BC) and create zero electric field intensity inside for a PEC sphere of radius a_i immersed in a medium with permittivity, ε, in the presence of a *static* uniform electric field intensity, E_0.

6.2 Integrate the surface charge density distribution of Problem 6.1 to show that the differential electric dipole moment, $d\vec{p} = \hat{a}_z dqd$, leads to a total *static* electric dipole moment, $\vec{p}_i = 4\pi\varepsilon a_i^3 E_0 \hat{a}_x$ for a PEC sphere.

QUESTION In a quasistatic approximation, would the induced charge density of Problem 6.1 for a sphere with conductivity, σ_{Cu}, be able to keep up with the external electric field intensity provided by a harmonically oscillating external field of frequency ω? For what range of frequencies would the quasistatic approximation apply?

6.3 Use the equation for electric field intensity in the far region of an electric dipole moment, $\vec{E}_{dipole} \cong (p_i/4\pi\varepsilon_0 R^3)[\hat{a}_R 2\cos\theta + \hat{a}_\theta \sin\theta]$, to find the electric field intensity in the far-field region of a slowly oscillating charge density distribution like that described in Problem 6.1.

6.4 For the instant in time shown in Figure 6.7, draw lines of electric field intensity from Problem 6.3 and add them to the uniform electric field intensity to confirm that

$$\vec{E}_{Inside\ PEC\ sphere}(R < a_i) = 0 \tag{6.13}$$

$$\vec{E}_{Outside\ PEC\ sphere}(R, \theta_p) = E_0 \hat{a}_x + E_0[2\cos\theta_p \hat{a}_R + \sin\theta_p \hat{a}_{\theta_p}](a_i/R)^3 \tag{6.14}$$

$$\vec{E}_{Outside\ PEC\ sphere}(R, \theta_p) = E_0 \left\{ \left[1 + 2\left(\frac{a_i}{R}\right)^3\right] \cos\theta_p \hat{a}_R - \left[1 - \left(\frac{a_i}{R}\right)^3\right] \sin\theta_p \hat{a}_{\theta_p} \right\} \tag{6.15}$$

$$\vec{E}_{Surface\ PEC\ sphere}(a_i, \theta_p) = 3E_0 \cos\theta_p \hat{a}_R \tag{6.16}$$

6.5 Confirm the electric scalar potential associated with a PEC sphere in external electric field intensity by showing that

$$\vec{E}_{Outside\ PEC\ sphere}(R, \theta_p) = -\vec{\nabla}\left[-E_0\left(R - a_i^2/R^2\right)\cos\theta_p\right] \tag{6.17}$$

yields the same result as Equation 6.14.

6.6 Use $\vec{H}_{Outside\ PEC\ sphere}(R,\ \theta_p) = (1/i\omega\mu)\vec{\nabla} \times \vec{E}_{Outside\ PEC\ sphere}$ to confirm that the surface charge density shown in Figure 6.7 yields

$$\vec{H}_{Outside\ PEC\ sphere}(R, \theta_p, t) = 0$$

CONCLUSIONS A PEC sphere in a uniform external electric field intensity, E_0, has

- Maximum surface electric field intensity, $3E_0$, at $\theta_p = 0$,
- Minimum surface electric field intensity, $-3E_0$, at $\theta_p = \pi$,
- Zero surface electric field intensity at $\theta_p = \pi/2$,
- Zero electric field intensity inside the sphere,
- External electric field intensity that is the sum of uniform and electric dipole field intensity,
- A scattered electric field intensity that falls off as a dipole field slowly oscillating in time,
- An electric scalar potential that behaves like a pure oscillating electric dipole.

PEC Sphere in a Uniform Magnetic Field Intensity

Incident uniform magnetic field intensity in the $\hat{a}_y$ direction also induces a magnetic dipole moment in a PEC sphere as shown in Figure 6.8.

The sphere will have corresponding surface current density induced:

$$\vec{J}_{l,\ Surface\ PEC\ sphere}(a_i) = [(3/2) H_0 \sin\theta_m] \hat{a}_{\phi_m} \tag{6.18}$$

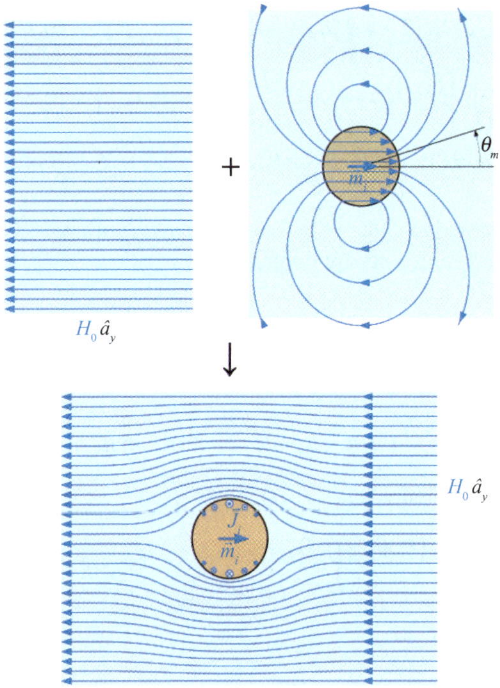

Figure 6.8 Rear view of the propagating magnetic field intensity (into the plane of the paper) as it induces a magnetic dipole moment, $\vec{m}_i = -2\pi a_i^3 H_0 \hat{a}_x$, on the surface of a perfect electrically conducting sphere.

PROBLEMS

6.7 Show that the surface current density distribution, Equation 6.18, in Figure 6.8 maintains the correct magnetic BC and creates zero magnetic field intensity inside a PEC sphere of radius a_i immersed in a medium with permittivity, ε, in the presence of a *static* uniform magnetic field intensity, H_0.

6.8 Integrate the surface current density distribution of Problem 6.7 to show that the differential magnetic dipole moment, $d\vec{m} \equiv \hat{a}_z dI\ S = \hat{a}_z dI\ \pi b^2$, leads to a total *static* magnetic dipole moment, $\vec{m}_{PEC\ sphere} = -2\pi a^3 H_0 \hat{a}_y$, for a PEC sphere.

6.9 Use the equation for the magnetic flux density in the far region of a magnetic dipole moment, $\vec{B}_{dipole} \approx (\mu_0/4\pi)\,(m/R^3)\,(\hat{a}_R 2\cos\theta + \hat{a}_\theta \sin\theta)$, to find the magnetic field intensity of a slowly, harmonically, oscillating current density distribution like that described in Problem 6.7.

6.10 For the instant in time shown in Figure 6.8, draw lines of magnetic field intensity from Problem 6.9 and add them to the uniform electric field intensity to confirm that

$$\vec{H}_{Inside\ PEC\ sphere}(R < a_i) = 0 \tag{6.19}$$

$$\vec{H}_{Outside\ PEC\ sphere}(R, \theta_m) = H_0 \hat{a}_y + (1/2) H_0 [2\cos\theta_m \hat{a}_R + \sin\theta_m \hat{a}_{\theta m}](a_i/R)^3 \quad (6.20)$$

$$\vec{H}_{Outside\ PEC}(R \geq a) = H_0 \left\{ \left[1 - \left(\frac{a_i}{R}\right)^3\right] \cos\theta_m \hat{a}_R - \left[1 + \frac{1}{2}\left(\frac{a_i}{R}\right)^3\right] \sin\theta_m \hat{a}_{\theta m} \right\} \quad (6.21)$$

$$\vec{H}_{Surface\ PEC\ sphere}(a, \theta_m) = (3/2) H_0 \sin\theta_m \hat{a}_{\theta m} \quad (6.22)$$

6.11 Show that we may use a magnetic scalar potential associated with a PEC sphere in external magnetic field intensity by confirming

$$\vec{H}_{Outside\ PEC\ sphere}(R, \theta_m) = -\vec{\nabla}\left[-H_0 \left(R - a_i^2/R^2\right)\cos\theta_m\right] \quad (6.23)$$

yields the same result as Equation 6.15.

6.12 Use $\vec{E}_{total,m}(R, \theta_m, \phi_m) = \frac{i}{\varepsilon\omega} \vec{\nabla} \times \vec{H}_{total,m}(R, \theta_m, \phi_m)$ to confirm that, for the surface current density shown in Figure 6.8,

$$\vec{E}_{Outside\ PEC\ sphere}(R, \theta_p, t) = 0$$

Conclusions

A PEC sphere in uniform external magnetic field intensity, H_0, has the following:

- Maximum surface magnetic field intensity, $3H_0/2$, at $\theta_p = \pi/2$,
- Minimum surface magnetic field intensity, 0 at $\theta_p = 0$ and π,
- Zero magnetic field intensity inside the sphere,
- External magnetic field intensity that is the sum of uniform and dipole field intensity,
- Scattered magnetic field intensity that falls off as a time oscillating dipole field,
- Magnetic scalar potential that behaves like a pure magnetic dipole moment.

In this section, we have assumed that PEC spheres in time-varying fields behave like their static counterparts. That assumption works for quasistatic solutions inside the conductor so the surface charge density and current density will be a limiting target when we consider finite conductivity in a *good* conductor. However, we have shown[11] that time-varying fields in a dielectric medium produced by charge or current densities cause a magnetic vector potential at a point in space that yields *both* electric and magnetic field intensities and it must be calculated with causality by assuring that responses to sources occur after the speed of light have permitted transit from a source location $\vec{x}'$ to a response location, $\vec{x}$. This required a "retarded time" analysis.

6.3 SPHERICAL CONDUCTORS IN TIME-VARYING FIELDS

Multipole Expansion

We have shown[12] that solutions for $\vec{H}$ and $\vec{E}$ due to a harmonic field in a propagating medium by source, $\vec{J}$, may be found by using the magnetic vector potential $\vec{A}(\vec{x}, \vec{x}', t) = (\mu/4\pi)\iiint_{V'}[\vec{J}(\vec{x}')e^{-jk|\vec{x}-\vec{x}'|}/|\vec{x} - \vec{x}'|]d^3x'e^{j\omega t}$ and then computing $\vec{H} = \vec{\nabla} \times \vec{A}/\mu$ and $\vec{E} = \vec{\nabla} \times \vec{H}/j\omega\varepsilon$. By invoking the Lorenz gauge, we also found that $\vec{E} = -\vec{\nabla}V - \partial\vec{A}/\partial t$ gives electric field intensity at a point in space even if the static term $\vec{\nabla}V$ is zero. In physics notation (letting $-j \rightarrow i$), these equations can be expressed as

$$\vec{A}(\vec{x}, \vec{x}', t) = (\mu/4\pi)\iiint_{V'}\left[\vec{J}(\vec{x}')e^{ik|\vec{x}-\vec{x}'|}/|\vec{x}-\vec{x}'|\right]d^3x'e^{-i\omega t} \tag{6.24}$$

$$\vec{H} = \vec{\nabla} \times \vec{A}/\mu \quad \text{and} \quad \vec{E} = i\vec{\nabla} \times \vec{H}/\omega\varepsilon \tag{6.25}$$

In the far-field region where $|\vec{x} - \vec{x}'| >> a_i$, as shown in Figure 6.9, we may approximate

$$|\vec{x} - \vec{x}'| \approx |\vec{x}| - \hat{n} \cdot \vec{x}' \tag{6.26}$$

and evaluate the magnetic vector potential as

$$\vec{A}(\vec{x}, \vec{x}', t) \approx (\mu/4\pi)\left(e^{ik|\vec{x}|}/|\vec{x}|\right)\iiint_{V'}\vec{J}(\vec{x}')e^{-ik\hat{n}\cdot\vec{x}'}d^3x'e^{-i\omega t} \tag{6.27}$$

$$\vec{A}(\vec{x}, \vec{x}', t) \approx (\mu/4\pi)\left(e^{ik|\vec{x}|}/|\vec{x}|\right)\sum_{l=0}^{\infty}\frac{(-ik)^l}{l!}\iiint_{V'}\vec{J}(\vec{x}')(\hat{n}\cdot\vec{x}')^l d^3x'e^{-i\omega t} \tag{6.28}$$

Equation 6.28 is the multipole expansion of the magnetic vector potential.

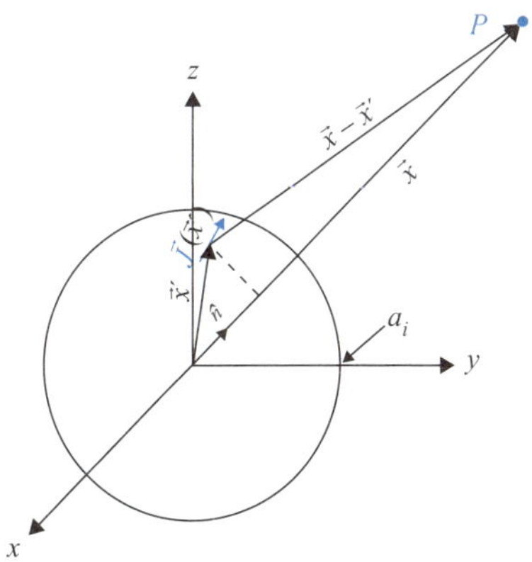

Figure 6.9 Response point P located at $\vec{x}$ due to source current density, $\vec{J}$, located at $\vec{x}'$.

Electric Dipole

The first term ($l = 0$) in series 6.28 gives the electric dipole radiation as

$$\vec{A}_{l=0}(\vec{x}, \vec{x}', t) \approx (\mu/4\pi)\left(e^{ik|\vec{x}|}/|\vec{x}|\right)\iiint_{V'} \vec{J}(\vec{x}')d^3x'e^{-i\omega t}, \tag{6.29}$$

which can be evaluated by using integration by parts and the equation of continuity:

$$\iiint_{V'} \vec{J}(\vec{x}')d^3x' = -\iiint_{V'} \vec{x}'\left(\vec{\nabla}'\cdot\vec{J}\right)d^3x' = -i\omega\iiint_{V'} \vec{x}'\rho(\vec{x}')d^3x' \tag{6.30}$$

with $\partial\rho/\partial t + \vec{\nabla}\cdot\vec{J} = 0$, which can be written as $i\omega\rho = \vec{\nabla}\cdot\vec{J}$.

Equation 6.30 thus results in

$$\vec{A}_{l=0}(\vec{x}, \vec{x}', t) \approx -i\omega\frac{\mu}{4\pi}\vec{p}\frac{e^{ik|\vec{x}|}}{|\vec{x}|}e^{-i\omega t}, \tag{6.31}$$

where

$$\vec{p} \equiv \iiint_{V'} \vec{x}'\rho(\vec{x}')d^3x' \tag{6.32}$$

is the definition of the electric dipole moment.

Using Equation 6.25 with 6.31, we thus deduce the magnetic and electric field intensity as

$$\vec{H}_{\vec{p}} = \frac{\omega k}{4\pi}\frac{e^{ik|\vec{x}|}}{|\vec{x}|}\left(1 - \frac{1}{ik|\vec{x}|}\right)(\hat{n}\times\vec{p})e^{-i\omega t} \tag{6.33}$$

and

$$\vec{E}_{\vec{p}} = \frac{1}{4\pi\varepsilon}\left\{k^2(\hat{n}\times\vec{p})\times\hat{n}\frac{e^{ik|\vec{x}|}}{|\vec{x}|} + [3\hat{n}(\hat{n}\cdot\vec{p}) - \vec{p}]\left(\frac{1}{|\vec{x}|^3} - \frac{ik}{|\vec{x}|^2}\right)e^{ik|\vec{x}|}\right\}e^{-i\omega t} \tag{6.34}$$

PROBLEM

6.13 Carry out the curl operations in Equation 6.25 with the magnetic vector potential given in Equation 6.31 to show Equations 6.33 and 6.34.

The second ($l = 1$) term in Equation 6.28 yields

$$\vec{A}_{l=1}(\vec{x}, \vec{x}', t) \approx -ik\frac{\mu}{4\pi}\frac{e^{ik|\vec{x}|}}{|\vec{x}|}\iiint_{V'} \vec{J}(\vec{x}')(\hat{n}\cdot\vec{x}')d^3x'e^{-i\omega t} \tag{6.35}$$

where

$$\iiint_{V'} \vec{J}(\vec{x}')(\hat{n}\cdot\vec{x}')d^3x' = (1/2)\iiint_{V'} [(\hat{n}\cdot\vec{x}')\vec{J} + (\hat{n}\cdot\vec{J})\vec{x}' + (\vec{x}'\times\vec{J})\times\hat{n}]d^3x' \quad (6.36)$$

PROBLEM

6.14 Verify the vector identity used in the integrand of Equation 6.36.

Magnetic Dipole

By using the definition of a magnetic dipole moment for the third term,

$$\vec{m} \equiv (1/2)\iiint_{V'} (\vec{x}'\times\vec{J})d^3x' \quad (6.37)$$

we may calculate the magnetic field intensity and electric field intensity from Equation 6.25 as

$$\vec{H}_{\vec{m}} = \frac{1}{4\pi}\left\{k^2(\hat{n}\times\vec{m})\times\hat{n}\frac{e^{ik|\vec{x}|}}{|\vec{x}|} + [3\hat{n}(\hat{n}\cdot\vec{m}) - \vec{m}]\left(\frac{1}{|\vec{x}|^3} - \frac{ik}{|\vec{x}|^2}\right)e^{ik|\vec{x}|}\right\}e^{-i\omega t} \quad (6.38)$$

and

$$\vec{E}_{\vec{m}} = -\frac{1}{4\pi\varepsilon c_2}k^2(\hat{n}\times\vec{m})\frac{e^{ik|\vec{x}|}}{|\vec{x}|}\left(1 - \frac{1}{ik|\vec{x}|}\right)e^{-i\omega t} \quad (6.39)$$

PROBLEM

6.15 Carry out the gradient operations in Equation 6.25, with the third term in the magnetic vector potential given in Equations 6.35 and 6.36 to show Equations 6.38 and 6.39.

Electric Quadrupole

The first two terms in Equation 6.36 may be evaluated through integration by parts and the equation of continuity $i\omega\rho = \vec{\nabla}\cdot\vec{J}$ to yield

$$(1/2)\iiint_{V'} [(\hat{n}\cdot\vec{x}')\vec{J} + (\hat{n}\cdot\vec{J})\vec{x}']d^3x' = (-i\omega/2)\iiint_{V'} \vec{x}'(\hat{n}\cdot\vec{x}')(\rho(\vec{x}'))d^3x' \quad (6.40)$$

PROBLEM

6.16 Use integration by parts to show 6.40.

By using Equation 6.40 for the first two terms in Equation 6.35, we can show

$$\vec{A}_{first\ 2} \approx -\omega k \frac{\mu}{8\pi}\frac{e^{ik|\vec{x}|}}{|\vec{x}|}\left(1-\frac{1}{ik|\vec{x}|}\right)\iiint_{V'} \vec{x}'(\hat{n}\cdot\vec{x}')\rho(\vec{x}')d^3x' e^{-i\omega t} \tag{6.41}$$

from which we may evaluate $\vec{H}_{\vec{Q}(\hat{n})}$ and $\vec{E}_{\vec{Q}(\hat{n})}$ in terms of an electric quadrupole tensor $\vec{Q}(\hat{n})$. Because we do not expect any multipole moments greater than dipole for a sphere in a uniform external electric and magnetic field intensity, we shall assume that this term and all higher order terms ($l = 2$ and greater) are negligible in Equation 6.28.

Conclusions

- Equations 6.33 and 6.34 give the magnetic and electric field intensities, $\vec{H}_{\vec{p}}$ and $\vec{E}_{\vec{p}}$, at a point $\vec{x}$ in space due to an electric dipole, $\vec{p}$, produced by a spherical charge distribution, $\rho(\vec{x}')$, at the origin. We see that $\vec{E}_{\vec{p}}$ has a radial component but $\vec{H}_p$ does not so we would call this radiation transverse magnetic (TM^r) because it is orthogonal to the radius vector in spherical coordinates.
- Equations 6.38 and 6.39 give the magnetic and electric field intensities, $\vec{H}_{\vec{m}}$ and $\vec{E}_{\vec{m}}$, at a point $\vec{x}$ in space because of a magnetic dipole, $\vec{m}$, produced by a spherical current distribution, $\vec{J}(\vec{x}')$, at the origin. We see that $\vec{H}_{\vec{m}}$ has a radial component but $\vec{E}_{\vec{m}}$ does not so we would call this radiation transverse electric (TE^r) because it is orthogonal to the radius vector in spherical coordinates.
- We do not expect quadrupole, octapole, hexadecapole, or higher-order moments in problems associated with spherical conductors in uniform external fields.

6.4 THE FAR-FIELD REGION

Equations 6.34 and 6.38 contain terms in the radial direction, but these terms fall off with distance $|\vec{x}|$ faster than the transverse radial term. Figure 6.10 shows the induced charge and current that contribute to the scattered radiation.

Figure 6.10 indicates that scattered electric and magnetic field intensity waves in the far-field region, $|\vec{x}| >> a_i$, are orthogonal to the radial vector because the radial field components have decreased to negligible amounts. We can calculate the total power radiated by scattering from the conducting sphere by integrating the Poynting vector dotted into a surface element over any closed surface that encloses the sphere (e.g., S_1 at the surface of the sphere). We can also choose a much larger sphere where the scattered fields have little radial component and get the same amount of total power radiated because scattered energy is conserved.

Using Equations 6.34 and 6.39 and keeping only the highest-order terms in $|\vec{x}|$, we can thus calculate the scattered electric field intensity in the far-field region, $|\vec{x}| >> a_i$, as

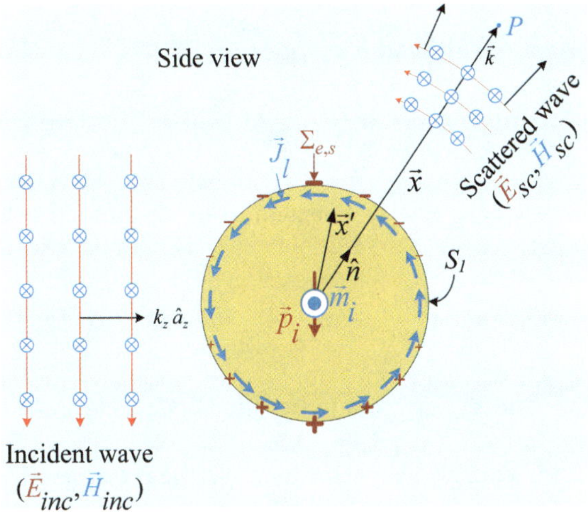

Figure 6.10 Scattering electric field intensity and magnetic field intensity due to a conducting sphere in a polarized incident field.

$$\vec{E}_{sc} = \frac{k^2}{4\pi\varepsilon}\left[(\hat{n}\times\vec{p}_i)\times\hat{n} - \hat{n}\times\frac{\vec{m}_i}{c_2}\right]\frac{e^{ik|\vec{x}|}}{|\vec{x}|}e^{-i\omega t} \tag{6.42}$$

The total scattered electric field intensity is, thus, a result of a vector sum of the electric dipole fields caused by an oscillating electric dipole and an oscillating magnetic dipole. These fields are orthogonal to the radius vector so they are called *spherically transverse electric* (normal to the radial direction in spherical coordinates) and are labeled TE^r. Using Equations 6.33 and 6.38 and keeping only the highest-order terms in $|\vec{x}|$, we can calculate the scattered magnetic field intensity in the far-field region, $|\vec{x}| >> a_i$, as

$$\vec{H}_{sc} = \frac{\omega k}{4\pi}\left[(\hat{n}\times\vec{p}_i) + \left(\hat{n}\times\frac{\vec{m}_i}{c_2}\right)\times\hat{n}\right]\frac{e^{ik|\vec{x}|}}{|\vec{x}|}e^{-i\omega t} \tag{6.43}$$

When quadrupole and higher moments and terms that die faster than the inverse value of $|\vec{x}|$ are ignored, the total scattered magnetic field intensity is thus a result of a vector sum of the magnetic dipole fields caused by an oscillating electric dipole and an oscillating magnetic dipole. These fields are orthogonal to the radius vector so they are called *spherically transverse magnetic* (normal to the radial direction in spherical coordinates) and are labeled TM^r. Because the electric and magnetic fields (in the far field region) are both orthogonal to the radial direction and are orthogonal to one another, they are called *spherically transverse electromagnetic radial* (TEM^r).

We can rewrite Equations 6.42 and 6.43 in terms of the base vectors $\hat{a}_{\theta_p}$ and $\hat{a}_{\phi_p}$ relative to the electric dipole moment and $\hat{a}_{\theta_m}$ and $\hat{a}_{\phi_m}$ relative to the magnetic dipole moment as

$$\vec{E}_{sc} = \frac{k^2}{4\pi\varepsilon}[-|\vec{p}_i|\sin\theta_p \hat{a}_{\theta_p} + |\vec{m}_i|\sin\theta_m \hat{a}_{\phi_m}/c_2]\frac{e^{ik|\vec{x}|}}{|\vec{x}|}e^{-i\omega t} \quad (6.44)$$

$$\vec{H}_{sc} = \frac{k^2}{4\pi\varepsilon}\frac{1}{\sqrt{\mu/\varepsilon}}[-|\vec{p}_i|\sin\theta_p \hat{a}_{\phi_p} - |\vec{m}_i|\sin\theta_m \hat{a}_{\theta_m}/c_2]\frac{e^{ik|\vec{x}|}}{|\vec{x}|}e^{-i\omega t} \quad (6.45)$$

Equation 6.44 is the same as $\sqrt{\mu/\varepsilon}\vec{H}_{sc} \times \hat{a}_R$ in Equation 6.45, confirming that $\vec{E}_{sc} = \sqrt{\mu/\varepsilon}\vec{H}_{sc} \times \hat{n}$.

PROBLEM

6.17 Confirm that $\sqrt{\mu/\varepsilon}\vec{H}_{sc} \times \hat{n}$, with $\vec{H}_{sc}$ as given in Equation 6.45, yields 6.44.

From Equations 6.44 and 6.45, we can compute the time average power density to be in the radial direction:

$$\vec{\mathcal{P}}_{Avg,sc} = \frac{1}{2}\mathrm{Re}[\vec{E}_{sc} \times \vec{H}^*_{sc}] = \frac{k^4}{2(4\pi\varepsilon)^2}\sqrt{\frac{\varepsilon}{\mu}}\left[2|\vec{p}_i|\sin\theta_p \frac{|\vec{m}_i|}{c_2}\sin\theta_m\right]\frac{\hat{a}_R}{|\vec{x}|^2} \quad (6.46)$$

Thus, on a large sphere (in the far region), the power scattered in the radial direction is 6.46 times the differential area $|\vec{x}|^2 d\Omega$. This result yields

$$\Delta\vec{\mathcal{P}}_{Avg,sc} = \frac{k^4}{2(4\pi\varepsilon)^2\sqrt{\mu/\varepsilon}}\left[2|\vec{p}_i|\sin\theta_p \frac{|\vec{m}_i|\sin\theta_m}{c_2}\right]\hat{a}_R d\Omega \quad (6.47)$$

and, using Equation 6.10 $\Delta P_{sc}(\theta, \phi)/\vec{\mathcal{P}}_{Avg} = d\sigma_{sc}(\theta, \phi)/d\Omega$ with $\vec{\mathcal{P}}_{Avg} = (1/2)E_0 H_0$ from Equation 6.5

$$\frac{d\sigma_{sc}(\theta, \phi)}{d\Omega} = \frac{k^4 c_2^2}{(4\pi)^2 H_0^2}[2|\vec{p}_i|\sin\theta_p |\vec{m}_i|\sin\theta_m/c_2] \quad (6.48)$$

Using Equation 6.12 for a PEC sphere, $\vec{p}_i = 4\pi a_i^3 \varepsilon_2 E_0 \hat{a}_x$, and $\vec{m}_{PEC\ sphere} = -2\pi a_i^3 H_0 \hat{a}_y$ from Problem 6.8, with $E_0 = \sqrt{\mu/\varepsilon}H_0$, we can express the differential cross section for scattered radiation from a PEC sphere as $d\sigma_{sc}(\theta, \phi)/d\Omega = k^4 a_i^6 \sin\theta_p \sin\theta_m$ From 6.11, we find that the total scattering cross section is

$$\sigma_{sc} = k^4 a_i^6 \int_0^{2\pi}\int_0^{\pi} \sin\theta_p \sin\theta_m \sin\theta d\theta d\phi \quad (6.49)$$

Note that the angle θ is just a variable of integration, and, because the integral is carried out over the entire sphere, we may choose it to be relative to either of the dipole directions. Assuming that all scattered power is lost from the incident wave, then the time averaged power lost per PEC sphere is

$$\Delta P_{Avg,i}/\vec{\mathcal{P}}_{Avg} = \sigma_{sc} = (10\pi/3)k^4 a_i^6 \quad (6.50)$$

PROBLEM

6.18 Use transformation relations[13] to rotate $\sin\theta_p$ or $\sin\theta_m$ into the other and then carry out the integral in Equation 6.49 to show Equation 6.50.

Scattered power is proportional to the fourth power of frequency because $k = \omega/c_2$, and the sixth power of radius is characteristic of Rayleigh[14] scattering because Lord Rayleigh was one of the first persons to quantify scattering from air variations by considering scattering from dielectric spheres with an electric dipole moment of the form

$$\vec{p}_i = 4\pi\varepsilon_0 [(\varepsilon_r - 1)/(\varepsilon_r + 2)] a_i^3 \vec{E}_{inc} \tag{6.51}$$

and no magnetic dipole moment, but his result yielded the characteristic $k^4 a_i^6$ dependence on the frequency and size of the scattering objects with a different numerical factor than Equation 6.50.

The quantity in Equation 6.50 can be used in Equation 6.11 as an additional loss per PEC sphere to that caused by the medium and fields that penetrate into the conducting trace and ground plane. By summing all of the losses due to all such spheres involved in surface roughness, we can find the contribution due to scattering by the snowballs. Below, we will discuss the interference problems associated with scattering from more than one sphere and the losses brought about from the dipole images that are created in the plane of the nearby trace or ground plane.

However, we chose to consider scattering from a PEC in this section so that no fields would penetrate the sphere, and, thus, there would be no contribution to losses due to absorption in the sphere. This gave us a value for a scattering cross section in Equation 6.50 that we can use in the limit as conductivity becomes very large in the following discussions. However, it turns out that, at frequencies below about 1 THz, the size of the copper snowballs we will encounter will give a larger contribution from absorption losses than from scattered losses, so we must take on the absorption problem associated with *good* rather than perfect spheres.

6.5 ELECTRODYNAMICS IN GOOD CONDUCTING SPHERES

In a flat *good* conductor, such as a perfectly flat trace, the magnetic field intensity penetrates the surface according to a skin depth formulation, as shown in Figure 6.11.

In Figure 6.11, the periodic magnetic field intensity on the surface is attenuated by an exponential envelope (dotted lines) as a result of conduction losses as it moves slowly (at velocity u_p) into the conductor. The magnetic field intensity inside the conductor propagates in the positive ξ direction at phase velocity

$$u_p = \omega\delta = c\sqrt{2}\sqrt{\omega\varepsilon_0/\sigma_{Cu}}, \tag{6.52}$$

where the quantity inside the square root is a figure of merit of the conductivity of a *good* conductor. For copper,

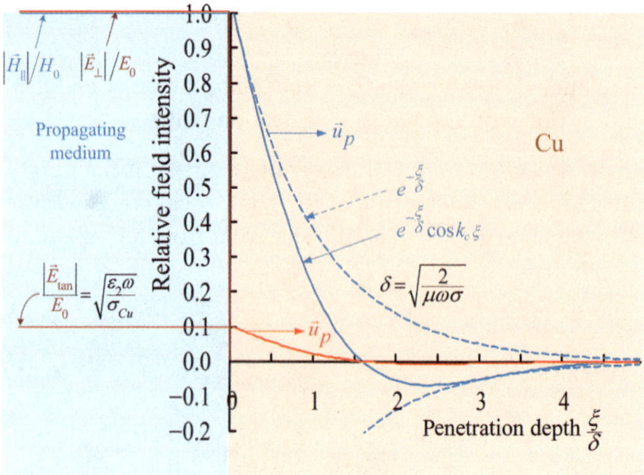

Figure 6.11 Tangential component of the magnetic field intensity and electric field intensity as a function of depth, ξ/δ, inside a flat good conductor.

$$\frac{\sigma_{Cu}}{\omega\varepsilon_0} = \begin{bmatrix} 1.04\times10^9 & \text{at} & 1\,\text{GHz} \\ 1.04\times10^8 & \text{at} & 10\,\text{GHz} \\ 1.04\times10^7 & \text{at} & 100\,\text{GHz} \end{bmatrix} \tag{6.53}$$

The magnetic field intensity immediately inside the conductor is oscillating in time with the tangential external field so that, by the time it reaches $\xi/\delta = \pi/2$, the field was that caused by the surface field when it had zero magnitude. For greater depths, for example $\xi/\delta = \pi$, the current magnetic field intensity was originally caused by a surface magnetic field intensity that had the opposite sign. Thus the current magnetic field intensity at $\xi/\delta = \pi$ has a negative value relative to the current magnetic field intensity at the surface. For a copper "snowball," we expect a similar behavior as shown at a snapshot in time in Figure 6.12.

In Figure 6.12, the wave front (dotted circle) produced by the Huygens construction of the inward propagating magnetic field intensity would be expected to be a sphere with radius, r, that decreases with time. However, this magnetic field intensity distribution will oscillate in inward distance, ξ, because the inward propagating wave is traveling slowly compared with the propagation of the external applied field; similar to the field oscillation shown in Figure 6.11. In Figure 6.12, the snapshot in time has been taken after the wave front has progressed a distance of about a half of a sphere radius, a_i, but we can see that the wave front will eventually reduce to zero radius after a time $\Delta t = a_i/u_p$. For greater times than this, the Huygens wave front construction will continue to propagate outward until it reaches the spherical surface. At time $2\Delta t = 2a_i/u_p$, what is left of the exponentially attenuated amplitude of the wave front will partially transmit back into the external medium across the boundary between a conductor and a dielectric and partially reflect from that boundary as concluded in Chapter 2.

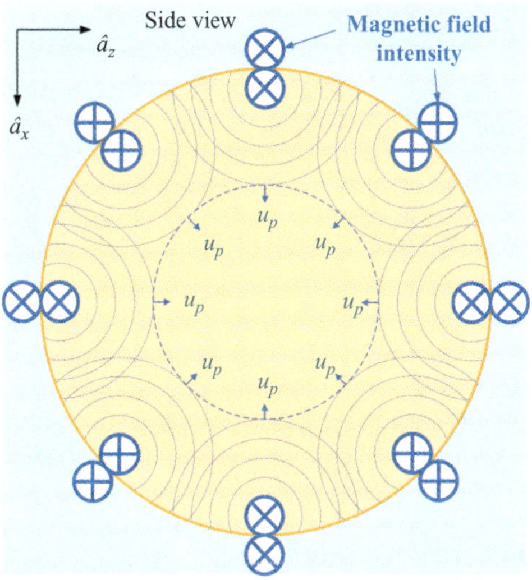

Figure 6.12 Huygens cross section "snapshot" in time of the inward propagating magnetic field intensity in a copper sphere as a result of a periodic uniform external magnetic field intensity tangent to the sphere.

We can reason that, if the diameter, $2a_i$, of the sphere is large compared with the skin depth at that frequency, there will be little amplitude remaining to reemerge back into the medium, so almost all of the power incident to the sphere will be absorbed by the copper. But, if the diameter $2a_i$ is comparable to the skin depth, we will expect some of the incident power to reemerge back into the external medium. Unless the reemerging wave front interferes constructively with the incident wave, the total incoming power will be lost to the incident wave. However, if the reemerging wave front interferes constructively, it will so with a time delay relative to the incident wave; that is, there will be a phase shift of the total wave that we would interpret as an inductive characteristic in a circuit model. Because we are modeling many snowballs of different sizes and the delay of the reemerging wave front is dependent on size, we would expect random interference with the incident wave from an assembly of spheres. We would thus conclude that the sum of the incoming power and the outgoing (scattered) power is phase shifted relative to the power of the incident wave. To find the lost power to a linear propagating wave, we need to determine how much power is outgoing from a conducting sphere and add that to the power that is incoming toward a conducting sphere. This can be accomplished by using the techniques of partial wave scattering analysis used by particle physicists and radar engineers. However, as we will discuss in Chapter 7, the total scattered power from a large number of conducting spheres can add constructively to the incident wave to produce a total intensity that is attenuated and phase shifted (delayed) relative to the incident wave. This will give us an electromagnetic wave foundation for resistance and inductance to compare with a circuit model interpretation.

6.6 SPHERICAL COORDINATE ANALYSIS

Although there is some concern that the vector form of Maxwell's equations does not adequately incorporate all of the concepts included in the original quaternion form, we basically need to assure that all harmonic electromagnetic problems satisfy the field equations for charge-free space that produce the **vector** Helmholtz Equations 6.54 where time-dependent terms are assumed to be of the form $e^{j\omega t}$:

$$\begin{aligned}
&\vec{\nabla}\cdot\vec{E}=\rho/\varepsilon && \text{so} \quad \vec{\nabla}\cdot\vec{E}_s=\rho_s/\varepsilon=0\\
&\vec{\nabla}\cdot\vec{H}=0 && \text{so} \quad \vec{\nabla}\cdot\vec{H}_s=0\\
&\vec{\nabla}\times\vec{E}=-\partial\vec{B}/\partial t && \text{or} \quad \vec{\nabla}\times\vec{E}_s=-j\omega\mu\vec{H}_s\\
&\vec{\nabla}\times\vec{H}=\vec{J}+\partial\vec{D}/\partial t && \text{or} \quad \vec{\nabla}\times\vec{H}_s=\sigma\vec{E}_s+j\omega\varepsilon\vec{E}_s
\end{aligned} \tag{6.54}$$

Further curl operations then turn the vector Maxwell equations into

$$\vec{\nabla}\times\vec{\nabla}\times\vec{E}_S=\vec{\nabla}\cdot(\vec{\nabla}\cdot\vec{E}_S)-\vec{\nabla}^2\vec{E}_S=-j\omega\mu\vec{\nabla}\times\vec{H}_S=-j\omega\mu\left(\sigma\vec{E}_S+j\omega\varepsilon\vec{E}_S\right)$$

$$\vec{\nabla}\times\vec{\nabla}\times\vec{H}_S=\vec{\nabla}\cdot(\vec{\nabla}\cdot\vec{H}_S)-\vec{\nabla}^2\vec{H}_S=\sigma\vec{\nabla}\times\vec{E}_s+j\omega\varepsilon\vec{\nabla}\times\vec{E}_S=-j\omega\mu\sigma\vec{H}_S+\omega^2\mu\varepsilon\vec{H}_S,$$

which can be written as the **vector** Helmholtz equations:

$$\vec{\nabla}^2\vec{E}_S+k^2\vec{E}_S=0 \quad \text{with } k^2=\omega^2\mu\varepsilon(1-j\sigma/\omega\varepsilon) \tag{6.55}$$

$$\vec{\nabla}^2\vec{H}_S+k^2\vec{H}_S=0 \quad \text{with } k^2=\omega^2\mu\varepsilon(1-j\sigma/\omega\varepsilon) \tag{6.56}$$

and letting $j\rightarrow -i$ with time described through $e^{-i\omega t}$ in physics notation,

$$\vec{\nabla}^2\vec{E}_S+k^2\vec{E}_S=0 \quad \text{with } k^2=\omega^2\mu\varepsilon(1+i\sigma/\omega\varepsilon) \tag{6.57}$$

$$\vec{\nabla}^2\vec{H}_S+k^2\vec{H}_S=0 \quad \text{with } k^2=\omega^2\mu\varepsilon(1+i\sigma/\omega\varepsilon) \tag{6.58}$$

Boundary Conditions

For BC specified with the geometry shown in Figure 6.13, the solutions for $\vec{H}$ and $\vec{E}$ outside the sphere may be written as

$$\begin{aligned}
\vec{E}&=E_0e^{i(k_2z-\omega t)}\hat{a}_x+\vec{E}_{sc}\\
\vec{H}&=H_0e^{i(k_2z-\omega t)}\hat{a}_y+\vec{H}_{sc}
\end{aligned}, \tag{6.59}$$

where the first terms on the right in Equation 6.59 for $\vec{H}$ and $\vec{E}$ are incident plane waves described in Cartesian coordinates propagating in the z-direction. Equation 6.59 is in the form of the zeroth and first Born approximation described earlier. Here, we want to specify boundary conditions:

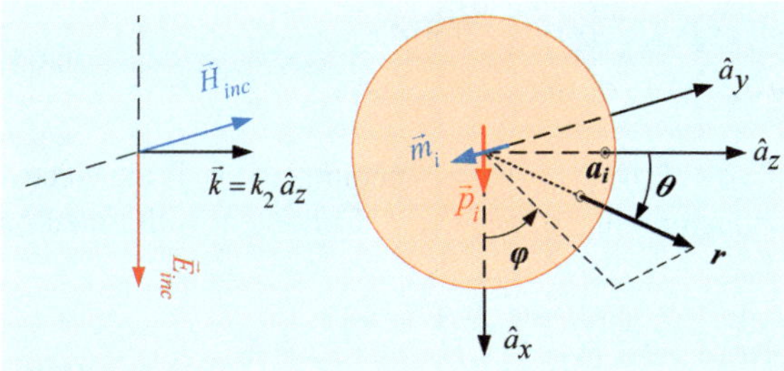

Figure 6.13 Helmholtz regions for the magnetic field intensity inside ($r \le a_i$) and outside ($r \ge a_i$) a *good* conducting sphere.

$$\vec{H}_{\parallel}(r = a_i +, \theta, \phi) = \vec{H}_{\parallel}(r = a_i -, \theta, \phi) \tag{6.60}$$

$$\vec{E}_{\parallel}(r = a_i +, \theta, \phi) = Z_s \hat{n} \times \vec{H}_{\parallel}(r = a_i +, \theta, \phi) \tag{6.61}$$

on the surface of the sphere with $Z_s = (1 - i)/\sigma_c\delta$ and that will be most easily carried out in spherical coordinates. To evaluate the normal components of $\vec{H}$ and $\vec{E}$ on the surface of the good conducting sphere, we can use $\vec{\nabla} \times \vec{E}(\vec{x}) = i\omega\mu\vec{H}(\vec{x})$ from Equation 6.54 to write

$$\vec{H}_{\perp} \simeq -\frac{(1+i)}{2}\delta\left[\vec{\nabla} \times (\hat{n} \times \vec{H}_{\parallel})\right]_{\perp} \tag{6.62}$$

To get a feel for how large this normal component will be, we can guess that the parallel magnetic field for a good conductor will approach the parallel field for a perfect conductor as the conductivity goes to infinity. For a PEC from Problem 6.6 gave $\vec{H}_{Outside\ PEC\ sphere,\vec{P}}(a_i, \theta_p) = 0$ and from Equation 6.22 $\vec{H}_{Surface\ PEC\ sphere,\vec{m}}(a_i, \theta_m) = (3/2)H_0 \sin\theta_m \hat{a}_{\theta_m}$ so

$$H_{\perp}(a_i, \theta_{\vec{m}}, \phi) \simeq \left[-\frac{(1+i)}{2}\delta\right]\frac{3}{2}H_0\left[\frac{1}{r^2\sin\theta_m}\begin{vmatrix} \hat{a}_r & \hat{a}_{\theta_m} r & \hat{a}_{\phi_m} r\sin\theta_m \\ \frac{\partial}{\partial r} & \frac{\partial}{\partial\theta_m} & \frac{\partial}{\partial\phi_m} \\ 0 & 0 & (r\sin^2\theta_m) \end{vmatrix}_{\perp}\right]_{r=a_i}$$

in which case

$$\vec{H}_{\perp}(a_i, \theta_{\vec{m}}, \phi) \simeq -(3/2)H_0(1+i)(\delta/a_i)\cos\theta_m \hat{a}_r \tag{6.63}$$

This value of the normal component goes to zero in the limit as $\sigma_c \to \infty$ and $\delta \to 0$ so it is consistent for a PEC and is what we would approximately expect for a *good* conductor.

As seen in Equations 6.57 and 6.58, the vector Helmholtz equations have a real value $k^2 = \omega^2\mu_0\varepsilon_2$ outside the sphere (in medium 2) and a complex value of k^2 inside the sphere:

$$k^2 = \begin{cases} (\omega^2/c^2)\varepsilon_{r,2} & \text{for } r > a_i \\ (\omega^2/c^2)[1 + i\sigma_c/\omega\varepsilon_0] & \text{for } r \le a_i \end{cases}, \tag{6.64}$$

where $\varepsilon_{r,2}$ is the relative permittivity of the propagating medium and σ_c is the electrical conductivity of the *good* conductor with $(\sigma_c/\omega\varepsilon_0) >> 1$. Consistent with Equation 1.30c, k for a *good* conductor is

$$k_c \approx \left(\frac{\omega}{c}\right)\left[\sqrt{\frac{\sigma_c}{\omega\varepsilon_0}}\frac{(1+i)}{\sqrt{2}}\right] = \frac{(1+i)}{\delta} \tag{6.65}$$

Radial Part of the Solution in Spherical Coordinates

To match the BC at $r = a$ in Figure 6.13, we need to express the plane wave in spherical coordinates. We previously discussed the solution to the **scalar** Helmholtz equation[15] as:

$$\psi(\vec{x}, t) = \begin{pmatrix} j_l(kr) \\ \eta_l(kr) \end{pmatrix} \begin{pmatrix} P_l^m(\cos\theta) \\ Q_l^m(\cos\theta) \end{pmatrix} \begin{pmatrix} e^{im\varphi} \\ e^{-im\varphi} \end{pmatrix} \begin{pmatrix} e^{j\omega t} \\ e^{-j\omega t} \end{pmatrix}$$

$j_l(kr)$ and $\eta_l(kr)$ were the spherical Bessel and Neumann functions and noted that the spherical Hankel functions of the first and second kind, $h_m^{(1)}(kr) = j_m(kr) + i\eta_m(kr)$ and $h_m^{(2)}(kr) = j_m(kr) - i\eta_m(kr)$, were an equally acceptable set of orthonormal functions. Either of these solution forms should thus be acceptable in the region outside a sphere. With complex values of k_c, however, it was noted that the modified spherical Bessel and McDonald functions $i_l(k_cr) \equiv i^{-l}j_l(ik_cr)$ and $k_l(k_cr) \equiv -i^l h_l^{(1)}(ik_cr)$ were often used because they assume that the quantity k_c is a complex number of the phase ¼ types, as indicated in Equation 6.65. The real and imaginary parts of the modified functions are called *ber*, *bei*, *ker*, and *kei* functions written as $i_m(kr) = ber_m(kr) + ibei_m(kr)$ and $k_m(kr) = ker_m(kr) + ikei_m(kr)$. In the following sections, the radial functions are written in the spherical Hankel function form (with complex k_c assumed inside the sphere), but it is understood that any other set (Bessel/Neumann, modified Bessel/McDonald, *ber-bei*/*ker-kei*) would be equally acceptable. Coefficients of the various functions are complex and determined by the BC at the surface of the sphere so they may be assumed to absorb the imaginary numbers i^l in relationships between normal and modified Bessel functions.

For reference, several of the **spherical Hankel functions** are listed below with properties that assist in obtaining higher-order functions:

$$\begin{aligned} h_0^{(1)}(x) &= (-i)(e^{ix}/x) & h_0^{(2)}(x) &= (i)(e^{-ix}/x) \\ h_1^{(1)}(x) &= (-i)^2(e^{ix}/x)(1+i/x) & h_1^{(2)}(x) &= (i)^2(e^{-ix}/x)(1-i/x) \\ h_2^{(1)}(x) &= (-i)^3(e^{ix}/x)(1+3i/x-3/x^2) & h_2^{(2)}(x) &= (i)^3(e^{-ix}/x)(1-3i/x-3/x^2) \\ &\vdots & &\vdots \end{aligned} \tag{6.66}$$

The so-called Wronskian of the spherical Hankel functions is

$$h_l^{(1)}(x)h_l^{(2)'}(x)-h_l^{(1)'}h_l^{(2)}(x)=-2i/x^2 \tag{6.67}$$

and useful recursion relations that let us expand the set Equation 6.61 to higher order are

$$\begin{aligned}(2l+1/x)h_l^{(j)}(x)&=h_{l-1}^{(j)}(x)+h_{l+1}^{(j)} \quad j=1\, or\, 2\\ \partial h_l^{(j)}(x)/\partial x&=\left[lh_{l-1}^{(j)}(x)-(l+1)h_{l+1}^{(j)}(x)\right]/(2l+1)\end{aligned} \tag{6.68}$$

The asymptotic values of Equation 6.61 (in the limit as $|x| = k_2r \to \infty$) are

$$h_l^{(1)}(k_2r)\underset{kr\to\infty}{\to}(-i)^{l+1}e^{ik_2r}/k_2r \quad \text{and} \quad h_l^{(2)}(k_2r)\underset{kr\to\infty}{\to}(i)^{l+1}e^{-ik_2r}/k_2r \tag{6.69}$$

and, when multiplied by the time harmonic, $e^{-i\omega t}$, $h_l^{(1)}(k_2r)e^{-i\omega t}$ and $h_l^{(2)}(k_2r)e^{-i\omega t}$ represent radial outgoing and incoming waves respectively in the far region.

Spherical Bessel and Neumann Functions

The real and imaginary parts of the spherical Hankel functions $h_m^{(1)}(kr) = j_m(kr) + i\eta_m(kr)$ and $h_m^{(2)}(kr) = j_m(kr) - i\eta_m(kr)$ are the spherical Bessel and Neumann functions:

$$\begin{aligned}
j_0(x)&=\frac{\sin x}{x} & \eta_0(x)&=-\frac{\cos x}{x}\\
j_1(x)&=\frac{\sin x}{x^2}-\frac{\cos x}{x} & \eta_1(x)&=-\frac{\cos x}{x^2}-\frac{\sin x}{x}\\
j_2(x)&=-\frac{\sin x}{x}\left(1-\frac{3}{x^2}\right)-3\frac{\cos x}{x^2} & \eta_2(x)&=\frac{\cos x}{x}\left(1-\frac{3}{x^2}\right)-3\frac{\sin x}{x^2}\\
&\vdots & &\vdots
\end{aligned} \tag{6.70}$$

with "Wronskian":

$$j_l(x)\eta_l'(x)-j_l'(x)\eta_l(x)=1/x^2 \tag{6.71}$$

and derivative relationships:

$$j_l(x)=(-1)^l x^l\left(\frac{1}{x}\frac{d}{dx}\right)^l\frac{\sin x}{x} \quad \text{and} \quad \eta_l(x)=-(-1)^l x^l\left(\frac{1}{x}\frac{d}{dx}\right)^l\frac{\cos x}{x} \tag{6.72}$$

Comparison of Equations 6.66–6.70 makes clear that the choice of spherical Hankel functions versus spherical Bessel and Neumann functions is like choosing to write

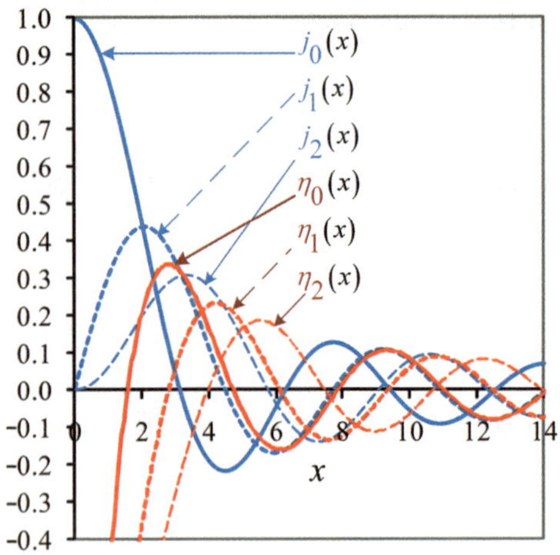

Figure 6.14 Behavior of the spherical Bessel and Neumann functions as a function of a real argument.

solutions for the harmonic oscillator differential equation as exponentials versus sine and cosine functions. Both forms are equally acceptable, but their arbitrary coefficients must be chosen correctly to make them identical. Sometimes, it is an advantage to mix the types of solutions when expanding other functions (e.g. in the case of writing the plane wave functions of Equation 6.59 in terms of spherical coordinate functions). The spherical Bessel and Neumann functions behave as shown in Figure 6.14 with a real argument:

An important characteristic to recognize is that the Neumann functions are singular for a zero argument. Because an implied BC is that all fields must be finite, the coefficient that multiplies the Neumann function must be zero if the solution applies for an argument at the origin (i.e., we must assure Neumann functions are not part of a solution for $r = 0$).

Modified Spherical Bessel and McDonald functions

The modified spherical Bessel and McDonald functions $i_l(kr) \equiv i^{-l}\, j_l(ikr)$ and $k_l(kr) \equiv -i^l\, h_l^{(1)}(ikr)$ are often used when the quantity k is a complex number of the phase $\frac{1}{4}$ type (such as for points inside a *good* conductor).

$$
\begin{aligned}
i_0(x) &= \frac{\sinh x}{x} & k_0(x) &= \frac{e^{-x}}{x} \\
i_1(x) &= \frac{\cosh x}{x} - \frac{\sinh x}{x^2} & k_1(x) &= \frac{e^{-x}}{x}\left[1+\frac{1}{x}\right] \\
i_2(x) &= \frac{\sinh x}{x}\left(1+\frac{3}{x^2}\right) - 3\frac{\cosh x}{x^2} & k_2(x) &= \frac{e^{-x}}{x}\left[1+\frac{3}{x}+\frac{3}{x^2}\right] \\
&\vdots & &\vdots
\end{aligned}
\tag{6.73}
$$

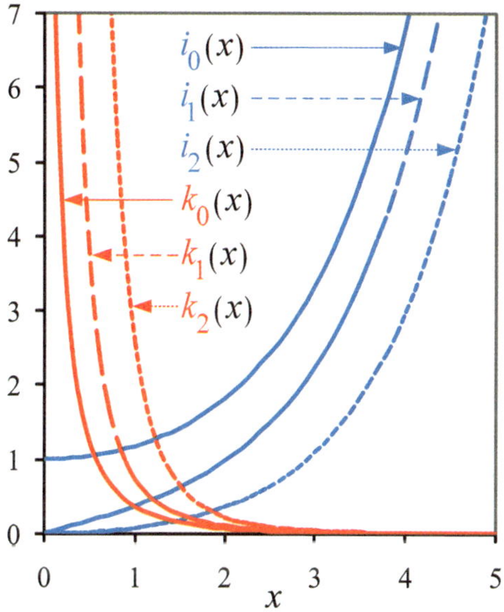

Figure 6.15 Behavior of the modified spherical Bessel and McDonald functions as a function of a real argument.

The "Wronskian" and derivative relationships of the spherical modified Bessel function are

$$i_l(x)k_l'(x) - i_l'(x)k_l(x) = -\frac{1}{x^2} \tag{6.74}$$

$$i_l(x) = x^l\left(\frac{1}{x}\frac{d}{dx}\right)^l \frac{\sinh x}{x} \quad \text{and} \quad k_l(x) = (-1)^l x^l \left(\frac{1}{x}\frac{d}{dx}\right)^l \frac{e^{-x}}{x} \tag{6.75}$$

The modified spherical Bessel and McDonald functions behave as shown in Figure 6.15 with a real argument.

An important characteristic to recognize is that the modified spherical McDonald functions are singular for a zero argument. Because an implied BC is that all fields must be finite, the coefficient that multiplies the modified spherical McDonald function must be zero if the solution applies for an argument at the origin (i.e., we must assure modified spherical McDonald functions are not part of a solution for $r = 0$). With a complex argument, these functions have a real and imaginary part.

SPHERICAL BER, BEI, KER, AND KEI FUNCTIONS

The real and imaginary parts of the modified spherical Bessel and McDonald functions are simply $i_m(kr) = ber_m(kr) + ibei_m(kr)$ and $k_m(kr) = ker_m(kr) + ikei_m(kr)$. Using $k_c \approx (1 + i)/\delta$, $x = k_c r = (1 + i)u$, with $u = r/\delta$ and $\sinh(u + iu) = \sinh(u)\cos u + i\cosh(u)\sin u$ $\cosh(u + iu) = \cosh(u)\cos u + i\sinh(u)\sin u$:

$$ber_0(u)+ibei_0(u)=\left[\frac{\sinh u\cos u+\cosh u\sin u}{2u}\right]+i\left[\frac{\cosh u\sin u-\sinh u\cos u}{2u}\right]$$

$$ber_1(u)+ibei_1(u)=\left[\frac{\cosh u\cos u+\sinh u\sin u}{2u}-\frac{\cosh u\sin u}{2u^2}\right]$$
$$+i\left[\frac{\sinh u\sin u-\cosh u\cos u}{2u}+\frac{\sinh u\cos u}{2u^2}\right]$$

$$ber_2(u)+ibei_2(u)=\left[\frac{(\sinh u\cos u+\cosh u\sin u)}{2u}-\frac{3\sinh u\sin u}{2u^2}\right.$$
$$\left.-\frac{3(\cosh u\sin u-\sinh u\cos u)}{4u^3}\right]$$
$$+i\left[\frac{(\cosh u\sin u-\sinh u\cos u)}{2u}+\frac{3\cosh u\cos u}{2u^2}\right.$$
$$\left.+\frac{3(\sinh u\cos u+\cosh u\sin u)}{4u^3}\right]$$

•

•

(6.76)

$$ker_0(u)+ikei_0(u)=\frac{e^{-u}}{2u}[\cos u-\sin u]-i\frac{e^{-u}}{2u}[\cos u+\sin u]$$

$$ker_1(u)+ikei_1(u)=\frac{e^{-u}}{2u^2}[u(\cos u-\sin u)-\sin u]-i\frac{e^{-u}}{2u^2}[u(\cos u+\sin u)+\cos u]$$

$$ker_2(u)+ikei_2(u)=\frac{e^{-u}}{2u^3}\left[u^2(\cos u-\sin u)-3u\sin u-\frac{3}{2}(\cos u+\sin u)\right]$$
$$-i\frac{e^{-u}}{2u^3}\left[u^2(\cos u+\sin u)+3u\cos u+\frac{3}{2}(\cos u-\sin u)\right]$$

•

•

(6.77)

We can see from these equations that the spherical *ker* and *kei* functions are all singular at the origin, $u = 0$, and hence cannot be a part of any solution that includes that point. The *ber* and *bei* functions behave as shown in Figure 6.16 with a real argument.

Angular Part of the Solution

The angular components may be written[16] in terms of the associated Legendre polynomials of the first, $P_l^m(\cos\theta)$, and second kind, $Q_l^m(\cos\theta)$, but the values of the second kind are singular for arguments $\cos\theta = \pm 1$, both of which are included in our scattering problem shown in Figure 6.13 so we cannot accept these solutions (i.e., the coefficient of these terms must be zero). For the angular part, this leaves only

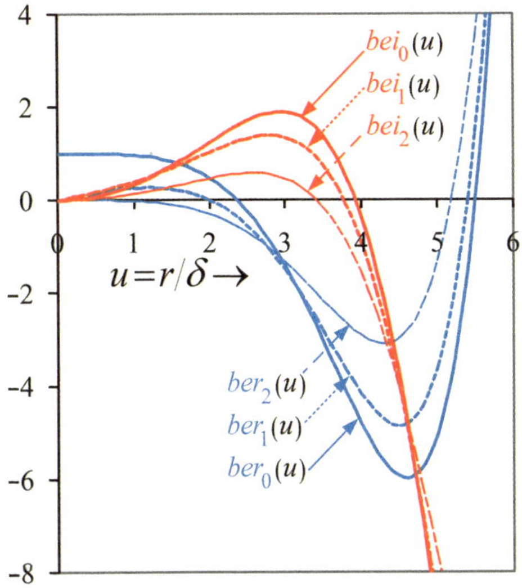

Figure 6.16 Behavior of the spherical *ber* and *bei* functions as a function of a real argument, $u = r/\delta$.

$$\sqrt{\frac{2l+1}{4\pi}\frac{(l-m)!}{(l+m)!}}P_l^m(\cos\theta)\begin{pmatrix} e^{im\varphi} \\ e^{-im\varphi} \end{pmatrix} = Y_l^m(\theta,\phi) \tag{6.78}$$

(the spherical harmonics) as angular components of the solution. The arbitrary normalizing constant in Equation 6.78 is chosen so that

$$\int_0^{2\pi}\int_0^{\pi} Y_{l'}^{m'*}(\theta,\phi)Y_l^m(\theta,\phi)\sin\theta d\theta d\phi = \delta_{l'l}\delta_{m'm} \tag{6.79}$$

This choice of normalizing constant makes the spherical harmonics an orthonormal set of functions as defined by Equation 6.79. The first few of the spherical harmonic functions are

$$\begin{aligned}
&&& Y_2^2(\theta,\varphi) = \sqrt{\frac{5}{96\pi}}3\sin^2\theta e^{i2\varphi} \\
& & Y_1^1(\theta,\varphi) = -\sqrt{\frac{3}{8\pi}}\sin\theta e^{i\varphi} \quad & Y_2^1(\theta,\varphi) = -\sqrt{\frac{5}{24\pi}}3\sin\theta\cos\theta e^{i\varphi} \\
Y_0^0(\theta,\varphi) = \frac{1}{\sqrt{4\pi}} \quad & & Y_1^0(\theta,\varphi) = -\sqrt{\frac{3}{4\pi}}\cos\theta \quad & Y_2^0(\theta,\varphi) = \sqrt{\frac{5}{4\pi}}\left(\frac{3}{2}\cos^2\theta - \frac{1}{2}\right) \\
& & Y_1^{-1}(\theta,\varphi) = +\sqrt{\frac{3}{8\pi}}\sin\theta e^{-i\varphi} \quad & Y_2^{-1}(\theta,\varphi) = \sqrt{\frac{5}{24\pi}}3\sin\theta\cos\theta e^{-i\varphi} \\
&&& Y_2^{-2}(\theta,\varphi) = \sqrt{\frac{5}{96\pi}}3\sin^2\theta e^{-i2\varphi}
\end{aligned} \tag{6.80}$$

and a useful property of the spherical harmonics is

$$Y_l^{-m}(\theta,\varphi)=(-1)^m\, Y_l^{*m}(\theta,\varphi) \tag{6.81}$$

Plane Wave in Spherical Coordinates

The plane wave, $e^{i(k_2z-\omega t)}$, of Equation 6.59 may be expressed in spherical coordinates by using the complete set of orthonormal functions $j_l(kR)$ and $\eta_l(kR)$ and $Y_l^m(\theta,\phi)$ above as

$$e^{i(k_2z-\omega t)}=\sum_{l=0}^{\infty}\sum_{m=-l}^{l} a_l\, j_l(k_2r)Y_l^m(\theta,\phi)e^{-i\omega t} \tag{6.82}$$

because we cannot permit either associated Legendre polynomials of the second kind or spherical Neumann functions in the solution because they are singular for values of $\cos\theta=\pm 1$ or $r=0$, both of which are needed in Figure 6.13. The unknown coefficients, a_l, can be found by writing $e^{i(k_2z-\omega t)}$ as $e^{i(k_2r\cos\theta-\omega t)}$ and noting that there is no ϕ dependence in this function. Thus, Equation 6.82 can be written as

$$e^{ik_2r\cos\theta}=\sum_{l=0}^{\infty} a_l\, j_l(k_2r)Y_l^0(\theta,\phi) \tag{6.83}$$

PROBLEM

6.19 Use the orthogonality relationship in Equation 6.79 to show that the coefficient,

$$a_l=i^l\sqrt{4\pi(2l+1)} \tag{6.84}$$

so that

$$e^{i(k_2z-\omega t)}=\sum_{l=0}^{\infty} i^l\sqrt{4\pi(2l+1)}\, j_l(k_2r)Y_l^0(\theta,\phi)e^{-i\omega t} \tag{6.85}$$

6.7 VECTOR HELMHOLTZ EQUATION SOLUTIONS

Solutions to Equations 6.57 and 6.58 in Cartesian coordinates are straightforward because the three components of the vector fields are separable, and the vector equations yield three scalar Helmholtz equations; that is,

$$\vec{E}_S=\sum_{l=0}^{\infty}\sum_{m=-l}^{l}\left[\left(\vec{A}_l^m\right)^{E,1} h_l^{(1)}(kr)+\left(\vec{A}_l^m\right)^{E,2} h_l^{(2)}(kr)\right]Y_l^m(\theta,\phi) \tag{6.86}$$

$$\vec{H}_S=\sum_{l=0}^{\infty}\sum_{m=-l}^{l}\left[\left(\vec{A}_l^m\right)^{H,1} h_l^{(1)}(kr)+\left(\vec{A}_l^m\right)^{H,2} h_l^{(2)}(kr)\right]Y_l^m(\theta,\phi), \tag{6.87}$$

where the vector coefficients of the spherical Hankel functions are arbitrary constant vectors that are chosen to match the boundary conditions.

However, the components of the fields in cylindrical or spherical coordinates are related to one another and do not separate so we can use a solution technique defined by Chandrasekhar and Kendall, Balanis and Harrington, or Bouwkamp and Casimir as reported by Jackson; the latter is used here, but all of the solutions *must* give the same answer. Here we will use the vector identities

$$\begin{aligned}\vec{\nabla}\left(\vec{r}\cdot\vec{E}_S\right)&=\vec{r}\cdot\left(\vec{\nabla}^2\vec{E}_s\right)+2\vec{\nabla}\cdot\vec{E}_S\\ \vec{\nabla}\left(\vec{r}\cdot\vec{H}_S\right)&=\vec{r}\cdot\left(\vec{\nabla}^2\vec{H}_s\right)+2\vec{\nabla}\cdot\vec{H}_S\end{aligned} \tag{6.88}$$

For charge-free space, we can then use Equations 6.57 and 6.58 in spherical coordinates to obtain

$$\vec{\nabla}^2\vec{E}_S+k^2\vec{E}_S=0 \quad \text{when } \vec{\nabla}\cdot E_S=0 \tag{6.89}$$

$$\vec{\nabla}^2\vec{H}_S+k^2\vec{H}_S=0 \quad \text{when } \vec{\nabla}\cdot\vec{H}_S=0 \tag{6.90}$$

Solving Equation 6.87 or 6.88 with the required relations $\vec{H}_S=(1/i\omega\mu)\,\vec{\nabla}\times\vec{E}_S$ or $\vec{E}_S=(i/\omega\varepsilon)\,\vec{\nabla}\times\vec{H}_S$ between electric and magnetic fields is equivalent to solving Maxwell's equations in Equation 6.54. If a dot product of these left equations is taken with $\vec{r}$

$$\left(\vec{\nabla}^2+k^2\right)\left(\vec{r}\cdot\vec{E}_S\right)=0 \tag{6.91}$$

$$\left(\vec{\nabla}^2+k^2\right)\left(\vec{r}\cdot\vec{H}_S\right)=0 \tag{6.92}$$

The terms $(\vec{r}\cdot\vec{E}_S)$ or $(\vec{r}\cdot\vec{H}_S)$ in Equations 6.91 and 6.92 are both scalar quantities so they are *scalar* Helmholtz equations in those quantities, for which we have found the answers above:

$$\left(\vec{r}\cdot\vec{E}_S\right)=\sum_{l=0}^{\infty}\sum_{m=-l}^{l}f_l(kr)Y_l^m(\theta,\phi) \tag{6.93}$$

$$\left(\vec{r}\cdot\vec{H}_S\right)=\sum_{l=0}^{\infty}\sum_{m=-l}^{l}g_l(kr)Y_l^m(\theta,\phi), \tag{6.94}$$

where $f_l(kr)$ and $g_l(kr)$ are a linear combination of two of the four functions $h_l^{(1)}(kr)$, $h_l^{(2)}(kr)$, $j_l(kr)$, $j\eta_l(kr)$, and values of k are complex for $r \le a_i$. Note that these functions are, in general, not zero so the coefficients of those terms will determine the magnitude of the radial component present in Equation 6.93 or 6.94. These equations are thus telling us that there is a radial component of $\vec{H}_S$ or $\vec{E}_S$, and hence they are **not** TEM^r. Note that these conclusions for a *good* conducting sphere are consistent with the conclusions we found for PEC spheres in section 6.3.

$$\vec{r}\cdot\sum_{l=0}^{\infty}\left[\frac{\partial h_l^{(1\text{ or }2)}}{\partial r}\sum_{m=-l}^{l}\left(\vec{A}_l^m\right)^{E\text{ or }M,1\text{ or }2}Y_l^m-\frac{ih_l^{(1\text{ or }2)}}{r}\vec{L}\times\sum_{m=-l}^{l}\left(\vec{A}_l^m\right)^{E\text{ or }M,1\text{ or }2}Y_l^m\right]=0 \tag{6.108}$$

The derivative term in Equation 6.108 can be replaced by use of the recursion relation 6.68 to include terms $h_{l-1}^{(1\text{ or }2)}(x)$ and $h_{l+1}^{(1\text{ or }2)}(x)$ that are linearly independent of $h_l^{(1\text{ or }2)}(x)$ in the square bracket. Thus, the coefficients of each of the terms in $h_l^{(1\text{ or }2)}$ (x) must go to zero independently so

$$\vec{r}\cdot\sum_{m=-l}^{l}\left(\vec{A}_l^m\right)^{E\text{ or }M,1\text{ or }2}Y_l^m=0 \tag{6.109}$$

$$\vec{r}\cdot\vec{L}\times\sum_{m=-l}^{l}\left(\vec{A}_l^m\right)^{E\text{ or }M,1\text{ or }2}Y_l^m=0 \tag{6.110}$$

Equation 6.109 assures that either the electric field intensity or magnetic field intensity is transverse to the radius vector, and Equation 6.110 provides a sufficient condition to determine a unique set of vector angular functions of order l, one for each value of m.

Comparing Equation 6.109 with 6.102, $\vec{r}\cdot\vec{L}=0$, we can see that an acceptable angular solution is equivalently given by a sum that uses scalar coefficients, $a(l, m)$:

$$\sum_{m'=-l}^{l}\left(\vec{A}_l^{m'}\right)^{E\text{ or }M}Y_l^{m'}=\sum_{m=-l}^{l}a_{E\text{ or }M}(l,m)\vec{L}Y_l^m \tag{6.111}$$

because the scalar sum on the right of the vector quantity, $\vec{L}Y_l^m$, also assures that that field is transverse to the radius vector. With the same substitution of a scalar sum in Equation 6.110 and using the commutation relationship that $\vec{L}\times\vec{L}=i\vec{L}$, we can see this equation is also satisfied.

Finally, components of the electric field intensity can be written as transverse to the radius vector (TE^r) as

$$\left(\vec{E}_l^m\right)^M=Zg_l(kr)\vec{L}Y_l^m(\theta,\phi) \tag{6.112}$$

with a corresponding magnetic field intensity

$$\left(\vec{H}_l^m\right)^M=-i\vec{\nabla}\times\left(\vec{E}_l^m\right)^M/kZ \tag{6.113}$$

Alternately, components of the magnetic field intensity can be written as transverse to the radius vector (TM^r) as

$$\left(\vec{H}_l^m\right)^E=f_l(kr)\vec{L}Y_l^m(\theta,\phi) \tag{6.114}$$

with a corresponding electric field intensity

$$\left(\vec{E}_l^m\right)^E = iZ\vec{\nabla}\times\left(\vec{H}_l^m\right)^E / k \qquad (6.115)$$

Equations 6.112–6.115 are the spherical components of the TE^r and TM^r fields that we earlier described in Equations 6.86 and 6.87 in Cartesian coordinates. The functions $f_l(kr)$ and $g_l(kr)$ are linear combinations of the spherical Bessel, Neumann, or Hankel functions. Just as we concluded in section 6.3 for a PEC, Equations 6.114 and 6.115 show that the electric multipole terms give rise to a transverse magnetic (TM^r) wave for a *good* conductor. Also consistent with the conclusions of section 6.3 for a PEC, Equations 6.112 and 6.113 show that the magnetic multipole terms give rise to a transverse electric (TE^r) wave for a *good* conductor.

Physicists prefer to use electric and magnetic multipole descriptions as opposed to the TE^r or TM^r descriptions because multipole fields come about from the distribution of charge density or current density in a source (like a *good* conducting sphere). In these descriptions, the vector spherical harmonic, $\vec{L}Y_l^m$ plays a central role so they often use the normalized vector spherical harmonic defined by

$$\vec{X}_l^m(\theta,\phi) \equiv \left(1/\sqrt{l(l+1)}\right)\vec{L}Y_l^m(\theta,\phi) \qquad (6.116)$$

because the vector spherical harmonics have the convenient orthonormality property:

$$\int_0^{2\pi}\int_0^{\pi}\left(\vec{X}_{l'}^{m'}\right)^* \cdot \vec{X}_l^m \sin\theta d\theta d\phi = \delta_{l,l'}\delta_{m,m'} \qquad (6.117)$$

and

$$\int_0^{2\pi}\int_0^{\pi}\left(\vec{X}_{l'}^{m'}\right)^* \cdot \left(\vec{r}\times\vec{X}_l^m\right)\sin\theta d\theta d\phi = 0 \qquad (6.118)$$

Note that the vector spherical harmonic for $l = 0$ is taken to be identically zero because solutions to the source free problem exist only in the static limit as $k \to 0$. Thus, in the following sums, the index l begins at 1.

The two sets of multipole fields form a complete set of vector solutions to Maxwell's equations so, by combining these two solution sets, we obtain the most general solution to Maxwell's equations in spherical coordinates as

$$\vec{E}_S = Z\sum_{l=1}^{\infty}\sum_{m=-l}^{l}\left[a_M(l,m)g_l(kr)\vec{X}_l^m + (i/k)a_E(l,m)\vec{\nabla}\times f_l(kr)\vec{X}_l^m\right] \qquad (6.119)$$

$$\vec{H}_S = \sum_{l=1}^{\infty}\sum_{m=-l}^{l}\left[a_E(l,m)f_l(kr)\vec{X}_l^m - (i/k)a_M(l,m)\vec{\nabla}\times g_l(kr)\vec{X}_l^m\right] \qquad (6.120)$$

These two equations give the spatial electric field intensity and magnetic field intensity in terms of coefficients, $a_E(l, m)$ and $a_M(l, m)$, which specify the amount of electric (l, m) and magnetic (l, m) multipole fields.

The scalars, $\vec{r}\cdot\vec{H}$ and $\vec{r}\cdot\vec{E}$ can be used to find the coefficients through

$$a_M(l,m))g_l(kr)=\left(k/\sqrt{l(l+1)}\right)\int_0^{2\pi}\int_0^{\pi}(Y_l^m)^*\vec{r}\cdot\vec{H}\sin\theta d\theta d\phi \tag{6.121}$$

$$Za_E(l,m))f_l(kr)=-\left(k/\sqrt{l(l+1)}\right)\int_0^{2\pi}\int_0^{\pi}(Y_l^m)^*\vec{r}\cdot\vec{E}\sin\theta d\theta d\phi \tag{6.122}$$

The radial functions, $f_l(kr)$ and $g_l(kr)$, are linear combinations of the spherical Bessel, Neumann, or Hankel functions as appropriate to their location in space. For example, inside the sphere or for a plane wave, we cannot have any component of a Neumann function because the point $r = 0$ is included in our solution so only the spherical Bessel function $j_l(k_c r)$ is permitted. We are also reminded from Equations 6.65 and 7.110 in *Maxwell's Equations* that $k_c \approx (1 + i)/\delta$ and $Z_c = [(1 - i)/\sigma_c\delta]$ inside a *good* conducting sphere are complex quantities. Outside the sphere, $k_2 = \omega/c_2$ and $Z_2 = \eta_2 = \sqrt{\mu_0/\varepsilon_2}$ are real and the point $r = 0$ is not included so we can, if we choose, write the solution as a linear combination of spherical Hankel functions of the first and second kind, $g_l(kr) = A_l^{(1)}h_l^{(1)}(kr) + A_l^{(2)}\ h_l^{(2)}(hr)$. Given their exponential character as shown in 6.66 when the Hankel functions are multiplied by $e^{-i\omega t}$, we can see that the $h_l^{(1)}(kr)$ and $h_l^{(2)}(kr)$ terms in $g_l(kr)$ correspond to an outgoing and an incoming wave, respectively.

6.9 SCATTERING OF ELECTROMAGNETIC WAVES

Linearly Polarized Incident Wave

Equation 6.85 gave the propagating wave $e^{i(k_2 z-\omega t)}$ in terms of scalar spherical harmonics, but we need to express this equation in terms of the vector spherical harmonics for boundary condition matching. This can be accomplished by using the Jackson formalism for a circularly polarized plane wave with helicity + for right- (– for left-) hand polarized waves incident along the z-axis:

$$\vec{E}_{\pm}(\vec{x}) = E_0(\hat{a}_x \pm i\hat{a}_y)e^{ik_2 z} \tag{6.123}$$

$$\vec{H}_{\pm}(\vec{x}) = (1/\eta_2)\hat{a}_z \times \vec{E}_{\pm}(\vec{x}) = \mp\sqrt{\varepsilon_2/\mu_2}\, i\vec{E}_{\pm}(\vec{x}) \tag{6.124}$$

An incident *linearly polarized* plane wave, in the x-direction, can then be written as

$$\vec{E}_{inc}(\vec{x}) = E_0\hat{a}_x e^{ikz} = \left(\vec{E}_+ + \vec{E}_-\right)/\sqrt{2} \tag{6.125}$$

Because the plane wave must be finite everywhere, including $r = 0$, we can write a multipole expansion for $\vec{E}_{inc}$ and $\vec{H}_{inc}$ only in terms of the spherical Bessel functions, $j_l(k_2 r)$:

$$\vec{E}_{inc} = E_0 \sum_{l=1}^{\infty} \sum_{m=-l}^{l} \left[a(l,m)\, j_l(k_2 r) \vec{X}_l^m + (i/k_2) b(l,m) \vec{\nabla} \times j_l(k_2 r) \vec{X}_l^m \right] \quad (6.126)$$

$$\eta_2 \vec{H}_{inc} = E_0 \sum_{l=1}^{\infty} \sum_{m=-l}^{l} \left[(-i/k_2) a(l,m) \vec{\nabla} \times j_l(k_2 r) \vec{X}_l^m + b(l,m)\, j_l(k_2 r) \vec{X}_l^m \right] \quad (6.127)$$

The orthogonality properties of the vector spherical harmonics identify coefficients as

$$a(l,m))\, j_l(kr) = \int_0^{2\pi} \int_0^{\pi} \left(\vec{X}_l^M\right)^* \cdot \vec{E}(\vec{x}) \sin\theta d\theta d\phi \quad (6.128)$$

$$b(l,m))\, j_l(kr) = \eta_2 \int_0^{2\pi} \int_0^{\pi} \left(\vec{X}_l^M\right)^* \cdot \vec{H}(\vec{x}) \sin\theta d\theta d\phi \quad (6.129)$$

Using Equations 6.128 and 6.129 in these equations yields

$$\vec{E}_{inc} = E_0 \sum_{l=1}^{\infty} i^l \sqrt{4\pi(2l+1)} \left\{ j_l(k_2 r) \vec{X}_l^+ + (1/k_2) \vec{\nabla} \times \left[j_l(k_2 r) \vec{X}_l^- \right] \right\} \quad (6.130)$$

$$\vec{H}_{inc} = H_0 \sum_{l=1}^{\infty} i^l \sqrt{4\pi(2l+1)} \left\{ (-i/k_2) \vec{\nabla} \times \left[j_l(k_2 r) \vec{X}_l^+ \right] - i j_l(k_2 r) \vec{X}_l^- \right\} \quad (6.131)$$

where $\vec{X}_l^+ = \left(\vec{X}_l^1 + X_l^{-1}\right)/\sqrt{2}$ and $\vec{X}_l^- = \left(\vec{X}_l^1 - X_l^{-1}\right)/\sqrt{2}$ (*linearly polarized* vector spherical harmonics) are orthogonal to one another, are transverse to the radial direction, and $E_0 = \eta_2 H_0$.

Fields Outside the Sphere

Consistent with our earlier analysis of scattering by PEC spheres, we can calculate the fields scattered by the incident electromagnetic wave on a *good* conducting sphere using the multipole fields produced by the induced electric and magnetic moments given in Equations 6.119 and 6.120. However, we know we will want to add the scattered fields to the incident fields so we will want to put them in the same format as Equations 6.130 and 6.131. Those scattered waves are outgoing waves at infinity so they are expressed as a spherical Hankel function $h_l^{(1)}(k_2 r)$:

$$\vec{E}_{sc} = E_0 \sum_{l=1}^{\infty} i^l \sqrt{\pi(2l+1)} \left\{ \alpha(l) h_l^{(1)}(k_2 r) \vec{X}_l^+ + (\beta(l)/k_2) \vec{\nabla} \times \left[h_l^{(1)}(k_2 r) \vec{X}_l^- \right] \right\} \quad (6.132)$$

$$\vec{H}_{sc} = H_0 \sum_{l=1}^{\infty} i^l \sqrt{\pi(2l+1)} \left\{ (-i\alpha(l)/k_2) \vec{\nabla} \times \left[h_l^{(1)}(k_2 r) \vec{X}_l^+ \right] - i\beta(l) h_l^{(1)}(k_2 r) \vec{X}_l^- \right\}, \quad (6.133)$$

where the unknown coefficients, $\alpha(l)$ and $\beta(l)$, are to be determined by the boundary conditions given in Equations 6.60, 6.61, and 6.62; $\alpha(l)$ giving the magnitude of the magnetic multipole moment (TE′ component) and $\beta(l)$ giving the magnitude of the electric multipole moment (TM′ component). To use these BC, we need to add the incident and scattered wave solutions to obtain the *total* solution outside the good conducting sphere as

$$\vec{E} = E_0 \sum_{l=1}^{\infty} i^l \sqrt{4\pi(2l+1)} \left\{ \begin{array}{l} \left[j_l(k_2 r) + \frac{\alpha(l)}{2} h_l^{(1)}(k_2 r) \right] \vec{X}_l^+ \\ + \frac{1}{k_2} \vec{\nabla} \times \left\{ \left[j_l(k_2 r) + \frac{\beta(l)}{2k_2} \vec{\nabla} \times h_l^{(1)}(k_2 r) \right] \vec{X}_l^- \right\} \end{array} \right\} \tag{6.134}$$

$$\vec{H} = H_0 \sum_{l=1}^{\infty} i^l \sqrt{4\pi(2l+1)} \left\{ \begin{array}{l} \frac{i}{k_2} \vec{\nabla} \times \left(\left[j_l(k_2 r) + \frac{\alpha(l)}{2} h_l^{(1)}(k_2 r) \right] \vec{X}_l^+ \right) \\ - i \left[j_l(k_2 r) + \frac{\beta(l)}{2} h_l^{(1)}(k_2 r) \right] \vec{X}_l^- \end{array} \right\} \tag{6.135}$$

If we take $\hat{a}_r = \vec{r}/|\vec{r}|$ to be directed outward normal, and because is $\vec{X}_l^m$ is transverse, then we can use the derivation for $\vec{\nabla} \times j_l(r)\vec{X}_l^{\pm 1}$ or $\vec{\nabla} \times h_l^{(1)}(r)\vec{X}_l^{\pm 1}$:

$$\nabla \times j_\ell(kr)\vec{X}_l^{\pm 1} = \frac{i\hat{a}_r \sqrt{\ell(\ell+1)}}{kr} j_\ell(kr) Y_l^{\pm 1} + \frac{1}{r}\frac{\partial}{\partial r}[r j_\ell(kr)]\hat{a}_r \times \vec{X}_l^{\pm 1} \tag{6.136}$$

$$\nabla \times h_\ell^{(1)}(kr)\vec{X}_l^{\pm 1} = \frac{i\hat{a}_r \sqrt{\ell(\ell+1)}}{kr} h_\ell^{(1)}(kr) Y_l^{\pm 1} + \frac{1}{r}\frac{\partial}{\partial r}[r h_\ell^{(1)}(kr)]\hat{a}_r \times \vec{X}_l^{\pm 1} \tag{6.137}$$

to show that

$$\vec{E} = E_0 \sum_{l=1}^{\infty} i^l \sqrt{4\pi(2l+1)} \left\{ \begin{array}{l} \left[j_l(x) + \frac{\alpha(l)}{2} h_l^{(1)}(x) \right] \vec{X}_l^+ \\ + \frac{1}{x}\frac{\partial}{\partial x} \left\{ x \left[j_l(x) + \frac{\beta(l)}{2} h_l^{(1)}(x) \right] \right\} \hat{a}_r \times \vec{X}_l^- \end{array} \right\} \tag{6.138}$$

$$\hat{a}_r \times \vec{H} = H_0 \sum_{l=1}^{\infty} i^l \sqrt{4\pi(2l+1)} \left\{ \begin{array}{l} \frac{i}{x}\frac{\partial}{\partial x} \left(x \left[j_l(x) + \frac{\alpha(l)}{2} h_l^{(1)}(x) \right] \right) \vec{X}_l^+ \\ - i \left[j_l(k_2 r) + \frac{\beta(l)}{2} h_l^{(1)}(x) \right] \hat{a}_r \times \vec{X}_l^- \end{array} \right\} \tag{6.139}$$

Now, if we impose the BC 6.61, $\vec{E}_{\|}(r = a_i+, \theta, \phi) = Z_s \hat{n} \times \vec{H}_{\|}(r = a_i+, \theta, \phi)$ with $Z_s = (1 - i)/\sigma\delta$ for $x = k_2 a_i$, we can equate the two series term by term:

$$\left[j_l(x) + \frac{\alpha(l)}{2} h_l^{(1)}(x) \right] = i(Z_s/\eta_2)\frac{1}{x}\frac{d}{dx}\left[x \left(j_l(x) + \frac{\alpha(l)}{2} h_l^{(1)}(x) \right) \right] \tag{6.140}$$

$$\left[j_l(x) + \frac{\beta(l)}{2} h_l^{(1)}(x) \right] = i(Z_s/\eta_2)\frac{1}{x}\frac{d}{dx}\left[x \left(j_l(x) + \frac{\beta(l)}{2} h_l^{(1)}(x) \right) \right] \tag{6.141}$$

If we use the identity $j_l(x) = (h_l^{(1)} + h_l^{(2)})/2$, we can solve for $\alpha(l)$ and $\beta(l)$:

$$\alpha(l) = -1 - \left[\frac{h_l^{(2)}(x) - i(Z_s/\eta_2)\frac{1}{x}\frac{d}{dx}\left(xh_l^{(2)}(x)\right)}{h_l^{(1)}(x) - i(Z_s/\eta_2)\frac{1}{x}\frac{d}{dx}\left(xh_l^{(1)}(x)\right)} \right]_{x=k_2a_i} \tag{6.142}$$

$$\beta(l) = -1 - \left[\frac{h_l^{(2)}(x) - i(\eta_2/Z_s)\frac{1}{x}\frac{d}{dx}\left(xh_l^{(2)}(x)\right)}{h_l^{(1)}(x) - i(\eta_2/Z_s)\frac{1}{x}\frac{d}{dx}\left(xh_l^{(1)}(x)\right)} \right]_{x=k_2a_i} \tag{6.143}$$

These coefficients, put back into the equation pair 6.132 and 6.133, give the *scattered* electromagnetic fields outside of a *good* conducting sphere. These coefficients, put back into the equation pair 6.134 and 6.135, give the *total* electromagnetic fields outside of a *good* conducting sphere.

For our case, in which $x = k_2a_i << 1$, we may use limiting properties of spherical Bessel and Neumann functions that

$$j_l(k_2a_i) \underset{k_2a\ll 1}{\rightarrow} \frac{(k_2a_i)^l}{(2l+1)!!}\left[1 - \frac{(k_2a_i)^2}{2(2l+3)} + \cdots\right] \tag{6.144}$$

$$\eta_l(k_2a_i) \underset{k_2a\ll 1}{\rightarrow} -\frac{(2l-1)!!}{(k_2a_i)^{l+1}}\left[1 - \frac{(k_2a_i)^2}{2(1-2l)} + \cdots\right] \tag{6.145}$$

so

$$\alpha(l) \approx \frac{-2i(k_2a_i)^{2l+1}}{(2l+1)[(2l-1)!!]^2}\left[\frac{(k_2a_i) - i(l+1)(Z_s/\eta_2)}{(k_2a_i) + il(Z_s/\eta_2)}\right] \tag{6.146}$$

$$\beta(l) \approx \frac{-2i(k_2a_i)^{2l+1}}{(2l+1)[(2l-1)!!]^2}\left[\frac{(k_2a_i) - i(l+1)(\eta_2/Z_s)}{(k_2a_i) + il(\eta_2/Z_s)}\right] \tag{6.147}$$

- For snowballs with radius of the order $a_i \sim 1$ μm or smaller, and for frequencies less than 100 GHz, $(k_2a_i) < 0.004$ so the limiting series above is satisfied for these applications.
- Because $(k_2a_i) << 1$, in Equations 6.146 and 6.147, we can see that each successive value of l leads to a value of the coefficients $\alpha(l)$ and $\beta(l)$ that are smaller than the previous one by $(k_2a_i)^2$ so that the dominant terms in the expansion Equations 6.132 and 6.133 will be those for $l = 1$.
- This is consistent with our previous conclusion that for spherically symmetric charge and current sources of radiation the dominant terms are dipole in nature.

Conclusion

The dominant radiation from scattering of linear polarized incident waves by *good* conducting spheres will be dominantly dipole in nature. The error in neglecting quadrupole and higher order multipole radiation will be less than 10^{-5} for snowballs with a radius of $a_i < 1$ μm and frequencies below 100 GHz.

Fields Inside the Sphere

When the incident linearly polarized plane waves scatter off a *good* spherical conductor, the magnetic field intensity penetrates into the conductor surface in a manner similar to that for flat conducting surfaces. However, because of the spherical symmetry the boundary conditions, (Equations 6.60–6.62), $\vec{H}_{\|}(r = a_i +, \theta, \phi) = \vec{H}_{\|}(r = a_i -, \theta, \phi)$, $\vec{E}_{\|}(r = a_i +, \theta, \phi) = Z_s \hat{a}_r \times \vec{H}_{\|}(r = a_i +, \theta, \phi)$, and $\vec{H}_{\perp} \simeq -[(1 + i)/2]\delta[\vec{\nabla} \times (\hat{a}_r \times \vec{H}_{\|})]_{\perp}$ must be satisfied with $Z_s = (1 - i)/\sigma_c\delta$ and $k_c \approx (1 + i)/\delta$. To evaluate the $\vec{H}$ and $\vec{E}$ fields inside the *good* conductor, we can expand the fields in terms of the complex spherical Bessel functions, $j_l(k_c r)$, or the complex modified spherical Bessel functions, $i_l(k_c r)$ (because $i_l(k_c r) \equiv (i)^{-l} j_l(ik_c r)$), or the real $ber_l(r/\delta)$ and $bei_l(r/\delta)$ functions, and, as found in Equations 6.119 and 6.120, the *linear polarized* vector spherical harmonics as

$$\vec{E}_c = E_0 \sum_{l=1}^{\infty} \{a_M(l,m)\, j_l(k_c r)\vec{X}_l^{+} + a_E(l,m)(i/k_c)\vec{\nabla} \times [j_l(k_c r)\vec{X}_l^{-}]\} \quad (6.148)$$

$$\vec{H}_c = H_0 \sum_{l=1}^{\infty} \{a_M(l,m)(-i/k_c)\vec{\nabla} \times [j_l(k_c r)\vec{X}_l^{+}] + a_E(l,m)\, j_l(k_c r)\vec{X}_l^{-}\} \quad (6.149)$$

The unknown coefficients can be found by evaluating Equations 6.148 and 6.149 at the boundary, $r = a$, and equating them to Equations 6.134 and 6.135 also evaluated at $r = a$. This leads to $a_M(l, 1) = a_M(l, -1)$, $a_M(l, 0) = 0$, $a_E(l, 1) = -a_E(l, -1)$ and $a_E(l, 0) = 0$ and

$$a_M(l,1) = \frac{i^l}{\sqrt{2}}\sqrt{4\pi(2l+1)}\frac{[j_l(k_2 a_i) + (1/2)\alpha(l)h_l^{(1)}(k_2 a_i)]}{j_l(k_c a_i)} \quad (6.150)$$

$$a_E(l,1) = \frac{i^{l-1}}{\sqrt{2}}\sqrt{4\pi(2l+1)}\left.\frac{\left(\dfrac{1}{k_2 r}\right)\dfrac{\partial}{\partial(k_2 r)}\{k_2 r[j_l(k_2 r) + (1/2)\beta(l)h_l^{(1)}(k_2 r)]\}}{\dfrac{1}{(k_c r)}\dfrac{\partial}{\partial(k_c r)}\{k_c r[j_l(k_c r)]\}}\right|_{r=a_i} \quad (6.151)$$

With these coefficients in Equations 6.148 and 6.149, we have found the electromagnetic fields inside a *good* conducting sphere. A picture of these fields (using $\vec{J}_c = \sigma_c\vec{E}_c$) is shown in Figure 6.17.

In Figure 6.17, the induced current density, $\vec{J}_c = \sigma_c\vec{E}_c$, and total magnetic field intensity are propagating inward at a relative slow phase velocity (compared with

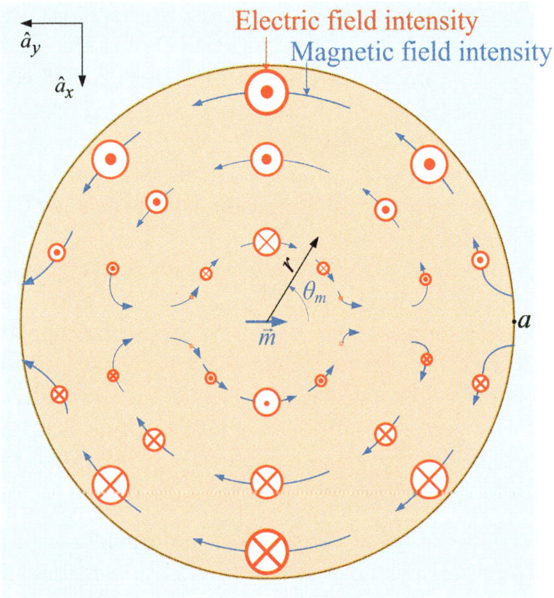

Figure 6.17 Snapshot of the induced electric field intensity, $\vec{E}_c$, and total magnetic field intensity, $\vec{H}_c$, induced inside a *good* conducting sphere in an externally incident polarized electromagnetic field.

the external phase velocity). As it propagates, the electromagnetic wave dies exponentially according to the skin depth, δ, loosing energy to heat in the conductor. The turnaround in the field lines at small radius, r, is a result of surface current density that was induced by the external field intensity a half cycle before this snapshot was taken. If the sphere has a small radius, a_i, then power can propagate all the way through the sphere and reemerge at a delayed time to contribute a phase-shifted component to the incident plane wave. The field intensity that penetrates the conducting sphere moves at slow speed $u_p(\omega) = \left(c/\sqrt{\varepsilon_r(\omega)}\right)\sqrt{2}\sqrt{\omega\varepsilon_0/\sigma_c}$ (see Equation 6.53) and has a complex wave vector in the negative radial direction, $\hat{a}_\xi = -\hat{a}_r$, and we recall that this wave vector shows that the electric field intensity leads the magnetic field intensity into the conductor with a phase angle of 45°. This means the tangential component of the electric field intensity induced at the outside surface (in the propagating medium) also leads the magnetic field intensity by 45°, as was confirmed by Equation 7.110 in *Maxwell's Equations*.

We cannot measure the instantaneous field intensities inside the sphere to confirm these functional forms but we can conduct a simulation of the field propagation with numerical solvers. Typically, numerically field solvers do not know anything about vector spherical harmonics, Bessel functions, or transverse solutions but are a brute force method of computing the field at a point due to the fields at adjacent points and they are set up to solve Maxwell's equations in an iterative process that converges to a solution (given enough memory and run time). Such iterations are normally conducted until the solutions converge to within a small chosen precision of the solution to the previous iteration; some statistical error (magnitude and direction) are typical of the final output. Those solvers are capable of determining relative phase of the fields to one another and to external stimuli. A numerical simulation of a 1-μm radius copper sphere in a linearly polarized external field (external electric field in the x-direction and external magnetic field in the y-direction) is shown for three different harmonic frequencies in Figure 6.18.

In Figure 6.18, points inside the sphere have been meshed with points outside the sphere under a copper trace, as shown in Figures 6.3 and 6.4. For clarity, the fields external to the sphere have been omitted, but the snapshot is taken at a point in time when the external magnetic field lines are from right to left as they go around the sphere much like the picture in Figure 6.8 for a PEC. Unlike the case for a PEC, the external magnetic field intensity penetrates into the surface of the sphere in phase, and the field intensity near the surface is the highest (mostly according to $\sin\theta_m$ for a dominantly magnetic dipole moment).

- At 1 GHz, the magnetic field intensity at the different mesh points at this instant in time is observed to be in phase with and parallel to the external magnetic field intensity and relatively uniform in magnitude (all the arrowheads are red), as expected for a large skin depth of $\delta \sim 2$ µm.
- At 5 GHz, the magnetic field intensity at the different mesh points at this instant in time is observed to be in phase with and parallel to the external magnetic field intensity, but the magnitude of the field near the center is smaller (the arrowheads become green there), as expected for a smaller skin depth of $\delta \sim 0.9$ µm. In addition, the magnitude of the magnetic field intensity

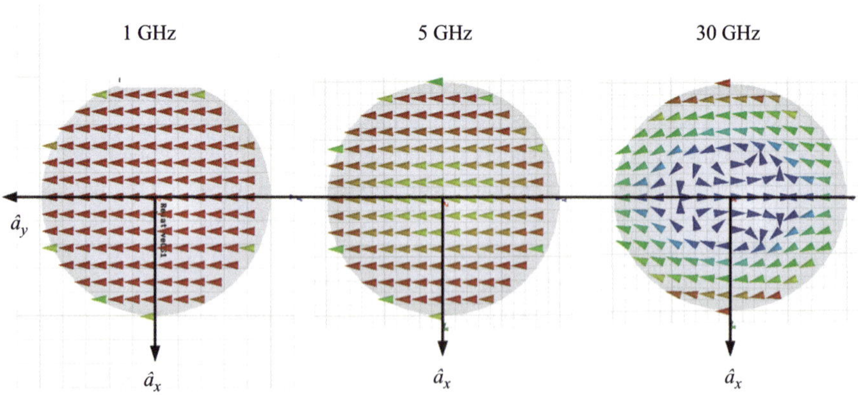

Figure 6.18 Cross section (as seen from the rear) of the magnitude (red, strongest; green, moderate; blue, weakest) and direction (arrowheads) of the magnetic field intensity at points inside a 1-µm copper sphere for three external frequencies. External magnetic field intensity lines to the spheres are from right to left in the three cases but are not shown.

is smaller for angles $\theta_m \sim 0$ and $\theta_m \sim \pi$, as expected for the magnetic dipole function $\sin\theta_m$.

- At 30 GHz, the magnetic field intensity at the different mesh points at this instant in time is observed to be in phase with and parallel to the external magnetic field intensity lines near the surface, but the magnitude of the field near the center is much smaller (the arrowheads become blue there), as expected for an even smaller skin depth of $\delta \sim 0.36$ μm. In addition, the direction of the magnetic field intensity is 180° out of phase with the external magnetic field intensity, as would be expected for magnetic surface fields that had diffused to the center from a half a cycle ago.
- Consistent results are produced by the numerical simulation for electric field intensity inside the good conducting 1-μm radius copper sphere but they are 45° out of phase.

Conclusion

Numerical simulation confirms that the angular and radial distributions of electric field intensity and magnetic field intensity follow the formulation given in Equations 6.148 and 6.149, with coefficients calculated using Equations 6.150 and 6.151.

To exhibit the dynamics of fields, Figure 6.19 shows the numerical simulation solutions for magnitude and direction of the magnetic field intensity at a set of mesh points inside a 1-μm radius copper sphere in a linearly polarized external field (external electric field intensity in the x-direction and external magnetic field intensity in the y-direction) for a 100-GHz harmonic frequency. In Figure 6.19, four snapshots in time are shown, upper left ($\varphi = 0$) when the external fields are from right to left (as was shown in Figure 6.8 for a PEC), upper right ($\varphi = 0.5\ \pi$) a quarter of a cycle later (0.25×10^{-11} s), lower left ($\varphi = \pi$) a half of a cycle later (0.50×10^{-11} s), and lower right ($\varphi = 1.5\ \pi$) three quarters of a cycle later (0.75×10^{-11} s).

- At 100 GHz the skin depth of copper is 0.2 μm so the mesh points near the surface of the sphere show a magnetic field intensity that is green; reduced in magnitude from that of the red external applied field (not shown).
- The magnetic field intensity at the different mesh points at this instant in time ($\varphi = 0$) is observed from the upper left figure to be in phase with and parallel to the external magnetic field intensity near the surface (albeit reduced in magnitude). At this instant of time, the magnitude of the field near the center is even smaller (the arrowheads become blue there), as expected for a skin depth of $\delta \sim 0.2$ μm. In addition, the direction of the magnetic field intensity is 180° out of phase with the external fields as would be expected for magnetic surface fields that had diffused to the center from a half a cycle ago.
- As time increases in the upper right, lower left, and lower right figures, the fields at mesh points near the surface move in phase with them, but the interior fields always remain out of phase due their diffusion to the center from earlier times.

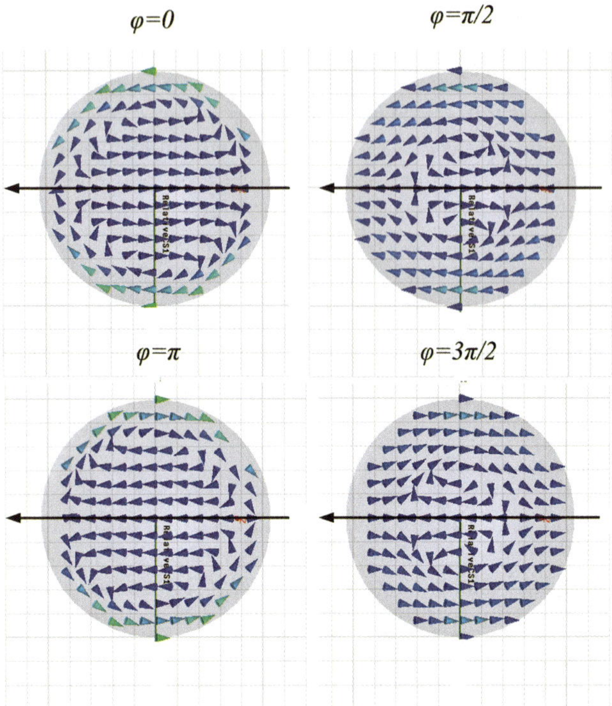

Figure 6.19 Cross section (as seen from the rear) of the magnitude (red, strongest; green, moderate; blue, weakest) and direction (arrowheads) of the magnetic field intensity at points inside a 1-μm copper sphere for 100 GHz at four different phases (snapshots at four different times for a harmonic external field after many cycles).

- The magnetic field intensity circulates in all of these figures (i.e., magnetic field lines close on themselves), as required for *no free* monopole sources or sinks of magnetic charge.

When intermediate phases are calculated and the internal fields are shown in a time sequence to emulate a motion picture, internal fields are seen to propagate toward the center of the sphere, as shown in Figure 6.12.

Conclusion

Magnetic field intensity inside a physically small sphere (in comparison with the external wavelength) periodically changes its direction in response to the external driving field but shows the effects of diffusion from the surface toward the interior. Because this sphere is large compared with the skin depth at this frequency, we also see exponential radial attenuation of the magnetic field intensity as it propagates inward.

6.10 POWER SCATTERED AND ABSORBED BY *GOOD* CONDUCTING SPHERES

Outgoing Power

In section 6.4, we found that the power radiated from the induced electric and magnetic multipole moments due to a sphere in a linearly polarized driving field could be calculated in the far region where the radial components had declined to the point that both electric and magnetic field intensities would be transverse. Equation 6.46 gave the time average power density (flux) in the radial direction $\vec{P}_{Avg,sc} = (1/2)\,\mathrm{Re}[\vec{E}_{sc} \times \vec{H}^*_{sc}]$. To get the time-averaged total power radiated, we can take the dot product of this Poynting vector with a differential area $|\vec{x}|^2 d\Omega \hat{a}_r$ and integrate over the solid angle:

$$\Delta P_{Avg,sc} = \int_0^{2\pi}\int_0^{\pi}(1/2)\mathrm{Re}[\vec{E}_{sc} \times \vec{H}^*_{sc} \cdot \hat{a}_r]|\vec{x}|^2 \sin\theta d\theta d\phi \tag{6.152}$$

and, using a vector triple product identity,

$$\Delta P_{Avg,sc} = \int_0^{2\pi}\int_0^{\pi}(1/2)\mathrm{Re}[\vec{E}_{sc} \cdot (\vec{H}^*_{sc} \times a_r)]|\vec{x}|^2 \sin\theta d\theta d\phi \tag{6.153}$$

Equations 6.132 and 6.133 give us the integrand quantities in Equation 6.153, and, if we use Equation 6.137, $\nabla \times h_\ell^{(1)}(kr)\vec{X}_l^{\pm 1} = \frac{i\hat{a}_r\sqrt{\ell(\ell+1)}}{kr} h_\ell^{(1)}(kr) Y_l^{\pm 1} + \frac{1}{r}\frac{\partial}{\partial r}\left[rh_\ell^{(1)}(kr)\right]\hat{a}_r \times \vec{X}_l^{\pm 1}$, we can see that the radial term $\hat{a}_r \times \hat{a}_r = 0$ leaves

$$\sigma_{sc} = \Delta P_{Avg,sc}/[(1/2)E_0 H_0] = \left(\pi/2k^2\right)\sum_{l=1}^{\infty}(2l+1)\left[|\alpha(l)|^2 + |\beta(l)|^2\right] \tag{6.154}$$

Because we use the radiated power of the induced multipole moments to calculate this ratio (and then refer to it as a cross section), outgoing power is called *scattered power*. Note that this quantity is always positive for all values of l. Note also that the units of σ are power divided by power per area (flux) so σ has units of m^2. In this sense, the scattering cross section has area units consistent with a physical two-dimensional scattering target.

Total Power

In Equations 6.134 and 6.135, we found that the *total* fields in the region exterior to the sphere power were the sum of the incident linearly polarized wave and the scattered wave from a sphere. In the far region, the radial components of this sum also decline so that both electric and magnetic field intensities are transverse. Using Equations 6.134 and 6.135 in Equation 6.46 and taking the dot product with the differential area, $|\vec{x}|^2 d\Omega \hat{a}_r$ then gives the *net* time-average power density *outgoing* in the radial direction. To get the time averaged *total* outgoing power, we thus take the dot product of the Poynting vector, $\vec{\mathcal{P}}_{Avg,total} = (1/2)\,\mathrm{Re}[\vec{E} \times \vec{H}^*]$, with a differential area $|\vec{x}|^2 d\Omega \hat{a}_r$ (pointed outward) and integrate over the solid angle using the same vector triple product identity to find the ***net*** *outgoing* power:

$$\Delta P_{Avg,\,total} = \int_0^{2\pi}\int_0^{\pi} (1/2)\mathrm{Re}\left[\vec{E}\cdot(\vec{H}^* \times \hat{a}_r)\right]|\vec{x}|^2 \sin\theta d\theta d\phi \tag{6.155}$$

Equation 6.155 is thus the *sum* of the incoming and outgoing components of the Poynting vector and yields

$$\sigma_{ext} = \Delta P_{Avg,total}/[(1/2)E_0 H_0] = -(\pi/k^2)\sum_{l=1}^{\infty}(2l+1)\mathrm{Re}[\alpha(l)+\beta(l)] \tag{6.156}$$

When Equation 6.156 is zero, the *incoming* power of the lth multipole moment equals the *outgoing* power of the lth multipole moment. Because we use the net outgoing power of the induced multipole moments to calculate this ratio (and then refer to it as a cross section), the quantity, σ_{ext}, is often called the *total cross section*, but this is a misnomer because, when the incoming power and outgoing power are equal, this cross section is zero even though power is going each way through the surface at $r = |\vec{x}|$. It is, thus, more appropriate to call this cross section the *extinction* cross section.

Incoming Power

If we interpret the incoming power to be the quantity that, when added to the outgoing power, gives the ***net*** outgoing power, then

$$\sigma_{\mathrm{incoming}} = \sigma_{ext} - \sigma_{sc} = (\pi/2k^2)\sum_{l=1}^{\infty}(2l+1)\left[2-|\alpha(l)+1|^2-|\beta(l)+1|^2\right] \tag{6.157}$$

The semantics are important in these paragraphs because Jackson interprets the incoming power as the absorbed power, and hence the quantity, σ_{abs}, is usually called the *absorption cross section*. This designation is appropriate when the incoming power is all absorbed by the *good* conducting sphere (as in the case when the radius, a_i, is much larger than the skin depth, δ). As we will discuss below, this is not always the case, especially when dealing with sub-μm radius spheres at low frequencies (where the skin depth, δ, can be larger than a_i).

PROBLEMS

6.22 Carry out the integrals in Equation 6.153 to show that the radiated power as a ratio to the incident power gives the scattering cross section in Equation 6.154.

6.23 Evaluate the quantities, $|\alpha(l) + 1|^2$ and $|\beta(l) + 1|^2$, in Equation 6.157 to show that they are the same as the difference between Equations 6.156 and 6.154.

The above derivation was made by using two assumptions:

- The radius of the sphere, a_i, is assumed small compared with the E&M wavelength, $\lambda_2 = 2\pi/k_2$.
- The radius of the sphere is assumed large compared with the skin depth, $\delta(\omega)$.

Applied to "snowballs" with a radius of $a_i \sim 10^{-6}$ m and a frequency of $f \sim 100$ GHz, $k_2 a_i = (2\pi f/c_2)a_i = 2\pi a_i/\lambda_2 \approx 4\times10^{-3}$, the first approximation, $a_i << \lambda_2$, is satisfied even at 100 GHz. The second approximation that the skin depth, δ (at 100 GHz) is about 0.2 μm, which **is small** compared* with $a_i = 1$ μm.

The first approximation shows that higher values of l in Equations 6.146 and 6.147 are successively smaller by $(k_2 a_i)^2$ so the $l = 1$ (dipole) terms dominate the 1 = 2 (quadrupole) and higher terms.

CONCLUSION We may use the cross-sectional analysis for scattered power at all frequencies but we may use the cross section for absorbed power only at high frequencies where the skin depth is small compared with the snowball radius.

Some Numbers

PEC

We can take the impedance of a perfect electric conductor to be zero (i.e., $Z_s = 0$). In this case, Equations 6.146 and 6.147 yield

$$\alpha(l) \approx \frac{-2i\,(k_2 a_i)^{2l+1}}{(2l+1)[(2l-1)!!]^2} \quad \text{so } \alpha(1) \approx \frac{-2i\,(k_2 a_i)^3}{3} \tag{6.158}$$

$$\beta(l) \approx \frac{-2i\,(k_2 a_i)^{2l+1}}{(2l+1)[(2l-1)!!]^2}\left[\frac{-(l+1)}{l}\right] \quad \text{so } \beta(1) \approx \frac{4i\,(k_2 a_i)^3}{3} \tag{6.159}$$

so
$$\sigma_{sc} \approx \left(\pi/2k^2\right)3\left[|\alpha(1)|^2 + |\beta(1)|^2\right] = (10\pi/3)\,k^4 a_i^6 \tag{6.160}$$

$$\sigma_{ext} = -\left(\pi/k^2\right)\sum_{l=1}^{\infty}(2l+1)\mathrm{Re}[\alpha(l)+\beta(l)] = 0 \tag{6.161}$$

$$\sigma_{\text{incoming}} = \sigma_{sc} \tag{6.162}$$

Note that the units of σ_{sc} in Equation 6.160 are m^2, in accordance with its label as a cross section.

Note that Equation 6.160 is the same as that found previously in Equation 6.50, confirming continuity.

Note the name σ_{incoming} seems more appropriate than σ_{abs} for this special case because no power should be absorbed by a PEC.

* δ (at 1 GHz) is about 2 μm, which **is not small** compared with $a_i = 1$ μm (we treat this condition in the following section).

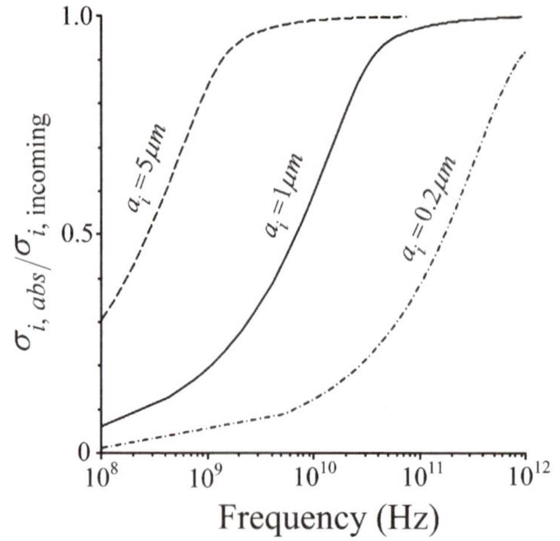

Figure 6.20 Ratio of the absorption cross section, $\sigma_{i,abs,c}$, for 5.0-, 1.0-, and 0.2-μm radius copper spheres at various frequencies compared with the absorption cross section, $\sigma_{i,abs}$ of a sphere with a relatively small skin depth.

$$\sigma_{i,abs}(1) \approx \frac{3\pi^2\sigma\eta_2\delta^4}{a_i}\frac{\left|j_1(k_2a_i)+(1/2)\alpha(1)h_1^{(1)}(k_2a_i)\right|^2}{\left|j_1(k_ca_i)\right|^2} \times\left\{\left(\frac{a_i}{\delta}\right)\left[\sin\left(\frac{2a_i}{\delta}\right)+\sinh\left(\frac{2a_i}{\delta}\right)\right]+\left[\cos\left(\frac{2a_i}{\delta}\right)-\cosh\left(\frac{2a_i}{\delta}\right)\right]\right\} \quad (6.172)$$

Comparing Equation 6.172 with 6.167, the absorption cross section of a small sphere with a large skin depth is smaller than the absorption cross section for small conducting spheres with a relatively small skin depth (Equation 6.167):

$$\frac{\sigma_{i,abs}(1)}{\sigma_{i,\text{incoming}}} \approx \pi\left(\frac{\sigma}{\omega\varepsilon_2}\right)\left(\frac{\delta^3}{a_i^3}\right)\frac{\left|j_1(k_2a_i)+(1/2)\alpha(1)h_1^{(1)}(k_2a_i)\right|^2}{\left|j_1(k_ca_i)\right|^2}\left[1+\frac{\delta}{a_i}+\frac{\delta^2}{2a_i^2}\right] \times\left\{\left(\frac{a_i}{\delta}\right)\left[\sin\left(\frac{2a_i}{\delta}\right)+\sinh\left(\frac{2a_i}{\delta}\right)\right]+\left[\cos\left(\frac{2a_i}{\delta}\right)-\cosh\left(\frac{2a_i}{\delta}\right)\right]\right\} \quad (6.173)$$

as shown in Figure 6.20.

6.11 APPLICATIONS OF FUNDAMENTAL SCATTERING

From these fundamental solutions for an isolated "snowball," we can calculate the effective magnetic dipole moment of an arbitrarily large, good conducting sphere, and its phase relative to a PEC sphere. In addition, we can calculate the power lost when an incident wave induces these dipoles. We can also calculate the additional loss on the nearby flat conducting trace as a result of the dipole and its image and we can evaluate the reduction in field at a "snowball" due to its first few neighbors. We can take into account the fact that the surface of snowballs has a lower density than that of pure copper and that there is a thin layer of solid solution alloy or impurity concentration in real spheres, as discussed in Chapter 4. Finally, a distribution of the snowballs in Figure 6.2 can be chosen to match the observed SEM pictures in Figure 6.1. These are considered second-order Born approximation effects and are discussed in Chapter 7 to estimate the effect of neglecting their contributions. In the following section, we will make the assumption that these are not large effects or that the errors tend to cancel one another in a random location of snowballs approximation. Perhaps it is serendipity, but we find the simple application of the fundamental scattering and absorption theory to multiple snowballs as calculated with Equation 6.167 for incoming power with small spheres match measured losses to a surprising accuracy for high- and low-profile rough surfaces to frequencies up to 50 GHz. However, the use of Equation 6.172 gives an even better description of measured losses. It is comforting that this application is a first-principles application of the theory that has no adjustable parameters and no data fitting and relies solely on measured material properties of copper and the propagating medium.

Scattering for *Multiple Good* Conducting Spheres

For two conducting spheres in a microstrip waveguide, we argue that the normal component of the electric field intensity is discontinuous at every point on their surface; that is, the normal component of the electric field intensity inside a PEC or a good conductor in the quasistatic approximation is zero. The external electric field intensity applied to the spheres will induce a surface charge density, $\Sigma_{e,s} = 3\varepsilon_2 E_0 \cos\theta_p$, on their surface where θ_p is the angle relative to the direction of the induced electric dipole moment ($\hat{a}_x$ in this case). The surface charge density does not depend on the radius of the sphere. and the area of common contact is equal for both spheres so the net charge cancels at their point of contact. As we see in Figure 6.21, the electric field intensity and magnetic field intensity at any point in space can thus be calculated by the principle of superposition.

In Figure 6.21, left, the external electric field intensity lines are not shown but they should be normal to the surface at every point. Induced conductor magnetic dipole moments are shown in Figure 6.21 right (lines of external magnetic field intensity are not shown). The induced magnetic dipole moment on each sphere is caused by surface currents that flow counterclockwise to the magnetic dipole moment direction. Surface currents on the two spheres also cancel one another and lead us to the conclusion that we may use the principle of superposition for two independent surface currents that will achieve the same results as a null concentration of charges and currents at their point of contact.

Using Auger electron spectroscopy, we have found that the outside surface of the snowballs on the copper samples used to manufacture our test boards is coated with a very thin (~0.01 μm) carbon, oxygen, chromium, sulfur, silicon layer over a thicker (~0.1 μm) CuZn layer, so it is possible that the individual spheres are not in good electrical contact. In addition, the CuZn (brass) layer causes a higher electrical resistivity that leads to a skin depth approximately double that of pure copper. Our power loss arguments are independent of a nonconducting surface because, once the EM wave is absorbed by the copper, the power is lost from the incident wave, but the presence of a brass coating will have a small effect[20] on the fractional power absorbed. As discussed in the next chapter, simulation of the scattered fields in such compound

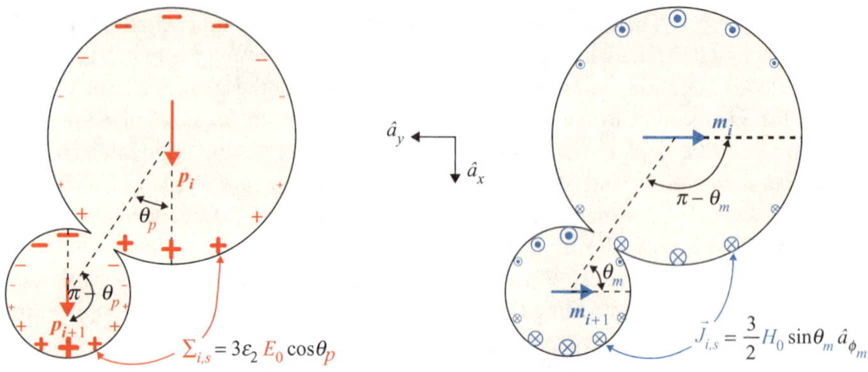

Figure 6.21 Surface charges and currents from contacting spheres in an external electric field intensity cancel, independent of their relative position to one another. $\vec{P}_i(\vec{m}_i)$ is the induced electric (magnetic) dipole moment for the *i*th sphere, and it is in the x $(-y)$ direction. θ_p (θ_m) is the angle with respect to that dipole moment.

spheres confirms that the screening or interference of fields between three and four adjacent snowballs may be treated as if they were independent scattering centers.

Surface Roughness Power Losses

Loss for a High Profile Surface

A methodology must be chosen to optimally represent the surface protrusions with stacked snowballs. A lower limit can be chosen by calculating the number of snowballs that will completely fit within a pyramid with a height of 5.8 μm and a base width of 9.4 μm, which are the root mean square RMS values of the protrusions measured on the high-profile copper samples using an optical profilometer. Using the superposition argument, we have replicated the structure seen in SEM photographs with an ensemble of spheres of various sizes to represent the geometry of the pyramid structures for high-profile samples.

The simplest ensemble is a set of uniform (less than a_i = 1-μm radius) spheres because that seems to be the dominant form of snowballs in either the high- or low-profile samples, so that is our first choice to compare with the measured data. For high-profile samples, we choose a set of hexagonal surface cells into which we may place a_i = 1-μm spheres to form a base upon which we stack up subsequent layers to form a pyramid structure with a height of *5.8 μm*. The structure is shown in Figure 6.22.

Eleven a_i = 1-μm radius snowballs fit easily inside a pyramid of width 9.4 μm in Figure 6.22 (the approximate distance between pyramids in Figure 6.2) and three layers form a stack of height that is less than 5.8 μm. Thrity-eight snowballs is the maximum number of snowballs that have part of their volume inside the pyramid and whose base snowballs touch the hexagonal boundary, as shown in Figure 6.23.

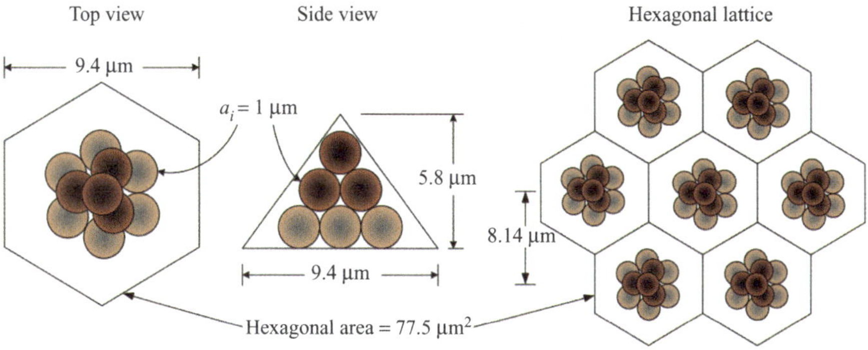

Figure 6.22 A basic 11-sphere model of a "snowball" stack in a hexagonal unit that can be replicated into a hexagonal structure to represent the surface protrusions on a rough surface.

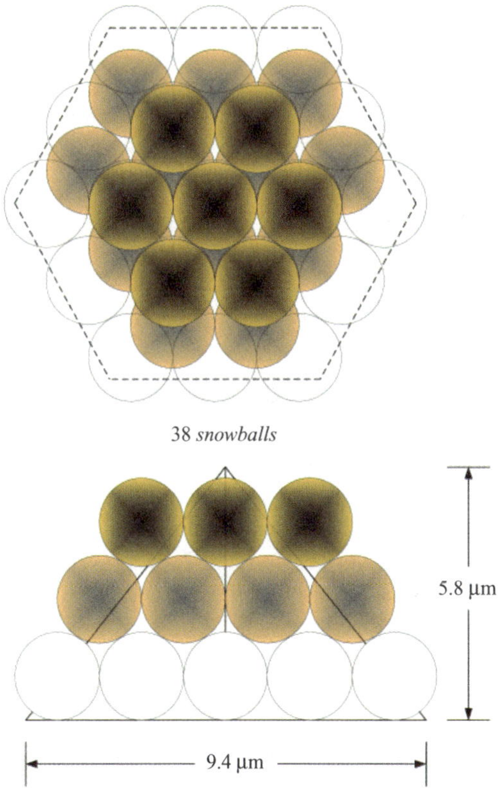

Figure 6.23 A basic 38-sphere model of a "snowball" stack in a hexagonal unit that can be replicated into a hexagonal structure to represent the surface protrusions on a rough surface.

A comparison of Figures 6.22 and 6.23 with Figure 6.1 reveals that 11 snowballs per pyramid are sparse and that 38 snowballs per pyramid are too large if the snowballs have a uniform radius of 1 μm. We do not determine the size or quantitative number of snowballs in a pyramid by this argument but rather use it to place reasonable limits on the numbers we determine through a fit below.

Power Losses Using the Snowball Model

We note that the relative surface area of N_i snowballs of radius a_i can be much larger than that of a flat trace, A. If the magnetic field intensity outside a very good conducting sphere behaves† like that outside a PEC, $H_{\|}(a_i, \theta_m) = 3H_0 \sin\theta_m/2$, the time-averaged power absorbed that will be the integral of the magnetic field intensity, $\Delta P_i/da = \mu_c\omega\delta|H_{\|}|^2/4$, over the surface of each sphere of radius a_i is

$$\Delta P_i = \frac{\mu_c\omega\delta}{4}\int_0^{\pi}\frac{9}{4}H_0^2\sin^2\theta_m(2\pi a_i\sin\theta_m)a_i d\theta_m = \frac{\mu_c\omega\delta}{4}H_0^2 6\pi a_i^2 \quad (6.174)$$

so the trace and the pyramid ensemble losses for N_i spheres of radius 1 μm at high frequencies are

$$\Delta P_{Flat} + N_1\Delta P_1 = \frac{1}{4}\mu_c\omega\delta H_0^2 A + N_1\frac{1}{4}\mu_c\omega\delta H_0^2 6\pi(1\,\mu\text{m})^2, \quad (6.175)$$

where A is the area (wl in Figure 4.2) of the trace. The ratio of maximum power lost at high frequencies for a rough versus a smooth trace is thus

$$K_{Max} = \frac{\Delta P_{Flat} + N_i\Delta P_i}{\Delta P_{Flat}} = 1 + \frac{N_1 6\pi(1\,\mu\text{m})^2}{A}$$
$$= \begin{cases} 3.7 & \text{for } 11 \quad 1\,\mu\text{m spheres/hex area} \\ 10.2 & \text{for } 38 \quad 1\,\mu\text{m spheres/hex area} \end{cases} \quad (6.176)$$

Equation 6.176 shows that, at high frequencies, the power loss for a flat plane plus a number of 1 μm radius spheres per hexagonal area can be **much** larger[21] than that of a flat surface‡ alone. With a wave incident power density, $\eta H_0^2/2$, the power losses produced by a copper surface in a PCB that is purposely roughened are thus larger than the relative power losses of a smooth surface by

† The magnetic flux excluded from a good conducting sphere will be close to that excluded from a PEC if the skin depth is small compared to the radius of the sphere (which is valid at very high frequencies).

‡ The largest loss factor of Morgan and Hammerstad was 2.0.

$$K(\omega) = \frac{\Delta P_{rough}}{\Delta P_{smooth}} \approx \frac{\frac{\mu_0 \omega \delta}{4} H_o^2 A + \sum_{i=1}^{j} N_i \sigma_{i,abs} \frac{\eta}{2} H_o^2}{\frac{\mu_o \omega \delta}{4} H_o^2 A} \tag{6.177}$$

For the frequency range 1–100 GHz, the product (ka_i) is a small quantity (smaller than 4×10^{-3} for a_i = 1 μm), and, because each successive multipole contribution to the cross section decreases by $(ka_i)^2$, we may neglect all but the dipole (l = 1) terms in the calculation of the total cross section, as in Equation 6.167:

$$K(\omega) = 1 + \sum_{i=1}^{j} \left(\frac{\sigma_{i,abs}(1)}{\sigma_{i,\text{incoming}}} \right) \left(\frac{6\pi a_i^2 N_i}{A} \right) \left[1 + \frac{\delta}{a_i} + \frac{\delta^2}{2a_i^2} \right]^{-1}, \tag{6.178}$$

where the sum is taken over all of the various size snowballs on a surface. In Equation 6.178, the important part of the losses is the total number of snowballs, N_i, with radius a_i per unit area, A, on the trace. The ratio $\sigma_{i,abs}$ (1)/ $\sigma_{i,\text{incoming}}$ is taken from Equation 6.173. For the special case stack-up of *uniform*, a_i = 1-μm snowballs, as shown in Figure 6.22 ($6\pi a_i^2 \cdot 11/77.5\ \mu\text{m}^2$) = 2.7 and for the higher density number of snowballs shown in Figure 6.23 ($6\pi a_i^2 \cdot 38/77.5\ \mu\text{m}^2$) = 9.2. The absolute loss of a line of (w/9.4 μm) hexagonal cells, (each separated by 8.14 μm as shown in Figure 6.22) may be calculated by using Equation 3.20:

$$\alpha_{snowballs} = \frac{(\Delta P_{snowballs} / 8.14\ \mu\text{m})}{2P_0} \tag{6.179}$$

expressed in Nepers per meter. If we choose to write *dB* per meter for a trace of width w:

$$\alpha_{snowballs} = \frac{10dB(w/9.4\ \mu\text{m})}{2(8.14\ \mu\text{m})} \sum_{i=1}^{j} \left(\frac{\sigma_{i,abs}(1)}{\sigma_{i,\text{incoming}}} \right) \left(\frac{6\pi a_i^2 N_i}{A} \right) \left[1 + \frac{\delta}{a_i} + \frac{\delta^2}{2a_i^2} \right]^{-1} \tag{6.180}$$

Comparison of S-Parameter Measurements to the Snowball Model

The primary output of a vector network analyzer measurement is the S_{21} parameter (called the insertion loss), which measures the *square root* of the transmitted power for a trace of length $z = l$ compared with the incident power at $z = 0$:

$$S_{21}(\omega) = \sqrt{\frac{P(l)}{P(0)}} = 10^{-\alpha_{dB} \frac{l}{20}} = \left(10^{-\alpha_{dielectric} \frac{l}{20}} \right) \left(10^{-\alpha_{flat} \frac{l}{20}} \right) \left(10^{-\alpha_{snowballs} \frac{l}{20}} \right), \tag{6.181}$$

which we can use as a metric to evaluate power losses as a function of frequency, ω, using snowball differential scattering analysis.

We can choose the number of spheres in a stack-up by using Equation 6.180 and various choices, N_i, of uniform a_i radius spheres to our measured losses. We can see from Figure 6.1 that there are several different sizes of snowballs in the real sample (especially if we include the 0.04-μm nodules to build individual snowballs), but a uniform assumption shows how the absorption changes with the mean radius (about 0.5 μm).

Loss for a High-Profile Microstrip

Figure 6.24 is a fit of the insertion loss using Equation 6.180 with three different average-sized snowballs per hexagonal area to that of the measured *high-profile* data. Here, losses due to the propagating medium and those due to the flat plane are included, as shown in Chapter 5.

Figure 6.25 shows a fit of the insertion loss using Equation 6.180 with 79 snowballs (of radius 0.5 μm) per hexagonal area to that of the measured high profile data between 50 MHz and 50 GHz.

Loss for a Low-Profile Microstrip

We show in Figure 6.26 a fit of the losses using Equation 6.167 with 42 snowballs (of radius 0.5 μm) per hexagonal area to that of the measured *low-profile* data. We have again included the losses due to the propagating medium[22] as we showed in the previous chapter.

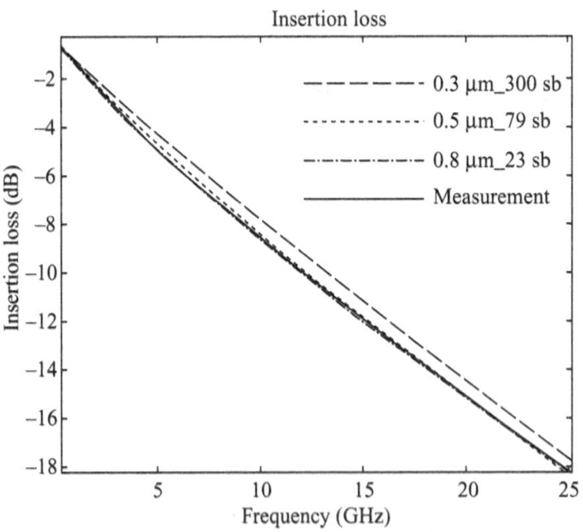

Figure 6.24 Measured insertion losses on a 7″ *high-profile* microstrip between 50 MHz and 25 GHz. Comparative loss trends are shown for three hundred 0.3-μm, seventy-nine 0.5-μm, twenty-three 0.8-μm radius snowballs.

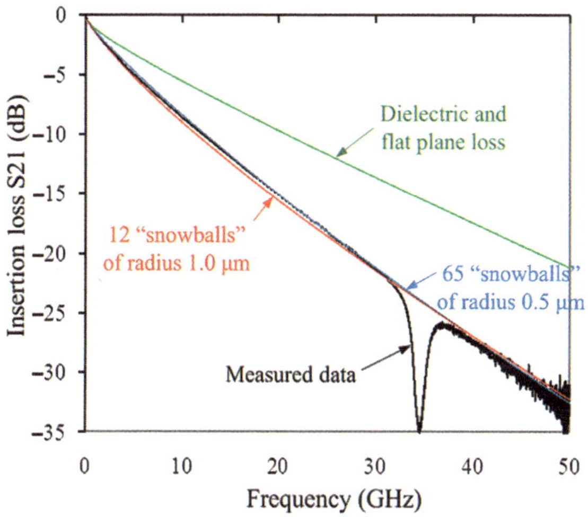

Figure 6.25 Measured insertion losses on a 7″ *high-profile* microstrip between 0 and 50 GHz.‡ A comparative loss trends for sixty-five 0.5-μm radius snowballs. Losses for twelve 1.0-μm radius snowballs and a Debye dielectric with a flat plane are shown for comparison.

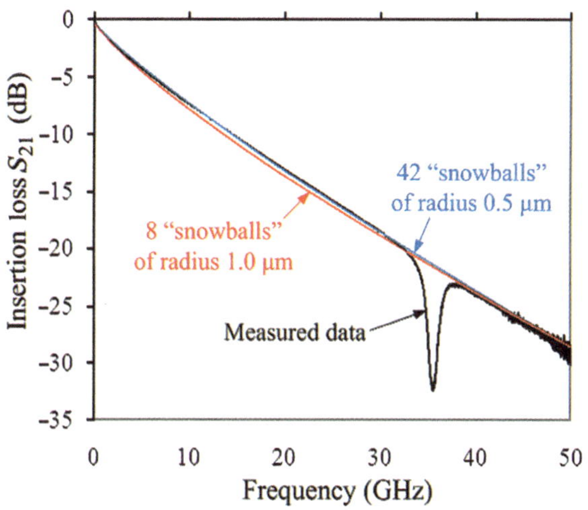

Figure 6.26 Measured transmission losses on a 7″ *low-profile* microstrip.‡ Loss trends for forty-two 0.5-μm snowballs and for eight 1.0-μm radius snowballs per hexagonal unit area on a flat plane are shown for comparison.

‡ The apparent resonance at about 36 GHz is considered in Chapter 7.

CONCLUSIONS

Measurements of transmission losses correlate[23] with electromagnetic absorption theory up to 50 GHz. Scattering and absorption of E&M waves by small copper "snowball spheres" assembled into *high-* or *low-profile* surface features are consistent with the rough surface preparation by PCB manufacturers, as shown by SEM photographs. The 3-D theoretical scattering and absorption model based on an exact solution to Maxwell's equations to the vector Helmholtz equations inside and exterior to individual spheres, neglecting interference between adjacent spheres, gives us an analytic expression for surface power losses of propagating E&M waves that is related to the cross section for absorption but relatively independent of how the "snowballs" are stacked-up[§]; in this first principles model, the *rms* value of surface roughness is unimportant.

We conclude from Equation 6.178:

$$K(\omega)=1+\sum_{i=1}^{j}\left(\frac{6\pi a_i^2 N_i}{A}\right)\left(\frac{\sigma_{i,abs}(1)}{\sigma_{i,\text{incoming}}}\right)\left[1+\frac{\delta}{a_i}+\frac{\delta^2}{2a_i^2}\right]^{-1}$$

Frequency dependence comes from a product of the two graphs in Figure 6.27. The main contributing factor to the power lost by a rough surface at high frequencies is proportional to the number of "snowball spheres" per unit area. We have assumed that uniform- (average-) sized snowballs adequately describe the power losses for *high-profile* copper and that the snowballs with the same size with a different surface density accurately describe the power losses for *low-profile* copper in our electrodeposited samples. However, the systematic difference between the "best" theoretical line in Figures 6.25 and 6.26 probably arises from the fact that a distribution of "snowball" sizes would be a more accurate approximation; the subject of a Master's thesis. While the additional rough surface loss depends upon the choice of the snowball radius in the Cu manufacturing process, it does not depend upon the manner in which the snowballs are stacked up. At low frequencies, two corrections to the power loss mechanism are required, as shown in Figure 6.27, when the radius of the "snowball spheres" are small compared with the skin depth. The definition of small is a statement of the snowball radius (which is fixed by the manufacturing process) relative to the skin depth (which depends upon the wave frequency); that is, a given snowball may be considered small at 1 GHz and large at 100 GHz.

As is shown in the next chapter on advanced signal integrity, screening of incident fields by near-neighbors is potentially compensated for by second-order corrections that can be masked by the number of "snowball spheres" per unit area. The measured and calculated values give a correlation within an accuracy

§ There is a second order correction in the power loss due to the location of a dipole in the neighborhood of a flat conducting surface that requires an image dipole above the interface. The correction depends on the distance of the dipoles from the interface and hence is a function of the stack-up.

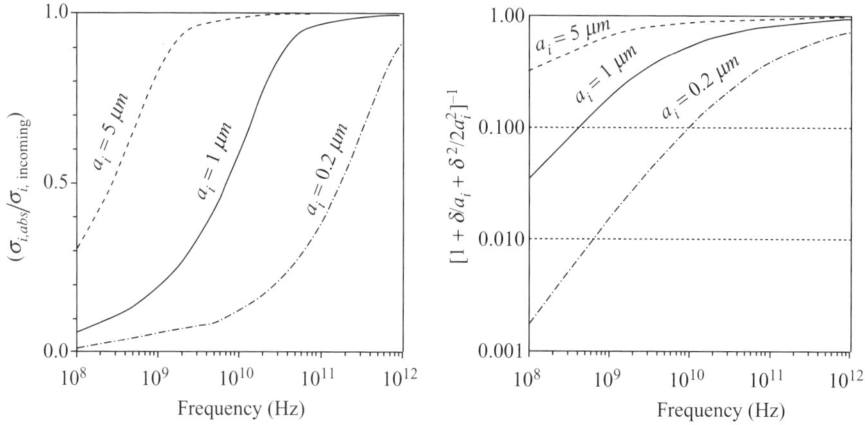

Figure 6.27 Frequency dependence of terms in the relative power, $K(\omega)$, for a rough versus a flat surface.

of 0.3 dB at all frequencies up to 50 GHz in numerical simulation codes. Within our approximations, the additional rough surface power losses are independent of the propagating medium of the microstrip. If existing PCB technology is to be used for increased data rates in the range of 10–100 GHz in the future, electrodeposition processes should be developed to produce less additional snowball area as a fraction of the trace area; that is, minimize the value of $K(\omega)$ through an optimal product of $(6\pi a_i^2 N_i/A)$ and stack-ups of smaller radii "snowball spheres."

ENDNOTES

1. M. Schlesinger and M. Paunovic, eds., *Modern Electroplating*, 4th ed. (New York: John Wiley and Sons, 2000).
2. S. Pytel, P. Huray, S. Hall, R. Mellitz, G. Brist, H. Meyer III, L. Walker, and M. Garland, "Physical and Chemical Analysis of Rough PWB Surfaces," Proceedings of the IEEE EPEP Conference, Oct. 29–31, 2007.
3. N. Ibl, in *Advances in Electrochemistry and Electrochemical Engineering*, vol. 2, C. S. Tobias, ed. (New York: Wiley, 1962).
4. S. Nakahara, *Acta Metall.* 36 (7) (1988): 1669.
5. S. P. Morgan Jr., "Effect of Surface Roughness on Eddy Current Losses at Microwave Frequencies," *Journal of Applied Physics* 20 (April 1949): 352–62.
6. P. G. Huray, S. Hall, S. G. Pytel, F. Oluwafemi, R. Mellitz, D. Hua, and P. Ye, "Fundamentals of a 3-D 'Snowball' Model for Surface Roughness Power Losses," Proceedings of the IEEE Conference on Signals and Propagation on Interconnects, May 14, 2007, Genoa, Italy.
7. P. G. Huray, Y. Chen, and R. Mellitz, "Physics of Scattering from Inclusions in FR-4 and Rough Trace Surfaces," Intel USC Dedication Poster, Feb. 25, 2005.
8. P. G. Huray, S. G. Pytel, R. I. Mellitz, S. H. Hall, "Dispersion Effects from Induced Dipoles," 10th Workshop on Signal Propagation on Interconnects, Berlin, Germany, May 9–12, 2006.
9. P. G. Huray, *Maxwell's Equations* (Hoboken, NJ: John Wiley & Sons, 2009), Equation 7.109a.

10. L. D. Landau, E. M. Lifshitz, and L. P. Pitaevskii, *Electrodynamics of Continuous Media*, 2nd ed. (Elsevier, 1982), p. 203.
11. Ibid., Chapter 7.
12. Ibid., Equation 7.86.
13. Ibid., Section 2.5.
14. Lord Rayleigh, *Philos. Mag.* 41 (1871): 107, 274.
15. Ibid., Equation 7.64.
16. Ibid., Equation 7.62.
17. C. J. Bouwkamp and H. B. G. Casimir, *Physica 20* (1954): 539.
18. John David Jackson, *Classical Electrodynamics*, 3rd ed. (John Wiley and Sons, 1999), section 10.4, p. 474.
19. Ibid., Equations 7.61a and 7.61b.
20. D. Aerne, "Electromagnetic Scattering from Multiple Copper Spheres Covered by a Thin Layer of Brass at 1 GHz, 10GHz and 100 GHz," thesis in partial fulfillment of the MS degree at the University of South Carolina, August 2008.
21. Hammerstad and Jensen, "Accurate Models for Microstrip Computer Aided Design," *IEEE MTT-S Int. Microwave Symposium Dig.*, May 1980, pp. 407–409.
22. S. G. Pytel, G. Barnes, D. Hua, A. Moonshiram, G. Brist, R. I. Mellitz, S. H. Hall, and P. G. Huray, "Dielectric Modeling and Characterization up to 40 GHz," *11th Annual IEEE SPI Proceedings*, May 13–16, 2007, Genoa, Italy.
23. F. Oluwafemi, "Surface Roughness and Its Impact on System Eye Margin," dissertation in partial fulfillment for the PhD degree at the University of South Carolina, Dec. 2007.

Chapter 7

Advanced Signal Integrity

LEARNING OBJECTIVES

- Use the second Born approximation to determine the magnitude of interference effects of scattered electromagnetic waves
- Model second-order scattering for multiple scattering centers to evaluate screening effects and find the magnitude of a Form Factor for a lattice of scattering centers
- Deduce the influence of a distribution of different size "snowballs" and show how they influence the second-order losses as a function of frequency
- Calculate the reduced electric and magnetic dipole moments due to the penetration of fields into *good* conductor scattering centers
- Evaluate the influence of alloy distributions in the surface layer of conductors as it pertains to electromagnetic losses
- Derive the influence of image electric and magnetic dipoles in the neighborhood of flat conducting surfaces
- Find the phase delay in pulses caused by surface roughness and lossy propagation media and show how the shape of a pulse is dispersed
- Evaluate inductance as a function of frequency due to scattering from rough surfaces in terms of the phase shift of the second-order wave compared with the incident wave

INTRODUCTION

In Chapter 6, the losses from electromagnetic scattering and absorption from conducting spheres was evaluated by using the first Born approximation. The electromagnetic power absorbed was derived from first principles by using only material properties and the solution to Maxwell's equations for spheres. The resulting power loss as a function of frequency was surprisingly accurate when compared with measured insertion losses from a Vector Network Analyzer (VNA)

The Foundations of Signal Integrity, by Paul G. Huray

up to 50 GHz. However, in the process of developing the scattering and loss theory, a number of approximations were made: for example, copper snowballs were treated like cue balls with a well-defined surface that is perfectly spherical, has the same density as pure copper, has no impurities in the surface layer, and has no alloy transition to pure copper. It was recognized that snowballs on a real surface are approximately made up of a distribution of different size spheres, but the losses were modeled with a single, average-size set of spheres that stacked up to form large pyramidlike "teeth" for good adhesion. A perturbation approximation (first Born approximation) was used to calculate the scattering from *good* conductors by comparing their results with a perfect electric conductor (PEC) as the conductivity increased. This is probably a good assumption, but it was later recognized that penetrating fields inside the *good* conductors propagate slowly inward with time so that interior fields can change direction. The model did account for the reduced "effective" magnetic moments of the good conducting sphere because the electromagnetic wave functions contained magnetic and electric coefficients that accurately describe the fields inside a sphere. It was recognized that electromagnetic fields can fully penetrate a *good* conductor if the radius is small compared with the skin depth so that incoming waves do not lose all of their energy to heat in the spheres but can emerge (after a delay) from the other side of a sphere as components of an outgoing wave. However, the model ignored this component to the resulting propagating wave. In Chapter 5, it was shown that the propagating medium is not homogeneous, yet the model used in Chapter 6 treated it as a uniform medium with only mildly frequency dependent losses; that is, the model did not take into account possible resonant frequencies because of voids or impurities in the material. In *Maxwell's Equations*, it was shown[1] that electric dipoles in the neighborhood of a flat conducting plane could use the method of images to calculate an additional induced electric charge on the surface. We also found that a magnetic dipole in the neighborhood of a flat conducting plane could induce an additional current in the plane, but that analysis was not completed, and there was no estimate of the magnitude or phase of that additional induced current in the analysis of total electromagnetic losses or dispersion; that analysis is provided in this chapter.

Perhaps the results of the theoretical losses from a first Born approximation to measurement were such a good fit because of a model that enjoyed the influence of compensating effects of a random phase approximation. In order to know the magnitude of each of the approximations noted above, each needs to be considered, and an estimate of its effect needs to be provided. Most of these effects can be evaluated through the second Born approximation, but some are complicated enough that the resulting boundary value problem is too difficult to solve; in that case, numerical solutions of Maxwell's equations are used to provide an estimate of the error magnitude. Measurements of highly specific material and geometric characteristics were made by using Fabry–Pérot absorption resonant cells, material-filled waveguides, and near-field scanners, but it was not shown how the results could be utilized to tailor transmission responses for future designs so, in a sense, this chapter is a guide to work that needs to be completed.

7.1 INDUCED SURFACE CHARGES AND CURRENTS

In *Maxwell's Equations*, the static electric scalar potential due to an electric dipole in the neighborhood of a conducting plane was shown[2] to be calculable by the method of images technique. In addition, arguments were made in Chapter 1 that time-varying fields in a *good* conductor could be treated like static fields as long as the frequency of oscillation was low enough to justify $(\sigma_c/\omega\varepsilon_0) << 1$; that is, the *quasistatic* condition was valid for copper for frequencies, $\omega = 2\pi f$, below 10^{15} Hz. It was also claimed that the magnetic field intensity in the neighborhood of a conducting plane could be found by the method of images technique. In this section, electric dipoles and magnetic dipoles and their corresponding images are used to calculate the electric and magnetic field intensities for horizontal and vertical orientations, as shown in Figure 7.1.

In Figures 6.7 and 6.8, it was found that an incident plane polarized electric field intensity, $\vec{E}_0\hat{a}_x$, and a corresponding magnetic field intensity, $\vec{H}_0\hat{a}_y$, propagating in the z-direction would induce an electric dipole moment, $\vec{p}$, like that shown in the far left of Figure 7.1 and a magnetic dipole moment, $\vec{m}$, like that shown in the far right of Figure 7.1.

Electric Dipole-Induced Surface Charge Density

For the case of induced electric dipole moments, Figure 7.2 shows the additional electric charge density induced on the surface of a flat conductor due to the presence of electric field intensity lines on the equipotential surface of the flat conductor. The induced surface charges due to the incident plane polarized electric field lines are intentionally omitted for clarity in Figure 7.2 even though they are likely larger than the additional charges indicated due to the dipole and its image. In *Maxwell's Equations*, we used[3] the negative gradient of the scalar electric potential to find the normal component of the electric field intensity and thereby calculate an induced surface charge density, $\Sigma_S = \varepsilon_0 E_n$.

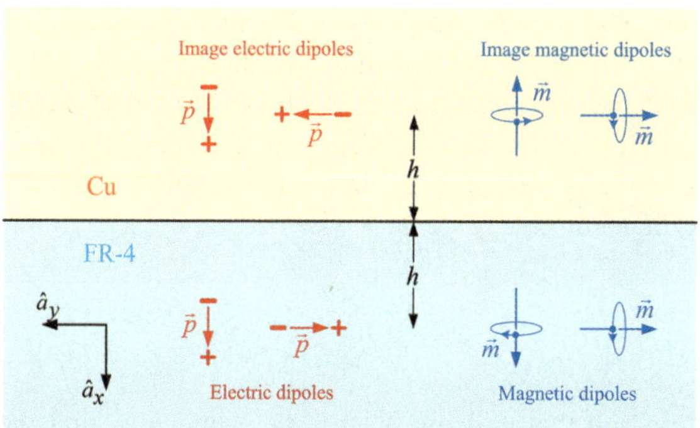

Figure 7.1 Electric and magnetic image dipole configurations used to calculate the equivalent electric and magnetic field intensities at the surface of a nearby flat conducting plane.

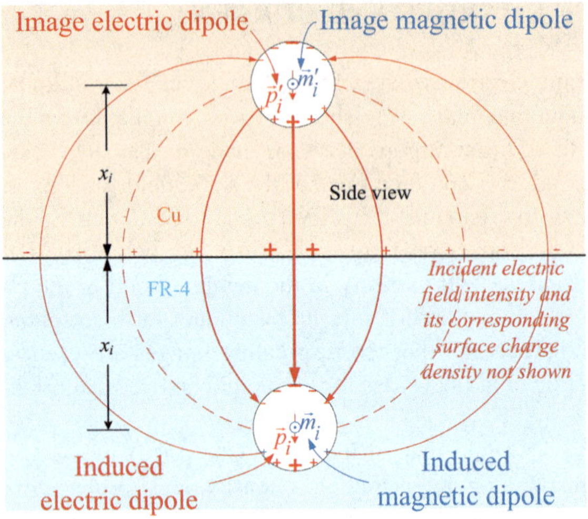

Figure 7.2 Electric field intensity lines as a result of a vertically oriented electric dipole located a distance x_i below a flat conducting plane. The electric field intensity is found by the equivalent method of images in which the potential at any point in space (below the conductor) can be found by calculating the scalar electric potential due to the electric dipole, $\vec{p}_i$, and its image dipole, $\vec{p}_i'$.

Note that, in Figure 7.2, the largest surface charge density is found directly between the two dipoles. This induced charge density is a second-order effect because it comes about as a result of the first Born approximation of a linearly polarized plane wave propagating in the z-direction as it first induces an electric dipole in the conducting sphere and that in turn induces an additional surface charge on the flat surface. It is important to note that this additional induced charge will not appear instantaneously on the flat conductor surface but will be forced to satisfy temporal causality; that is, the charge will appear only after the speed of the E&M wave in medium 2 propagates to the surface (a time-retarded electric potential). By this time, the incident wave will have moved in the z-direction by the same amount. The retarded response will cause the maximum additional surface charge density to lag behind the inducing incident wave. Had the incident wave (that induced the dipole moment), $\vec{p}_i$, been shown, it would thus lie to the right in this figure by an amount equal to the distance x_i (because the speed of propagation is the same in all directions in the propagating medium), a phase shift of 45° from the incident charge density.

PROBLEMS

7.1 Neglecting time retardation and the inducing electric field intensity, find the electric potential as a result of an electric dipole and its image at a point halfway between them, and calculate the electric field by means of the gradient normal to the surface at all points on the interface plane.

7.2 Calculate the total electric field intensity due to the incident electric field and the induced dipole and its image to show the time-retarded electric potential equations. Show that the point of maximum electric field gradient (and, thus, maximum charge density) lags the inducing electric field by an angle of 45°.

Figure 7.2 shows the field lines as a result of a static dipole moment and its image. Note that, unlike the case for a monopole charge and its image where the induced charge density was monatonic and extended to infinity, the additional charge density caused by an electric dipole moment and its image changes sign at a radius of x_i from the midpoint. This is a result of the fact that, for the case of an electric dipole charge and its image, there will be a set of points on a sphere of radius x_i for which the electric potential is zero. The additional electric charge density caused by the incident wave will thus lie at this crossover point on the interface surface.

Power losses due to the additional dipole-image-induced charge density will be negligible because charge density moves transverse to the direction of propagation in the form of a wave of charge. While this moving wave of additional charge constitutes an additional current (that in turn can induce a third-order component of adjacent magnetic field intensity), it does not scatter from impurities, voids, or lattice distortions like conventional longitudinal current density and hence produces little power loss. Only that loss due to acceleration of the charges in a transverse motion and the scattering that would occur from that microscopic displacement would contribute to the loss.

PROBLEM

7.3 Find the energy required to displace conduction electrons transversely by an amount consistent with the production of surface charge density for both the incident polarized electric field intensity and for the additional induced electric dipoles and show that it is negligible compared with the loss of energy that occurs from penetration of the magnetic field intensity (eddy currents).

Magnetic Dipole-Induced Surface Current Density

For the case of induced magnetic dipole moments, Figure 7.3 shows the additional electric current density induced on the surface of a flat conductor as a result of the presence of tangential magnetic field lines on the surface of the flat conductor. The induced surface current density due to the incident plane polarized magnetic field intensity lines are intentionally omitted for clarity in Figure 7.3 even though they are likely larger than the additional current density indicated due to the magnetic dipole moment and its image. In *Maxwell's Equations*, the normal component of the magnetic field intensity at a conducting/dielectric interface was found[4] to be approximately zero, and the tangential component was approximately continuous. It was also shown in *Maxwell's Equations* that an electric field intensity was induced inside the conductor (lagging the magnetic field intensity by 45°) and that this field intensity could cause a current density $\vec{J} = \sigma\vec{E} = (\partial H_{1y}/\partial x)\hat{a}_z$ to be induced inside a conductor in the quasistatic approximation.

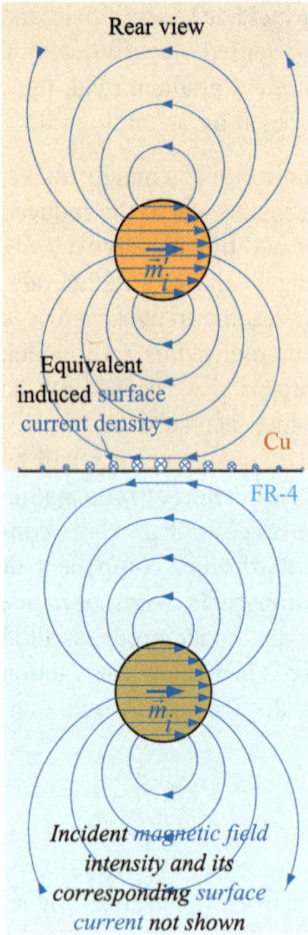

Figure 7.3 Magnetic field intensity lines due to a horizontally oriented magnetic dipole moment located a distance x_i below a flat conducting plane. The total magnetic field intensity is found by the equivalent method of images in which the magnetic field intensity at any point in space (below the conductor) is found by calculating the vector sum of the field intensity due to the magnetic dipole, $\vec{m}_i$, and its image dipole, $\vec{m}_i'$.

We can then use Figure 7.4 to calculate the vector sum of the two magnetic field intensity at a point on the surface of the interface for the distance, ρ, from the vertical between the magnetic dipole and its image. The time-retarded response of the additional surface current produced by a magnetic dipole and its image would lag the surface current that supported the incident wave (along the direction of propagation) at an angle of 45° in a side view.

Contrary to our normal color scheme, Figure 7.4 uses blue to correspond to the magnetic field intensity due to the induced dipole, and magenta to correspond to the magnetic field intensity due to its corresponding image dipole. In two dimensons, the vector sum calculation is straightforward. From this calculation, we can also find the additional surface current density caused by the induced magnetic dipole by calculating the derivative of the vector sum of the magnetic field intensity with respect to the vertical (x) variable as $\vec{J}_z = \sigma\vec{E}_z = (\partial H_{1y}/\partial x)\hat{a}_z$. This additional current density should be added to the incident current density not shown in any of these drawings, but it should be noted that the additional amount of induced dipole current will lag the incident surface current by an amount in time that corresponds to the distance, x_i, divided by the propagation velocity. This produces a dispersion of

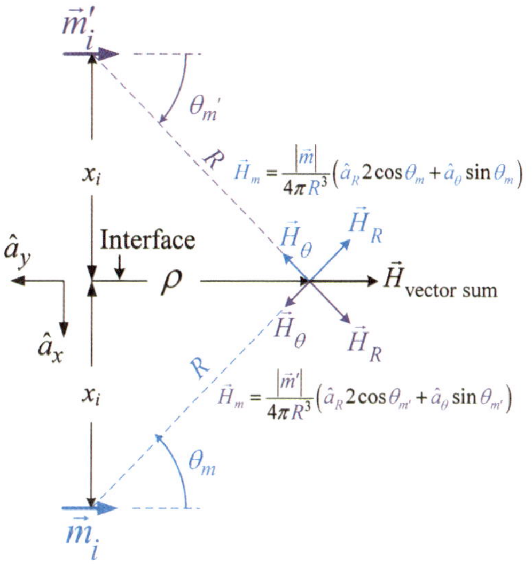

Figure 7.4 Vector sum of the magnetic field intensity due to a magnetic dipole below a conducting plane and its magnetic dipole image above the interface, as seen from the rear view of propagation. The incident magnetic field intensity has been omitted from this drawing.

the incident signal because of the lagging additional current density as a function of the location of the dipole below the interface; the lower the dipole, the more the dispersion.

Magnetic Dipole-Induced Surface Current Density in Three Dimensions (3-D)

The additional current density induced on the *good* conductor interface surface in 3-D can be calculated by using a series of coordinate transformations shown in Figure 7.5.

To accomplish this, we first transform the functional form in spherical coordinates into Cartesian coordinates[5] for rotation about the $\hat{a}_m = \hat{a}_1$ axis:

$$\begin{pmatrix} H_1 \\ H_2 \\ H_3 \end{pmatrix} = \begin{pmatrix} \cos\theta_m & -\sin\theta_m & 0 \\ \sin\theta_m \cos\varphi_m & \cos\theta_m \cos\varphi_m & -\sin\varphi_m \\ \sin\theta_m \sin\varphi_m & \cos\theta_m \sin\varphi_m & \cos\varphi_m \end{pmatrix} \begin{pmatrix} H_{R_m} \\ H_{\theta_m} \\ H_{\varphi_m} \end{pmatrix} \tag{7.1}$$

Then we can make a transformation to cylindrical coordinates[6] for rotation about the $\hat{a}_3$ axis:

$$\begin{pmatrix} H_\rho \\ H_\varphi \\ H_z \end{pmatrix} = \begin{pmatrix} \cos\varphi & \sin\varphi & 0 \\ -\sin\varphi & \cos\varphi & 0 \\ 0 & 0 & 1 \end{pmatrix} \begin{pmatrix} H_1 \\ H_2 \\ H_3 \end{pmatrix} \tag{7.2}$$

(e.g., for $a_i = x_i$), the additional magnetic field intensity due to current at the top of the sphere will occur immediately, while the induced additional magnetic field intensity due to currents lower on the sphere will occur after a delayed time. A differential time-retardation analysis of this delay will give the dispersion of the incident wave and its perturbed contribution as a result of the induced current in a second Born approximation.

Incremental Power Lost Due to Additional Induced Current on a Flat Plane

As shown in section 6.1, the total power lost by the penetration of field on the flat plane is the power density:

$$\vec{\mathcal{P}}_{Avg,c} = (1/2)\mathrm{Re}[\vec{E}_c \times \vec{H}_c^*] = (1/2)\mathrm{Re}[(1+j)/\sigma\delta]\left|\vec{H}_c(0)\right|^2 = (1/2\sigma\delta)\left|\vec{H}_{medium}(0)\right|^2$$

multiplied by a surface element and integrated over the plane. If the magnetic field intensity on the flat plane is modified by an additional amount as a result of the induced surface current, then, because the ρ and φ components are real, the total magnetic field intensity is

$$\left|\vec{H}_{total}(x_i, \rho, \varphi)\right|^2 = \left|H_0 e^{i\omega x_i/u_p}\,\hat{a}_y + H_{B,\rho}\,\hat{a}_\rho + H_{B,\varphi}\,\hat{a}_\varphi\right|^2$$

in calculating the total power lost to the flat conductor. We may calculate the differential power lost, $\Delta P_{differential}$, to the flat conductor by taking the difference between $\vec{\mathcal{P}}_{Avg,c}$ with and without the additional induced current as

$$\Delta P_{differential} = \int_0^\infty \int_0^{2\pi} (1/2\sigma\delta)\left[\left|\vec{H}_{total}(x_i, \rho, \varphi)\right|^2 - \left|\vec{H}_0\,\hat{a}_y\right|^2\right]\rho\, d\varphi\, d\rho \tag{7.8}$$

We can calculate the additional fractional power lost to the flat conductor, $f(x_i)$, for N_i dipoles on a flat plane of area $(w \cdot l)$ by taking the ratio of Equation 7.8 to the power lost on the flat plane with no inducing dipoles,

$$f(x_i) = \frac{(1/2\sigma\delta)N_i}{(1/2\sigma\delta)H_0^2(w\cdot l)} \int_0^\infty \int_0^{2\pi} \left[\left|\vec{H}_{total}(x_i, \rho, \varphi)\right|^2 - \left|\vec{H}_0\,\hat{a}_y\right|^2\right]\rho\, d\varphi\, d\rho \tag{7.9}$$

where the first and last terms in the integrand cancel over the integrated plane and the remaining integrand terms are

$$\begin{aligned} &2\cos(\omega x_i/u_p)H_0^2 \frac{a^3\rho^2}{(x_i^2+\rho^2)^{\frac{5}{2}}}\left\{-\cos^2\varphi\left[2-\left(\frac{x_i}{\rho}\right)^2\right]+\sin^2\varphi\left[6\cos^2\varphi-1-\left(\frac{x_i}{\rho}\right)^2\right]\right\} \\ &+ H_0^2 \frac{a^6\rho^4}{(x_i^2+\rho^2)^5}\left\{\cos^2\varphi\left[2-\left(\frac{x_i}{\rho}\right)^2\right]^2+\sin^2\varphi\left[6\cos^2\varphi-1-\left(\frac{x_i}{\rho}\right)^2\right]^2\right\} \end{aligned} \tag{7.10}$$

By using standard integrals from tables (e.g., Dwight[7] 203.05, 121.9, 123.9, and 125.9),

$$f(x_i) = \left(\frac{N_i \pi a_i^2}{w \cdot l}\right)\left[-\frac{2\cos(\omega x_i/u_p)a_i}{x_i} + \frac{5a_i^4}{16x_i^4}\right] \tag{7.11}$$

Equation 7.11 gives the fractional change in power lost on the flat, *good* conducting interface trace of width w and length l as a result of additional current induction on that plane caused by N_i dipole PEC spheres of radius a_i located a distance x_i below the plane.

PROBLEM

7.4 Use the additional magnetic field intensity terms given in Equation 7.7 to construct the power loss integral of Equation 7.8 and carry out the individual integral terms to produce Equation 7.11.

This expression can be applied to the set of $N_i = 8$ spheres shown in Figure 7.6. In this figure, the spheres all have the same radius a_i and are located at a distance x_i below the copper trace of width w and length l. The blue lines in the figure relate to the induced current density on the surface of the spheres and on the surface of the trace plane due to the incident electromagnetic wave and the additional amount of current induced in the plane by the nearby magnetic dipole moments. Note that the current density is induced transversely on all conducting surfaces so that the currents seen are a transverse wave of charge that propagates in the direction of the arrows.

Note that the spheres in Figure 7.6 are physically located in a geometric hexagonal close-packed (hcp) arrangement. The term $(N_i \pi a_i^2/w \cdot l)$ in Equation 7.11 is, thus, the ratio of the cross-sectional area of the spheres compared with the total area of the trace and the ratio $(a_i/x_i) \leq 1$. In the *special case* of a field of equal-radius

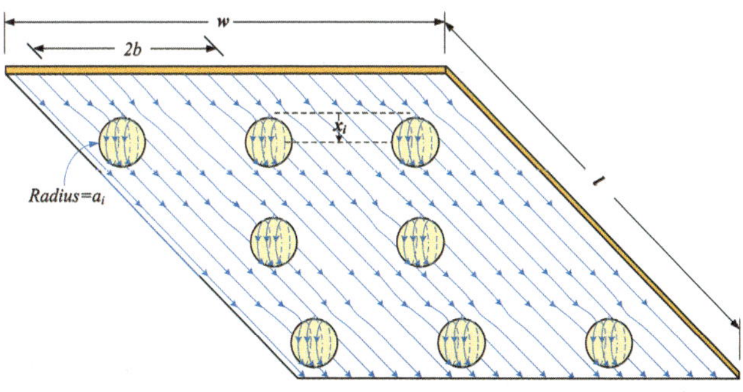

Figure 7.6 Eight equal radius, a_i, PEC spheres located the same distance, x_i, below a copper trace of width w and length l as an incident polarized plane electromagnetic wave (not shown) propagates from upper left to lower right. Blue lines indicate the direction of the transverse-induced current density wave on both the trace surface and the sphere surfaces.

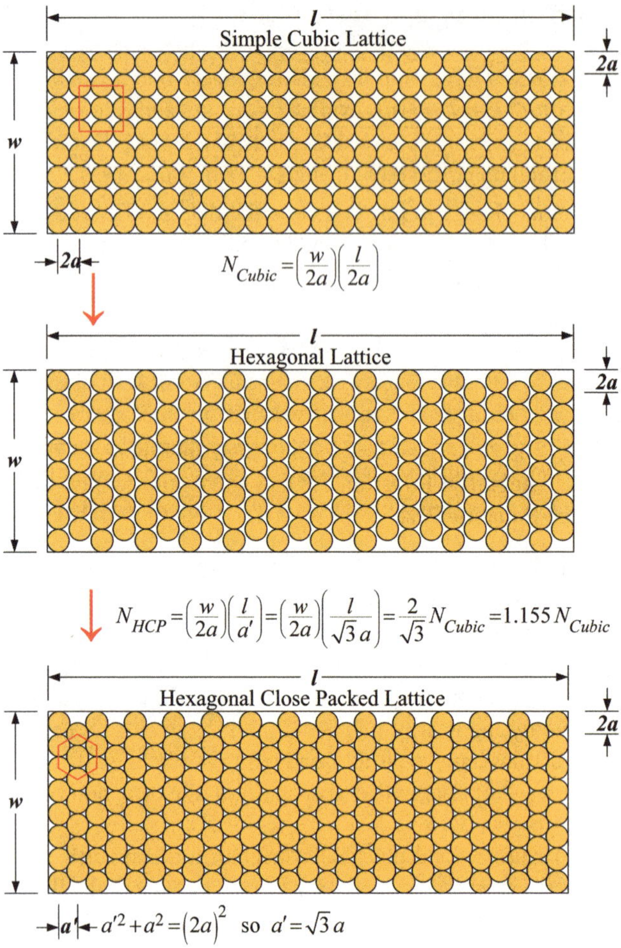

Figure 7.7 Normal view of equal size copper snowballs on the surface of a flat copper trace for comparing the packing factor of hexagonal close packed versus a simple cubic lattice arrangement: (top) a simple cubic lattice arrangement, (middle) alternate rows shifted by a, (bottom) a close-packed hexagonal field.

copper snowballs lying on the flat plane (the first layer of snowballs), the ratio $(a_i/x_i) = 1$ and the spheres would be expected to fill the space to the extent possible, as shown in cross section in Figure 7.7.

Figure 7.7 is an exaggeration (snowballs would be much smaller in cross section, and there would be many more of them on a typical copper trace) so that packing in simple cubic can be compared to the close packing in hexagonal geometric arrangement. In Figure 7.7, all of the snowballs are assumed to be of the same radius, ***a***, and that they lie ***on*** the copper plane; that is, $x_i = a_i = \boldsymbol{a}$. The top sketch shows how closely the spheres could be if they touch in a simple cubic arrangement. After an intermediate shift of every other row of the simple cubic arrangement, the middle sketch shows that there are empty spaces between the circles. If the space is closed up so that the circles touch, the near neighbors to any given sphere are in the form of a hexagonal arrangement (as shown by the red hexagonal unit). In the bottom arrangement, 15% more spheres can be packed into the same area, $(w \cdot l)$. The hcp arrangement leads to

$$f(x_i) = \left(\frac{\pi}{2\sqrt{3}}\right)\left[-2\cos(k_2 a_i) + \frac{5}{16}\right] \tag{7.12}$$

Because $k_2 a_i << 1$, $\cos(k_2 a_i) \approx 1$ for frequencies below 100 GHz. This is a somewhat surprising result for it says that the losses on the flat plane of a good conducting trace are substantially reduced by the presence of the induced dipoles below the plane. It might be said that, in this analysis, the presence of the PEC spheres screens the flat conductor from the propagating electromagnetic wave. A criticism of the analysis includes the assumptions made that all of the spheres are of the same size and that the magnetic field intensity caused by a *good* copper snowball magnetic dipole is the same as that produced by a PEC. Chapter 6 showed that magnetic field intensity penetrates the surface of a *good* conducting sphere and decays according to a modified spherical Bessel function inside the sphere (changing sign because the E&M fields propagate inward at a relatively slow velocity). This penetration will reduce the magnetic dipole moment of a real conducting sphere so that is the topic considered next.

However, this analysis also shows that there is a retarded current density that lags the incident current density in a perfectly flat waveguide due to the size and location of the good conducting spheres. If a single row of uniform spheres delays a fraction of the current density, then there will be an overlap of the original pulse and a smaller delayed pulse. Those two pulses will encounter the next row of spheres and produce three overlapping pulses with diminishing amplitudes. At the output end of a transmission line, a sequence of similar delayed pulses that overlap the original pulse shape would create a dispersed waveform that should be predictable. That analysis has not been carried out.

7.2 REDUCED MAGNETIC DIPOLE MOMENT DUE TO FIELD PENETRATION

In Equations 6.148 and 6.149, the electric and magnetic field intensities inside a good conducting sphere in a linearly polarized incident electromagnetic wave were found to be

$$\vec{E}_c = E_0 \sum_{l=1}^{\infty} \sum_{m=-l}^{l} \left\{a_M(l,m) i_l(k_c r) \vec{X}_l^m + a_E(l,m)(i/k_c)\vec{\nabla} \times \left[j_l(k_c r)\vec{X}_l^m\right]\right\}$$

$$\vec{H}_c = H_0 \sum_{l=1}^{\infty} \sum_{m=-l}^{l} \left\{a_M(l,m)(-i/k_c)\vec{\nabla} \times \left[j_l(k_c r)\vec{X}_l^m\right] + a_E(l,m) j_l(k_c r)\vec{X}_l^m\right\}$$

and because

$$a_M(l,1) = a_M(l,-1),\ a_M(l,0) = 0,\ a_E(l,1) = -a_E(l,-1) \text{ and } a_E(l,0) = 0$$

$$\vec{E}_c = E_0 \sum_{l=1}^{\infty} \{ a_M(l) j_l(k_c r) \vec{X}_l^+ + a_E(l)(i/k_c)\vec{\nabla}\times[j_l(k_c r)\vec{X}_l^-] \} \tag{7.13}$$

$$\vec{H}_c = H_0 \sum_{l=1}^{\infty} \{ a_M(l)(-i/k_c)\vec{\nabla}\times[j_l(k_c r)\vec{X}_l^+] + a_E(l) j_l(k_c r)\vec{X}_l^- \}, \tag{7.14}$$

where the coefficients $a_M(l) = a_M(l,1)$ and $a_E(l) = a_E(l,1)$ were found by evaluating Equations 6.148 and 6.134 for $\vec{E}_c$ and Equations 6.149 and 6.135 for $\vec{H}_c$. For the μm-size spheres used in surface roughness "anchor modules," Equations 6.146 and 6.147 showed that only the dipole terms, $l = 1$, contribute significantly to the sum in Equation 7.13 and that the higher-order multipole terms are negligible, so

$$\vec{E}_c \approx E_0 \{ a_M(1) j_1(k_c r) \vec{X}_1^+ + a_E(1)(i/k_c)\vec{\nabla}\times[j_1(k_c r)\vec{X}_l^-] \} \tag{7.15}$$

$$\vec{H}_c \approx H_0 \{ a_M(1)(-i/k_c)\vec{\nabla}\times[j_1(k_c r)\vec{X}_1^+] + a_E(1) j_1(k_c r)\vec{X}_1^- \} \tag{7.16}$$

where

$$a_M(1) = i\sqrt{6\pi}\,[j_1(k_2 a_i) + (1/2)\alpha(1) h_1^{(1)}(k_2 a_i)] / j_1(k_c a_i) \tag{7.17}$$

and

$$a_E(1) = \sqrt{6\pi}\, \frac{\left(\dfrac{1}{k_2 r}\right)\dfrac{\partial}{\partial(k_2 r)}\{k_2 r[j_1(k_2 r) + (1/2)\beta(1) h_1^{(1)}(k_2 r)]\}}{\dfrac{1}{(k_c r)}\dfrac{\partial}{\partial(k_c r)}\{k_c r[j_1(k_c r)]\}}\Bigg|_{r=a_i} \tag{7.18}$$

We can use $\alpha(1)$ and $\beta(1)$ from Equations 6.146 and 6.147 to find these two coefficients as a function of frequency. These coefficients are plotted in Figure 7.8.

These coefficients have been evaluated for various frequencies and for three spheres of different radii as shown in Figure 7.8. Note that the vertical scales are very different so that these figures show that the magnitude of coefficient $a_E(1)$ is much smaller than the magnitude of $a_M(1)$ for frequencies below 100 GHz so $a_E(1)$ can be approximately neglected. Thus,

$$\vec{E}_c \approx E_0 [a_M(1) j_1(k_c r) \vec{X}_1^+] \tag{7.19}$$

$$\vec{H}_c \approx H_0 \{ a_M(1)(-i/k_c)\vec{\nabla}\times[j_1(k_c r)\vec{X}_1^+] \} \tag{7.20}$$

and we can use Equation 6.136 to find

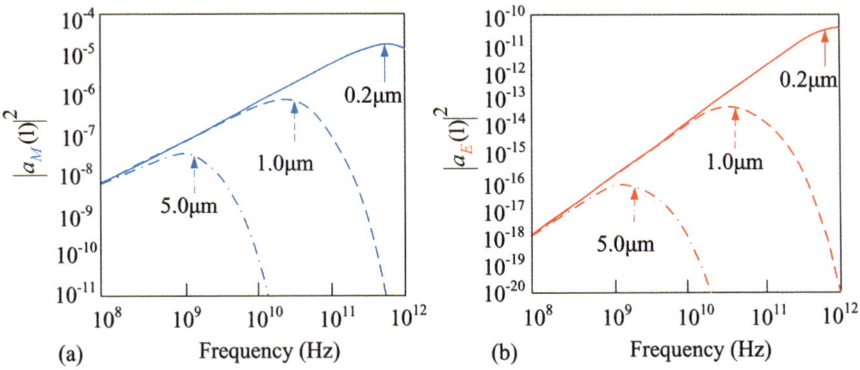

Figure 7.8 (a) Magnitude of coefficient $|a_M(1)|^2$ as a function of frequency for three spheres of different radii; (b) magnitude of coefficient $|a_E(1)|^2$ as a function of frequency for three spheres of different radii.

$$\vec{H}_c = H_0 a_M(1)(-i/k_c)\left\{\frac{i\hat{a}_r}{k_c r} j_1(k_c r)\left[-2\sqrt{\frac{3}{8\pi}}\sin\theta\sin\phi\right] + \frac{1}{r}\frac{\partial}{\partial r}[rj_1(k_c r)]\hat{a}_r \times \vec{X}_1^+\right\} \tag{7.21}$$

The first term in Equation 7.21 is in the radial direction, and the second is transverse radial. Furthermore, $\hat{a}_r \times \vec{X}_1^+$ is orthogonal to $\vec{X}_1^+$ so the transverse radial terms in Equation 7.19 and 7.20 are orthogonal to one another. The radial term can be transformed into the coordinates of the dipole moment by choosing $\theta = \pi/2$ in which case the radial component depends on $\sin\phi$ relative to the x-axis or $\cos\theta_m$ relative to the negative y-axis. The radial magnetic field intensity thus penetrates the sphere at $\theta_m = 0$ and exits from the sphere mostly at $\theta_m = \pi$, as shown in Figure 6.17, and the $\cos\theta_m$ dependence is consistent with Equation 6.63.

The general relations for $\vec{X}_l^+$ and Equation 6.116 may be used for values of the transverse radial terms in Cartesian coordinates through

$$\vec{X}_l^+ = (1/\sqrt{2})[\vec{X}_l^1 + \vec{X}_l^{-1}] = (1/\sqrt{2}\sqrt{l(l+1)})\{\vec{L}[Y_l^1 + Y_l^{-1}] + \vec{L}[Y_l^1 - Y_l^{-1}]\} \tag{7.22}$$

where

$$\vec{L}Y_l^m = [L_x\hat{a}_x + L_y\hat{a}_y + L_x\hat{a}_z]Y_l^m \tag{7.23}$$

and

$$\begin{aligned} L_x Y_l^m &= (1/2)\sqrt{(l-m)(l+m+1)}Y_l^{m+1} + (1/2)\sqrt{(l+m)(l-m+1)}Y_l^{m-1} \\ L_y Y_l^m &= (1/2i)\sqrt{(l-m)(l+m+1)}Y_l^{m+1} - (1/2i)\sqrt{(l+m)(l-m+1)}Y_l^{m-1} \\ L_z Y_l^m &= mY_l^m \end{aligned} \tag{7.24}$$

For the specific ($l = 1$) transverse radial term $\vec{X}_1^+$,

$$\vec{X}_1^+ = (1/2)\{\vec{L}[Y_1^1 + Y_1^{-1}] + \vec{L}[Y_1^1 - Y_1^{-1}]\} \tag{7.25}$$

so

$$\vec{X}_1^+ = \{[L_x\hat{a}_x + L_y\hat{a}_y + L_z\hat{a}_z]Y_1^1\} = (1/2)\sqrt{2}Y_1^0\hat{a}_x - (1/2i)\sqrt{2}Y_1^0\hat{a}_y + Y_1^1\hat{a}_z \tag{7.26}$$

or

$$\vec{X}_1^+ = \frac{1}{\sqrt{2}}\left(-\sqrt{\frac{3}{4\pi}}\cos\theta\right)\hat{a}_x + \frac{i}{\sqrt{2}}\left(-\sqrt{\frac{3}{4\pi}}\cos\theta\right)\hat{a}_y + \left(-\sqrt{\frac{3}{8\pi}}\sin\theta e^{i\phi}\right)\hat{a}_z \tag{7.27}$$

and, setting $\theta = \pi/2$, this quantity is expressed in the plane perpendicular to the propagation direction:

$$\vec{X}_1^+ = \sqrt{3/8\pi}\,(\cos\phi + i\sin\phi)\hat{a}_z \tag{7.28}$$

In which case Equation 7.19 becomes

$$\vec{E}_c \approx -\sqrt{3/8\pi}E_0 a_M(1)\,j_1(k_c r)(\cos\phi + i\sin\phi)\hat{a}_z \tag{7.29}$$

Note that, in this plane, $\hat{a}_{\phi_m} = \hat{a}_z$ and the real part of 7.29, $\cos\phi = \sin\theta_m$, is equivalent to the tangential part of the previous rear-view picture 6.17, as shown in Figure 7.9.

Fields inside the sphere in Figure 7.9 are the same as those of Figure 6.17 if we neglect the radial components in that figure. Equation 7.28 and Figure 7.9 show that the current density, $\vec{J}_c = \sigma\vec{E}_c$, inside a good conducting sphere is approximately transverse to the radial direction. Note that the currents inside the sphere actually turn around because they are propagating slowly (relative to the propagation velocity in the propagating medium) inward. Thus, those fields near the center of Figure 7.9 are a result of the magnetic field intensity that was on the surface half a cycle ago. Note also that the fields on a *good* conductor with $k_c = (1 + i)/\delta$ reduce to those of a PEC in the limit as $\delta \rightarrow 0$. In that special case, all of the current density flows on the surface of the PEC and is of the form, $\vec{J}_{l,PEC}(a) = (3/2)H_0\sin\theta_{m\hat{a}\phi_m}$, as was shown in Equation 6.18.

Electric field intensity inside a good conducting sphere can also be found for the cross section perpendicular to the induced magnetic dipole moment by setting $\phi = \pi$ in Equation 7.27 to obtain

$$\vec{E}_c \approx E_0 a_M(1)\,j_1(k_c r)\sqrt{3/8\pi}\left[-\cos\theta\hat{a}_x - \sin\theta\hat{a}_z - i\cos\theta\hat{a}_y\right] \tag{7.30}$$

The choice $\phi = \pi$ locates points above the y–z plane, as shown in Figure 7.10. We can also see that the angle $\phi_m = \theta + \pi/2$ so that, above the y–z plane, the base

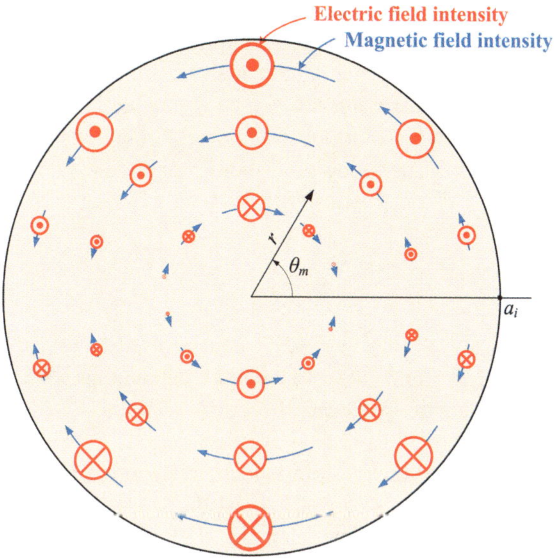

Figure 7.9 Tangential components of the induced magnetic field intensity, $\vec{H}_c(r, \theta_m)$, and corresponding electric field intensity, $\vec{E}_c(r, \theta_m)$ (phase shifted by 45°), inside a spherical *good* conductor due to an externally applied linear polarized electromagnetic field.

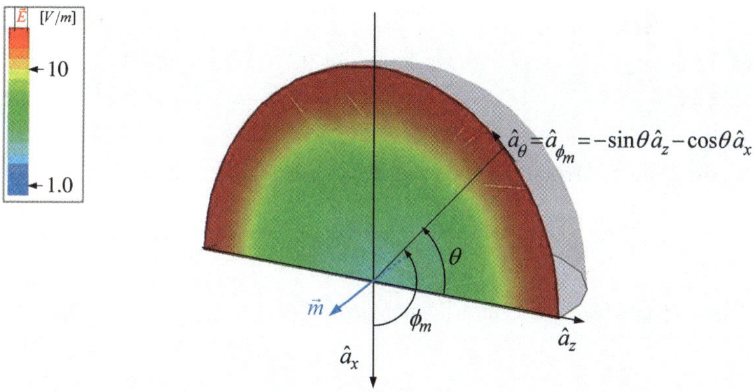

Figure 7.10 Spherical coordinate angles relative to the propagation direction ($\hat{a}_Z$) and the direction of the induced magnetic moment ($-\hat{a}_y$) superimposed on an Ansoft 3D Ful-wave Electromagnetic Field (HFSS) solution for the magnitude of the electric field intensity inside a *good* conducting sphere.

vector $\hat{a}_\theta = \hat{a}_{\phi_m} = -\cos\theta\hat{a}_x - \sin\theta a_z$. Thus, the real part of the electric field intensity in Equation 7.30 is in the $\hat{a}_{\phi_m}$ direction and is constant with respect to the angle ϕ_m. In Equation 7.30, the magnitude of the electric field intensity varies with radius inside the conductor as $j_1(k_c r)$. It is satisfying to see in Figure 7.10 that the superimposed Ansoft 3D Full-wave Electromagnetic Field (HFSS) numerical solution of the magnitude of the electric field intensity on a cross section in the middle of the sphere (in the x–z plane) shows no variation with angle ϕ_m but it does show a variation in the radial direction that is consistent with a spherical Bessel function of the first kind. Once again, the electric field intensity interior to a good conducting sphere has produced a result consistent with analytic solutions to Maxwell's equations. Note that, at this frequency, the field has not turned around in the center as it would for higher frequencies.

Reduced Dipole Moment

The magnetic dipole moment, $\vec{m} \equiv (1/2)\iiint_{V'}(\vec{x}' \times \vec{J}_c)d^3x'$, can be used to find the effective magnetic dipole moment of the total current inside the conducting sphere. To compute the magnetic dipole moment, we need the cross product $(\vec{x}' \times \vec{J}_c) = (\vec{r} \times \vec{J}_c)$ shown in Figure 7.11.

From Figure 7.11, it is clear that only the $-y$ components of the cross product survive the integral because the x-components cancel one another so that the contribution to the magnetic moment in the volume element $(r\sin\theta_m d\phi)(rd\theta_m)(dr)$ shown is given by using $\vec{J}_c = \sigma\vec{E}_c$ from Equation 7.30:

$$\left(\vec{r}\times\vec{J}_c\right)_y \hat{a}_y \approx -\sigma\sqrt{3/8\pi}E_0 a_M(1)\, j_1(k_c r)(\sin\theta_m + i\cos\theta_m)(r\sin\theta_m)\hat{a}_y \quad (7.31)$$

The total effective magnetic moment due to the sphere is the integral over the volume of the sphere:

$$\vec{m} \approx -\sigma\sqrt{3/32\pi}E_0 a_M(1)\iiint_{V'} j_1(k_c r)(\sin\theta_m + i\cos\theta_m)r^3\sin^2\theta_m\, dr\, d\theta_m\, d\phi_m\, \hat{a}_y \quad (7.32)$$

and because there is no ϕ_m dependence of the integrand $\int_0^{2\pi} d\phi_m = 2\pi$ or

$$\vec{m} \approx -\sqrt{3\pi/8}\,\sigma E_0 a_M(1)\int_0^{\pi}\int_0^{a} j_1(k_c r)\left(\sin^3\theta_m + i\sin^2\theta_m\cos\theta_m\right)r^3\, dr\, d\theta_m\, \hat{a}_y \quad (7.33)$$

and because $\int_0^{\pi}\sin^2\theta_m\cos\theta_m d\theta_m = 0$ and $\int_0^{\pi}\sin^3\theta_m d\theta_m = 4/3$:

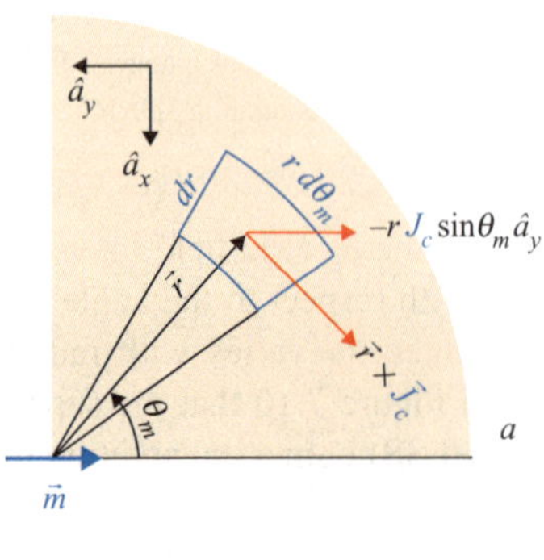

Figure 7.11 Geometric details of the cross product used to calculate the magnetic dipole moment of a good conducting sphere of radius a.

$$\vec{m} \approx -\sqrt{2\pi/3}\sigma E_0 a_M(1)\int_0^a [j_1(k_c r)]r^3 dr\ \hat{a}_y \tag{7.34}$$

For Equation 7.34 to be evaluated, the spherical Bessel function, $j_1(x) = \cos x/x - \sin x/x^2$, may be used with $x = k_c r = (1 + i)r/\delta = (1 + i)u$. In that case,

$$\vec{m} \approx -\sqrt{2\pi/3}k_c^{-4}\sigma E_0 a_M(1)\int_0^{k_c a} [j_1(x)]x^3 dx\ \hat{a}_y \tag{7.35}$$

$$\vec{m} \approx -\sqrt{2\pi/3}k_c^{-4}\sigma E_0 a_M(1)\int_0^{k_c a} \left[x^2\cos(x) - x\sin(x)\right]dx\ \hat{a}_y \tag{7.36}$$

$$\vec{m} \approx \sqrt{\pi/24}\delta^4\sigma E_0 a_M(1)\int_0^{a/\delta} \left[(1+i)^3 u^2\cos(u+iu) - (1+i)^2 u\sin(u+iu)\right]du\ \hat{a}_y \tag{7.37}$$

and $\cos(u+iu) = \cos u\cosh u - i\sin u\sinh u$ and $\sin(u+iu) = \sin u\cosh u + i\cos u\sinh u$ so

$$\vec{m} \approx \sqrt{\frac{\pi}{6}}\delta^4\sigma E_0 a_M(1)\int_0^{a/\delta} \left\{\begin{matrix}\left[u^2\sin u\sinh u - u^2\cos u\cosh u + u\cos u\sinh u\right] \\ +i\left[u^2\cos u\cosh u + u^2\sin u\sinh u - u\sin u\cosh u\right]\end{matrix}\right\}du\ \hat{a}_y \tag{7.38}$$

Expressing the electric field intensity, E_0, in the propagating medium in terms of the magnetic field intensity, H_0, in the propagating medium, $E_0 = \eta_2 H_0$ and using the surface impedance of the conductor $Z_s = (1 + i)/\sigma\delta$ the reduced magnetic dipole moment is

$$\begin{aligned}\vec{m} \approx -\sqrt{\frac{\pi}{3}}\delta^3\frac{(1+i)}{\sqrt{2}}\frac{\eta_2}{Z_S}H_0\hat{a}_y a_M(1) \\ \int_0^{a/\delta}\left\{\begin{matrix}\left[u^2(\cos u\cosh u - \sin u\sinh u) - u\cos u\sinh u\right] \\ -i\left[u^2(\cos u\cosh u + \sin u\sinh u) - u\sin u\cosh u\right]\end{matrix}\right\}du\end{aligned} \tag{7.39}$$

where $u = r/\delta$. Expressing the reduced magnetic dipole moment, $\vec{m}$, in terms of the magnetic dipole moment of a PEC, $\vec{m}_{PEC} = -2\pi a^3 H_0\hat{a}_y$, we see that the term $(1+i)/\sqrt{2} = e^{i\pi/4}$ in Equation 7.39 implies that the reduced magnetic dipole moment of a *good* conductor is 45° out of phase with the magnetic dipole moment of a PEC due to the fact that the induced electric field intensity inside a good conductor leads the applied external field by 45°. Comparing the reduced magnetic dipole moment to that of a PEC

$$\begin{aligned}\frac{\vec{m}}{\vec{m}_{PEC}} \approx \sqrt{\frac{1}{12\pi}}\frac{\delta^3}{a^3}e^{i\frac{\pi}{4}}\frac{\eta_2}{Z_S}a_M(1) \\ \int_0^{a/\delta}\left\{\begin{matrix}\left[u^2(\cos u\cosh u - \sin u\sinh u) - u\cos u\sinh u\right] \\ -i\left[u^2(\cos u\cosh u + \sin u\sinh u) - u\sin u\cosh u\right]\end{matrix}\right\}du\end{aligned} \tag{7.40}$$

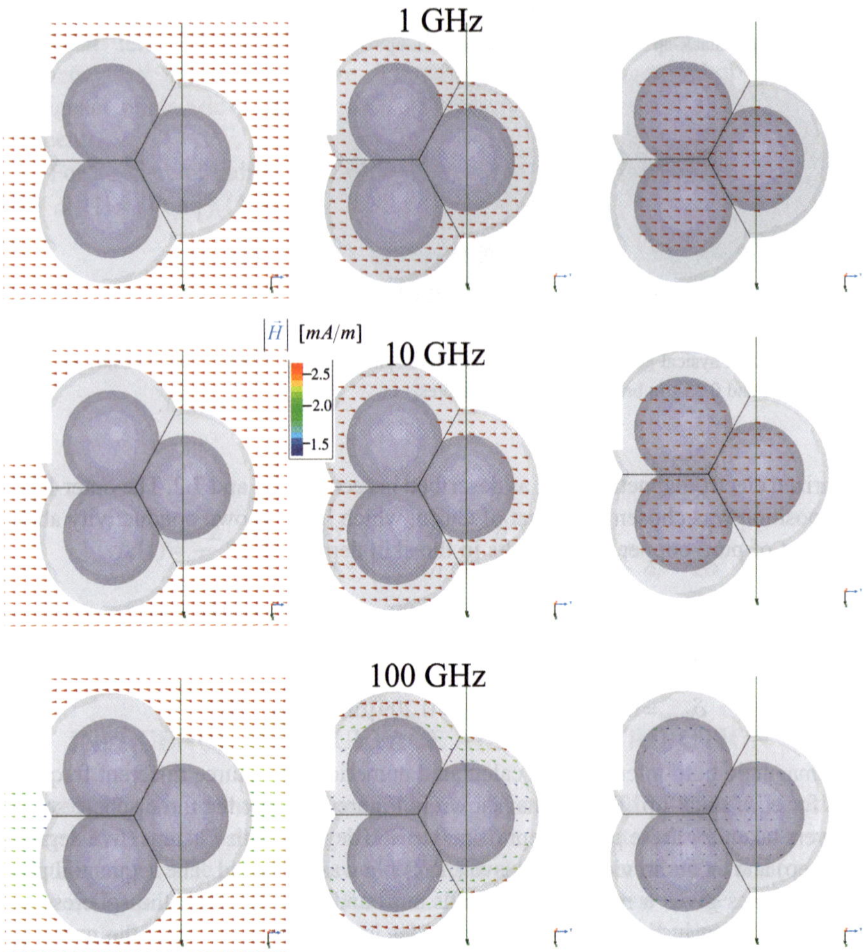

Figure 7.13 Cross section of the magnetic field intensity distribution in three regions of a of a three-copper-brass-sphere stack-up. All copper spheres have a radius of $a = 0.75$ μm, with an outer layer of 0.25-μm brass (CuZn).

at that instant when the external field lines are a maximum in their right to left direction. These field directions are separate from those outside the stack-up (left column) or inside the pure copper core (right column) because the overlap of the arrows at the boundaries is confusing. The upper-middle plot shows that the external field lines are largely unaffected by the presence of the cores or the other spheres at 1 GHz. There is only a slight decrease in the magnetic field intensity in the brass layer at 10 GHz, but we see that the external field is significantly reduced (the arrows become green) in the lower-middle cross section taken at 100 GHz. It is difficult to see in this column of plots, but the directions of the arrows also change at 100 GHz. It can

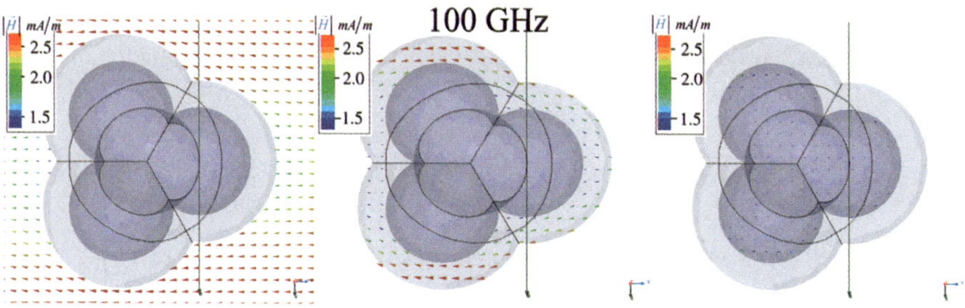

Figure 7.14 Magnetic field intensity distribution in three regions of a of a four-copper-brass-sphere stack-up. All copper spheres have a radius of $a = 0.75$ µm, with an outer layer of 0.2-µm brass (CuZn).

also be seen that the flow of the arrows around the copper cores has the same symmetry where spheres touch as for those surfaces near the exterior.

The three plots in the right column of Figure 7.13 show the magnetic field intensity and directions inside the copper core at three different applied frequencies at that instant when the external field lines are a maximum in their right to left direction. The upper-right plot shows that the external field lines are largely unaffected by the presence of the brass shell or the other spheres at 1 GHz. There is a slight decrease in the magnetic field intensity in the brass layer at 10 GHz (the arrows become green), and the swirling effect of a slow, inward propagating wave is seen in the lower-right cross section taken at 100 GHz. It can also be seen that the flow of the arrows around the copper cores is slightly different at the point of contact of the spheres as contrasted with the surface on the exterior; this last point is thus showing where the uniform field, first Born approximation begins to fail.

For the influence of a multicore stackup of spheres to be studied, another set of numerical solutions was made with four copper cores, as shown in Figure 7.12. Again, the four copper cores physically touch one another, and the brass outer layer covers this ensemble (now in a 3-D pyramid configuration). One snapshot of the fields outside, one in the brass layer, and one in the copper core are shown in Figure 7.14.

In Figure 7.14, the worst-case scenario at 100 GHz is considered. The results are similar to those of the three-copper-brass-sphere stackup, but there is a more pronounced asymmetry between the locations where the copper spheres touch and those locations near the outer edges.

7.4 SCREENING OF NEIGHBORING SNOWBALLS AND FORM FACTORS

Fields outside a Good Conducting Sphere

Section 7.1 found the fields on a flat conducting plane by considering a worst case scenario of scattering by PEC spheres, but the fields scattered by the incident electromagnetic wave on a *good* conducting sphere were calculated by using the multipole fields produced by the induced electric moment and magnetic moment given in the following equations:

$$\vec{E}_{sc} = E_0 \sum_{l=1}^{\infty} i^l \sqrt{\pi(2l+1)} \{\alpha(l) h_l^{(1)}(k_2 r)\vec{X}_l^+ + (\beta(l)/k_2)\vec{\nabla} \times [h_l^{(1)}(k_2 r)\vec{X}_l^-]\} \quad (6.132)$$

$$\vec{H}_{sc} = H_0 \sum_{l=1}^{\infty} i^l \sqrt{\pi(2l+1)} \{(-i\alpha(l)/k_2)\vec{\nabla} \times [h_l^{(1)}(k_2 r)\vec{X}_l^+] - i\beta(l) h_l^{(1)}(k_2 r)\vec{X}_l^-\}, \quad (6.133)$$

where the unknown coefficients $\alpha(l)$ and $\beta(l)$ were determined by the boundary conditions and were given in Equations 6.164 and 6.165, $\alpha(l)$ giving the magnitude of the magnetic multipole moment (transverse electric component) and $\beta(l)$ giving the magnitude of the electric multipole moment (transverse magnetic component). In that analysis, the dipoles terms were found to be dominant, so that

$$\alpha(1) = \frac{-2}{3}(k_2 a_i)^3 \left\{ \frac{3\delta}{2a_i} + \left[1 - \frac{\delta}{2a_i} - \left(\frac{\delta}{a_i}\right)^2\right] i \right\} \left[1 + \frac{\delta}{a_i} + \frac{\delta^2}{2a_i^2}\right]^{-1} \quad (6.164)$$

$$\beta(1) = \frac{-2}{3}(k_2 a_i)^3 \left\{ \frac{3(k_2^2 a_i \delta) + [k_2^4 a_i^2 \delta^2 + k_2^2 a_i \delta - 4] i}{(k_2^2 a_i \delta - 1)^2 + 1} \right\} \quad (6.165)$$

For the case of μm-radius spheres and frequencies in the range $1\,\text{GHz} \le f \le 100\,\text{GHz}$, the skin depth of copper has been found to be between $2\,\mu\text{m} \ge \delta \ge 0.2\,\mu\text{m}$ in which case the terms $k_2 a_i \ll 1$ and $k_2\delta \ll 1$ so that Equation 6.165 becomes $\beta(1) \approx 4i(k_2 a_i)^3/3$, which is the same value as that of a PEC (e.g., let $\delta \to 0$ in Equation 6.165). In the same way, Equation 6.164 becomes $\alpha(1) \approx -2i(k_2 a_i)^3/3$ in the limit as $\delta \to 0$.

Conclusion

Electromagnetic fields outside a good conducting sphere are approximately the same as electromagnetic fields outside of a PEC; the approximation made in section 7.1 was valid for spheres with radius $a_i \le 1\,\mu\text{m}$ and frequencies $1\,\text{GHz} \le f \le 100\,\text{GHz}$.

Screening of a Snowball Because of Another Good Conducting Sphere in the Same Plane

The magnetic field intensity outside a PEC was previously calculated in Equation 7.6 so if $x_i = 0$ the field intensity *in the plane* of a good conducting sphere is

$$\begin{pmatrix} H_{B,\rho} \\ H_{B,\varphi} \\ H_{B,z} \end{pmatrix} = \frac{H_0}{2}\frac{a_i^3}{\rho^3} \begin{pmatrix} 2\cos\varphi \\ \sin\varphi[6\cos^2\varphi - 1] \\ 0 \end{pmatrix} \quad (7.43)$$

and evaluated at a nearest neighbor distance, $a_i/\rho = 1/2$, so the differential fractional power change at a nearest neighbor, f_{nn}, is given by

$$f_{nn} = \left[\left| H_0 e^{i\omega x_i/u_p} \hat{a}_y + H_{B,\rho} \hat{a}_\rho + H_{B,\varphi} \hat{a}_\varphi \right|^2 - \left| \vec{H}_0 \hat{a}_y \right|^2 \right] \Big/ \left| H_0^2 \right| \tag{7.44}$$

and, if time retardation in the forward and backward directions are neglected, $\cos(\omega 2a_i/u_p) \approx 1$, and

$$f_{nn} = \frac{1}{16}\left[\sin^2\varphi\left(6\cos^2\varphi - 1\right) - 2\cos^2\varphi\right] + \frac{1}{256}\left[\sin^2\varphi\left(6\cos^2\varphi - 1\right)^2 + 4\cos^2\varphi\right] \tag{7.45}$$

The fractional change in power at a nearest neighbor in the plane of the dipole is shown plotted in Figure 7.15 as a function of the angle of the nearest neighbor relative to the direction of the induced magnetic dipole moment, $\vec{m}$.

Figure 7.15 shows that, if a nearest neighbor atom is located along (or perpendicular to) the axis of the induced magnetic dipole, $\theta_m = 0$ (or $\theta_m = \pi/2$), the square of the magnetic field intensity is *less* than it would be if the neighbor were absent. Only at angles near $\theta_m = \pi/4$, $3\pi/4$, $5\pi/4$, or $7\pi/4$ would there be an increase in the square of the magnetic field intensity (+1.6%). The blue dotted line in Figure 7.15 shows the fractional loss at a random location for a neighboring snowball.

Conclusion

The square of the magnetic field intensity at a particular snowball due to a randomly located neighbor snowball is, on the average, 3.44% *less* than it would be if there

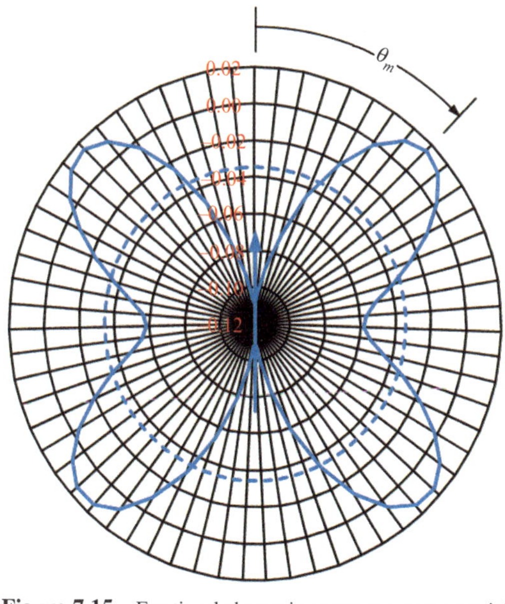

Figure 7.15 Fractional change in power at a nearest neighbor snowball as a function of angle, θ_m, relative to the induced magnetic dipole moment due to the external magnetic field intensity from a good conducting sphere.

were no neighbor. The snowballs thus *screen* the field intensity from their neighbors by 3.44% each. In this analysis, the influence of six neighbors in a randomly oriented hexagonal near-neighbor configuration would be a 20.6% decrease in the field.

Extensions of the Screening Concept

- Several approximations have been made in the above analysis of a second-order Born approximation, and the results are significant, perhaps beyond the bounds of a normal Born approximation. However, the analysis was conducted only for snowballs in the plane of nearest neighbors. What about the snowballs slightly above or below those in the plane? What about the snowballs in a typical stack-up of a pyramid "tooth" structure? Such an analysis would constitute a good research project if randomness of the incident wave direction is included.
- In applications of similar effects, some theorists would say that the total field intensity decrease is so large that the entire incident field intensity must be *renormalized* to account for all of the scattering effects and maintain conservation of charge in the medium.
- If the neighboring atoms are stacked up in a chosen periodic array, they can be located at points of higher or lower field intensity, as shown by Figure 7.15. In that case, second-neighbor (or higher) screening influences should be considered due to the multiplicative effects of larger number of neighbors. This *Form Factor* analysis due to the periodic structure of neighboring snowballs is discussed in Jackson[10] and in Kittel.[11]
- Chapter 6 focuses on snowballs of an average size even though scanning electron microscope images in Chapter 4 showed that a distribution of snowball sizes was more likely in real-world samples. In calculating the power loss due to several different average-size snowballs, it was concluded that the losses were not very sensitive to smaller sizes due to compensation by a larger number of smaller spheres and hence surface areas that could account for the same characteristic loss factors. However, it should be possible to make a geometric analysis of a typical snowball stack-up to justify a distribution of sizes.

7.5 PULSE PHASE DELAY AND SIGNAL DISPERSION

In Chapter 3, we showed that the circuit model for a low-loss transmission line gives

$$\begin{gathered}\gamma_{LL} \approx j\omega\sqrt{LC}\left(1+\frac{1}{2}\frac{R}{j\omega L}+\frac{1}{8}\frac{R^2}{\omega^2 L^2}\right)\left(1+\frac{1}{2}\frac{G}{j\omega C}+\frac{1}{8}\frac{G^2}{\omega^2 C^2}\right) \text{ so} \\ \alpha_{LL} \approx \frac{1}{2}\left(\frac{R}{\sqrt{L/C}}+\frac{G}{\sqrt{C/L}}\right) \text{ and } \beta_{LL} \approx \omega\sqrt{LC}\left[1+\frac{1}{8\omega^2}\left(\frac{R}{L}-\frac{G}{C}\right)^2\right]\end{gathered} \tag{3.10}$$

and from Equation 3.10, we saw that low loss (α_{LL} = small) implies that R must be small compared with $\sqrt{L/C}$, and G must be small compared with $\sqrt{C/L}$. We also found a general expression for phase velocity:

$$u_{p,LL} = \omega/\beta_{LL} \approx \frac{1}{\sqrt{LC}} \frac{1}{\left[1+\left(1/8\omega^2\right)\left(R/L-G/C\right)^2\right]}, \tag{3.11}$$

which leads to signal dispersion. The amount of signal dispersion can be determined from the time delay per inch, TD_{in},:

$$TD_{in,LL} = \frac{1\,in}{u_p} = (1\,in)\sqrt{LC}\left[1+\frac{1}{8\omega^2}\left(\frac{R}{L}-\frac{G}{C}\right)^2\right] \tag{3.12}$$

so if $\sqrt{LC} = \sqrt{\mu\varepsilon} = \sqrt{\mu_r\varepsilon_r}/c$ for nonmagnetic materials ($\mu_r = 1$),

$$TD_{in,LL} = (1\text{in})(0.0254\,\text{m/in})\left(\sqrt{\varepsilon_{r,eff}}/c\right)\left[1+\left(1/8\omega^2\right)\left(R/L\quad G/C\right)^2\right], \tag{3.13}$$

where we have let the term $\sqrt{\varepsilon_r} \Rightarrow \sqrt{\varepsilon_{r,eff}}$ to permit an effective permittivity in microstrip applications, where the electric field lines are partly in the propagating medium (e.g., FR-4) and partly in air. Furthermore, for $\varepsilon_{r,\,eff} \sim 3.055$, as predicted for a typical microstrip of our dimensions, the constants in front of the square bracket were found to be 148 ps, so

$$TD_{in,LL}(\text{microstrip}) \approx 148\text{ps}\left[1+\left(1/8\omega^2\right)\left(R/L-G/C\right)^2\right] \tag{3.14}$$

We also expanded the impedance of a transmission line to show

$$Z_0 = \sqrt{\frac{L}{C}}\left(1+\frac{1}{2}\frac{R}{j\omega L}+\frac{1}{8}\frac{R^2}{\omega^2 L^2}+\cdots\right)\left(1-\frac{1}{2}\frac{G}{j\omega C}-\frac{1}{8}\frac{G^2}{\omega^2 C^2}-\cdots\right)$$

and, if we again use the definition of low loss to be $R \le \omega L$ and $G \le \omega C$,

$$Z_{0,LL} \approx \sqrt{\frac{L}{C}}\left[1+\left(1/8\omega^2\right)\left(R/L-G/C\right)\left(R/L+3G/C\right)-j/2\omega\left(R/L-G/C\right)\right] \tag{3.15}$$

Conclusions

- For a low-loss transmission line in the circuit model, the propagation constant $\gamma = \alpha + j\beta = \sqrt{(R+j\omega L)(G+j\omega C)}$ will (in general) depend upon ω, and this will lead to a frequency dependent phase velocity, u_p.
- As different frequency components of a signal pulse propagate along the transmission line (with different velocities), the signal pulse will attenuate

and broaden (i.e., it will suffer distortion and dispersion). A low-loss transmission line (with nonzero R and/or G) is therefore dispersive.

The quantities α_{LL}, β_{LL}, $u_{p,LL}$, TD_{LL}, and Z_{LL} may be evaluated as a function of frequency in two ways:

1. Use electromagnetic software to compute tabular *w*-element values of the circuit model values of the paramaters *R*, *L*, *G*, and *C* at different frequencies of each of the quantities to numerically evaluate the Equations 3.10–3.15. Chapter 8 shows how software tools can be used to find tabular *w*-elements and to plot the quantities as a function of frequency. This method relies upon a computer code so users should always ask, "What assumptions were made in the generation of this computer code?" "Did it take into account impurities in the propagating medium, did it include surface roughness, and if so, was it carried out by using some empirical basis like a Hammerstad fit that has no physical basis?" These questions should prompt students or practitioners to use care in applying unknown codes to avoid incorrect predictions.
2. Use the frequency dependence of *R*, *L*, *G*, and *C* predicted by a causal, physical mechanism directly in the circuit model expressions 3.10 to 3.15 to find an analytic expression for the quantities α_{LL}, β_{LL}, $u_{p,LL}$, TD_{LL}, and Z_{LL}. This book has at one or more places considered the physical basis of all of these dependent variables.

For the ability of technique 1 to predict the attenuation, delay, and dispersion of a pulse to be appreciated, Figure 7.16 shows the simulated setup of a 15-GHz pulse sent into the measured s-parameter data of the 7-inch channel and the transmission line models developed from the snowball methodology from Figure 4.4. As can be observed, the snowball model shows good correlation to the measured data in terms of amplitude and width but arrives 4 ps earlier than the measured data. This delay corresponds to the length of the launch structure on the measured board, which was not included in the snowball model.

CHAPTER CONCLUSIONS

Second-order perturbation effects have been used to predict the following:

- A fractional change in power is lost on the flat, *good* conducting interface trace of width w and length l as a result of additional current induction on that plane caused by N_i dipole PEC spheres of radius a_i located a distance x_i below the plane (Equation 7.11). A retarded current density that depends on the size and location of good conducting spheres lags the incident current density of a perfectly flat waveguide. At the output end of a transmission line, a sequence of delayed pulses combines with the attenuated original pulse shape to create a dispersed pulse as a result of surface roughness.

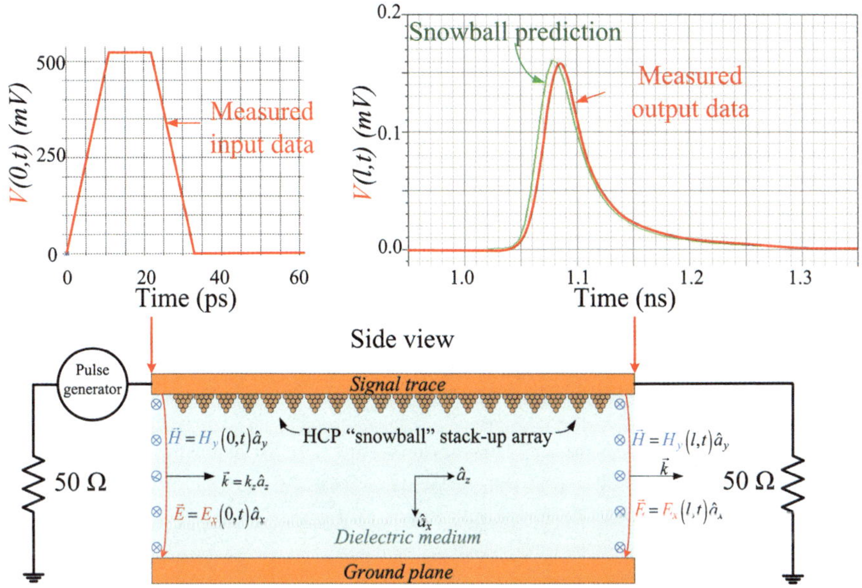

Figure 7.16 Measured data from an Agilent time domain reflectometer (TDR) at the input, $V(0,t)$, and output, $V(l,t)$, of a 7-inch transmission line with a high-profile surface roughness for a microstrip used in Chapter 6. The time domain voltage predicted by a snowball scattering model is superimposed on the output data.

- The magnitude of the magnetic dipole moment for a *good* conductor compared with the magnitude of the magnetic dipole moment of a PEC (Equation 7.40) is reduced because of the lower phase velocity of electric field intensity and magnetic field intensity that penetrates into good conductor-scattering centers.
- Alloy distributions in the surface layer of conductors cause a variable depth of electrical resistivity that influences electromagnetic field penetration skin depth and thereby influences the magnitude of electromagnetic propagation losses.
- Scattered electromagnetic fields from snowballs increase or decrease the magnitude of the incident power at nearest neighbor snowballs by a small amount (Figure 7.15). The degree of the increase/decrease depends upon orientation, and this in turn results in a form factor consequence for how snowballs are stacked up.
- Pulse output shape for a transmission line can be accurately predicted by the snowball model by using R, L, G, and C parameters chosen to match frequency-dependent power loss. An analytic form of the time-dependent pulse shape can be found in terms of Fourier components with a formulation that includes medium and roughness parameters.

ENDNOTES

1. P. G. Huray, *Maxwell's Equations* (Hoboken, NJ: John Wiley & Sons, 2009), Chapter 4.
2. Ibid., Figure 4.12.
3. Ibid., Example 4.1, Equation 4.55.
4. Ibid., Table 7.7.
5. Ibid., Equation 2.56.
6. Ibid., Equation 2.51.
7. H. B. Dwight, *Table of Integrals and Other Mathematical Data,* 4th ed. (New York: Macmillan, 1961).
8. Adefisayo (Fisayo) Adepetun, Thesis, in partial fulfillment for the degree of Master's of Science in Electrical Engineering at the University of South Carolina (2009).
9. David Aerne Thesis, in partial fulfillment for the degree of Master's of Science in Electrical Engineering at the University of South Carolina (2009).
10. J. D. Jackson, *Classical Electrodynamics*, 3rd ed. (Hoboken, NJ: John Wiley, 1999), sec. 10.1D, p. 461.
11. C. Kittel, *Introduction to Solid State Physics*, 7th ed. (Hoboken, NJ: John Wiley, 1996), Chapter 2, p. 44.

Chapter 8

Signal Integrity Simulations

LEARNING OBJECTIVES

- Define common simulation terms, techniques, and tool types
- Review transient analysis techniques
- Learn fast convolution and statistical methods
- Review applications of quasistatic field solvers
- Review applications of three-dimensional finite element field solvers
- Show how numerical solution tools are used by Signal Integrity engineers to make decisions about high-speed circuit parameters

INTRODUCTION

The foundations of Signal Integrity stand upon solutions to Maxwell's equations. In *Maxwell's Equations*,[1] the mathematical meaning of the original equations was developed along with some definitions of what constituted quasistatic solutions within *good* conductors and dielectric materials. The propagation of electromagnetic waves was then studied in several media that were defined in terms of good, medium, and poor conductors, and characteristics of those waves were considered at the boundary between materials. In the current text, applications were studied in terms of common types of transmission lines, including those involved in the fabrication of printed circuit boards. Real-world materials were introduced after which analysis of propagation was conducted for complex dielectric materials and rough surfaces. The first Born approximation permitted an approximate solution to material features such as spherical snowballs on the surface, and theoretical calculations were made and compared with measurements with good correlation. A second-order Born approximation permitted us to consider effects of multiple scattering and absorption on realistic geometries, and a scale of magnitudes was developed to show how good

The Foundations of Signal Integrity, by Paul G. Huray

Figure 8.1, the data voltage fell to zero some of the time but was slightly higher than zero at other times because of some irregular voltage source such as cross talk from other parts of the circuit. The data voltage pulses rose as high as 3.2 V for part of the cycle and only to 3.0 V for other parts of the cycle. The data pulses (light blue and dark blue) were not exactly in the same time phase with the strobe voltage pulses, as one rose slightly earlier than the other. If the rise and fall times are statistically variable (jitter in the timing) then we see the case of a two data voltage measurement in Figure 8.1. To achieve a more statistically significant set of results, we could superimpose additional voltage signals on the two-pulse set of Figure 8.1. With many such superimposed measurements, the width of the lines would overlap to such an extent that they would appear to be a blurred rising and falling set of some finite width, as shown in Figure 8.2. Here, the clock strobe voltage pulses of Figure 8.1 have been omitted. These so-called eye diagrams can be measured on an oscilloscope for real-world voltages on a high-speed circuit or simulated by using numerical analysis that incorporates cross talk and jitter from a complex circuit among its parameters. When more statistical data are superimposed in a plot like Figure 8.2, the width of the crossing points (~10 ps centered around 50 and 150 ps) will grow, and the various low- and high-voltage blurred lines will get fatter; the eye is said to be closing in such cases. For the data shown in Figure 8.2, the gray diamond of width $\Delta UI = 0.50$ and height ΔEH (eye height) = 700 mV is a timing/voltage regime where the signal has not intruded for the set of measurements or simulations shown. Thus, any circuit decision (such as the trigger of a data count) made within this regime is certain to be reliable; that is, the data signal has integrity.

In Figure 8.2, a bit rate of 100 ps is shown, and the UI is equivalent to the bit rate. A 0.5 UI results in an eye width of 50 ps, as shown by the width of the gray diamond.[4] The timing required for a reliable operating channel is 50 ps in this example. In addition, for the receiver threshold, the EH is also specified to be between EH_{upper} = 350 mV and EH_{lower} = −350 mV, resulting in ΔEH = 700 mV. Specification of the receiver parameters in this method is generally referred to as defining an eye mask. Eye masks are commonly specified at test point locations for both the transmitter and the receiver. Numerical simulations are employed to predict

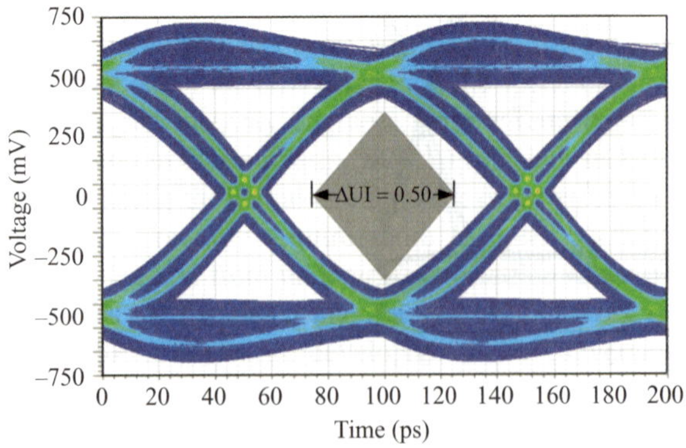

Figure 8.2 Unit interval plot where the eye width is $\Delta UI = 0.50$ with eye height between EH_{min} = −350 mV, and EH_{max} = 350 mV.

the eye width and EH, and measurements of a prototype board are then made to confirm or validate those predictions. Depending on the standard, the mask may or may not be specified for termination into a standardized load (say, 50 Ω).

Another aspect of modern signaling includes the ability to understand information from the frequency domain. Commonly, bus standards specify channel loss, cross talk, phase lock loop bandwidth, and filter information using an s-parameter matrix where the maximum insertion loss (IL) or minimum return loss (RL) is specified at a given frequency (see Figures 6.24, 6.25, 6.26 as examples). Other less known derivations such as insertion loss deviation[5] are sometimes specified to provide the reader with a deeper insight into the channels characteristics to ensure proper operation. These data are obtained through simulation and measurement and are then used in circuit simulators to determine if the system will pass a set of required margin specifications.

Historically, numerical analysis performed in the time domain used transient simulations, but, for higher-speed applications, inter-symbol interference (ISI) has become a dominant factor of the overall margin in an operating system. ISI analysis sometimes requires prohibitively long simulations into the million to billions of bits. The requirement has necessitated the need for simulation methodologies such as fast convolution and statistical techniques in lieu of repeated transient simulations because these types of simulations can quickly provide millions to billions bits of data to the user. These techniques make the assumption that the system is linear time invariant (LTI) because of the use of superposition and convolution and will be discussed in the next section.

One of the fundamental building blocks within the signal integrity engineer's simulation toolbox is the tabular transmission line, often referred to as the tabular *w*-element. The *w*-element quickly emulates the effects of loss in transmission line structures up to at least 30 GHz.[6] This element can reproduce the effects of frequency-dependent dielectrics and surface roughness and then be translated into a distributed network of passive resistor, inductor, conductor, and capacitor (RLGC) components, but the user of this technique must be warned that there is no physical basis for the *w*-element; it is a parameterization of the outcomes.

Initial modeling of the transmission line requires a two-dimensional (2-D) electrostatic field solver in which the transmission line geometry may be accurately described. Software tools such as Q3D Extractor™ by Ansoft, a subsidiary of ANSYS, Inc. of Pittsburg, PA incorporate a 2-D electrostatic field solver that can accurately extract RLGC parameters for transmission line structures and then generate the required "net list" for the tabular *w*-element that may be used in a SPICE sub-circuit model (Figure 8.3).

Real systems such as vias and connectors often include three-dimensional (3-D) geometries that cannot be modeled by using simple 1-D and 2-D techniques (Figure 8.4). For the electromagnetic interactions to be more accurately captured, a more complete physical geometry must be modeled. This requires the use of more complex simulation tools that model and numerically solve Maxwell's equations. Within industry and academia, the finite element method (FEM) has been proven to provide the robustness required to accurately predict the electromagnetic fields

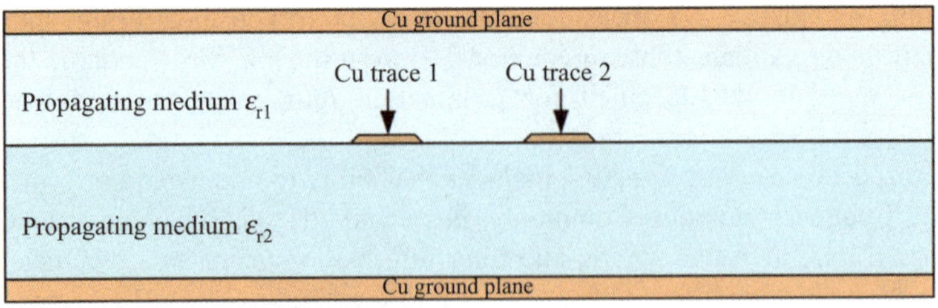

Figure 8.3 Example of differential stripline geometry drawn in Q3D™ employing a two-dimensional electrostatic field solver used for exporting to tabular *w*-element tools.

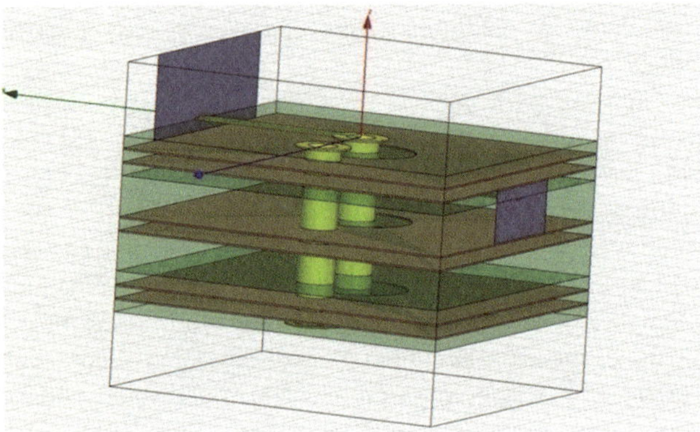

Figure 8.4 Example of signal and ground via in a multi-layer PWB drawn in HFSS™.

within arbitrary 3-D shapes primarily because of the tools' ability to mesh arbitrary shapes using tetrahedrons. High Frequency Structure Simulator (HFSS™, also by Ansoft, Inc.) uses the FEM technique along with an adaptive meshing algorithm, which automatically creates the required mesh based on several user inputs such as the frequency band of interest and the maximum deviation allowed between sweeps of s-parameters. Once solved, this information is generally exported as s-parameters by using the Touchstone[7] file format developed by the Agilent Corporation of Santa Clara, CA to be used in circuit simulation.

Simulation tools that combine several modeling techniques are becoming more common within the signal integrity discipline. The primary motivation is to simulate complex structures that have large geometries compared with the wavelength (e.g., server board structures operating at 5–10 Gb/s). The resulting hybrid simulation tools have enabled simulation of entire printed wiring boards (PWBs) in an acceptable run time (less than 1 day). Typically, these types of simulation tools are utilized post layout and extracted into the simulation environment from computer aided drawings (CAD). Another interesting benefit of these hybrid simulation tools is the ability to analyze interactions between a system's signal integrity, power integrity, and electromagnetic compatibility/interference. Previously, these features were

(a) Physical system (b) SIwave™ model

Figure 8.5 (a) Xilinx Virtex-4 FPGA physical test board; (b) extracted test board using the SIwave™ software hybrid simulation tool by Ansoft at Pittsburg, PA.

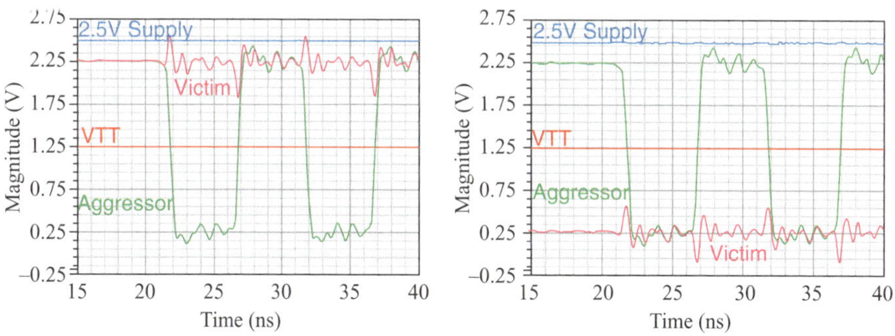

Figure 8.6 Circuit simulation showing the effects of simultaneous switching noise from a 32 bit SDRAM memory interface using the extracted hybrid model from SIwave™ within the Nexxim™ circuit simulation environment by Asoft Corporation of Pittsburg, PA.

analyzed independently, if at all. The hybrid results are often exported as SPICE "net lists" or Touchstone s-parameter files for use with circuit simulators. A common problem that necessitates this type of analysis occurs with memory applications where simultaneous switching noise is a problem caused by multiple drivers switching off the same power and ground sources at the same time. Noise is often introduced into the power delivery system that may have significant impact on timing margins and silicon reliability.

An example[8] of the effects simultaneous switching output (SSO) noise can have on signals is shown below in Figures 8.5 and 8.6. This example shows the Xilinx, Inc. of San Jose, CA Virtex-4 FPGA test board and its extracted model using a hybrid simulation tool. The results of the 32-bit synchronous dynamic random access memory (SDRAM) interface show significant coupling from aggressors onto a victim signal and slight coupling onto a 2.5-V power supply. The extracted model was simulated in the time domain by using a circuit simulation tool.

Conclusion

Different modeling tools using 2-D, 3-D methods, and hybrid techniques provide the pieces required to perform parameterized system-level simulations within the time domain using a circuit simulator.

8.2 CIRCUIT SIMULATION

As discussed in sections 3.7 and 3.8, PSpice was developed by the Electronics Research Laboratory of the University of California at Berkeley and made available to the public in 1975.[9] Donald O. Pederson could be called the father of SPICE[10] because of his research and commitment that information in the public domain should be used by the engineering community. Many electrical engineering undergraduate programs require the usage of a SPICE simulator, while some use PSpice or Multisim created by the National Instruments Electronics Workbench group of Austin, TX, which have been available at no cost in student versions.

As described in section 8.1, eye diagrams are useful tools to help answer one of the fundamental questions of signal integrity: "If a sequence of ones and zeros is transmitted into a channel, separated in time by a specified UI, what are the chances of *correctly* detecting the data at a given point within the channel?" As mentioned above, an eye diagram takes the waveform generated at a selected location within the channel and overlays multiple copies of the data, each shifted in time by UI. A simple example of an eye diagram is shown in Figure 8.7 to demonstrate its usefulness. Here, a 50 Ω, single-ended driver is feeding a 5.5-inch stripline (parameterized as a tabular *w*-element) on FR-4 with $\varepsilon_r = 3.321$ and $\tan\delta = 0.0231$ at 1 GHz using a resistor capacitor (RC) load for the receiver with $R_{load} = 50\,\Omega$ and $C_{load} = 5\,pF$, as was shown in section 3.8.

Fundamentally, the purpose of an eye diagram is to indicate the allowable window for reliably distinguishing bits from one another at the receiver end. There is a nonzero eye width required because receive jitter makes the time location of the bit imprecise, and setup and hold times also increase the width of the margin window. The required height of the window is usually specified by the noise margin

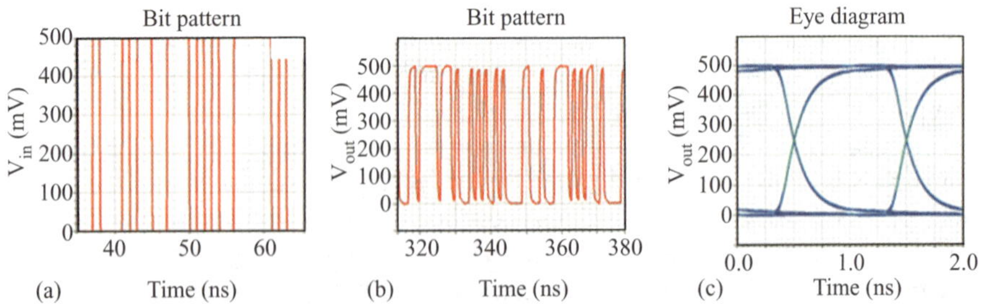

Figure 8.7 Creation of an "eye diagram": (a) the input bit pattern, (b) the output bit pattern at the RC load, and (c) the eye diagram at the RC load.

of the receivers. The required width is often determined by the performance of clock data recovery, phase locked loop, and phase interpolator characteristics.

The number of times a waveform "violates" or crosses into a given margin window is an important metric because it predicts a signal logic error. This is the bit-error rate, or BER, which may be determined by

$$\text{BER} = \frac{\text{Number of Errors}}{\text{Number of Bits}} \tag{8.1}$$

Traditionally, eye diagrams were measured in the lab, or parametrically simulated in the way described above: The waveforms were overlaid with multiple time-shifted copies to form an output like that shown in Figure 8.2. However, as speeds increased, particularly in the case of high-speed serial links, the acceptable BER required a significant decrease (often to a level of 10^{-9}–10^{-12}). This decrease, in turn, led to a great increase in the number of bits (10^{9}–10^{12}) that were needed to be simulated in order to predict the BER. Such lengthy transient simulations are impractical and have led to predictive techniques that rely upon the assumption of linear time invariance.[11]

8.3 TRANSIENT SPICE SIMULATION

SPICE transient analysis combines traditional circuit analysis methods learned in undergraduate Electrical Engineering courses with microprocessors that have the ability to perform millions to billions of numeric calculations per second. This allows for an extremely accurate analysis of complicated nonlinear circuits that represent actual circuit behavior. A basic and brief overview of SPICE transient analysis is shown in the flow chart of Figure 8.8 to help the reader understand the differences and limitations between techniques such as fast convolution analysis, statistical transient analysis, and peak distortion analysis. Readers are referred to *Computer Design Aids for VLSI Circuits*[12] and *The Designer's Guide to SPICE & SPECTRE®*[13] for a more in-depth discussion on SPICE analysis.

The first step to performing a SPICE transient analysis is the creation of the "net list" whether performed by means of a graphic user interface or by a text editor using the SPICE programming language.

The graphical creation of the schematic, Figure 8.9, produces a SPICE syntax "net list," as described in section 3.7, that is then parsed to create a matrix of unknown equations, as shown in Figure 8.10.

After the SPICE "net list" is created, a direct current (DC) solution is computed. The "net list" is comprised of nonlinear differential equations assuming capacitor currents, and inductor voltages are zero (this is acceptable because the derivative of a constant is zero).

$$i_L = L\frac{di(t)}{dt} \text{ and } v_C = C\frac{dv(t)}{dt} \tag{8.2}$$

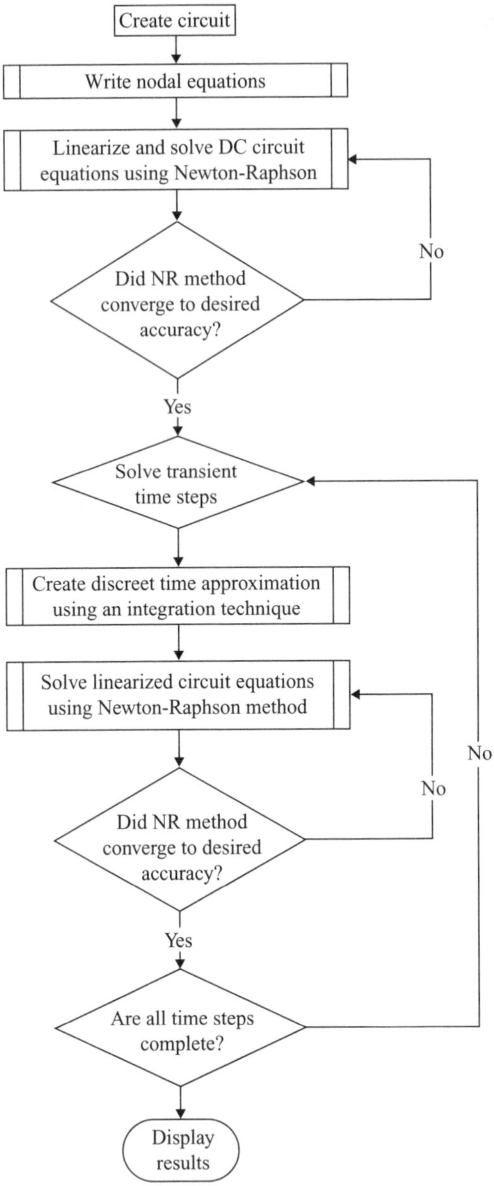

Figure 8.8 Flowchart detailing the steps required to produce transient analysis results.

The nonlinear differential nodal equations are discretized by using the Newton–Raphson (NR) method to form a series of linear algebraic equations that are generally solved by using a form of Gaussian elimination (GE). The NR method repeats itself until the convergence criteria set forth by the user are met and a DC operating point has been established.

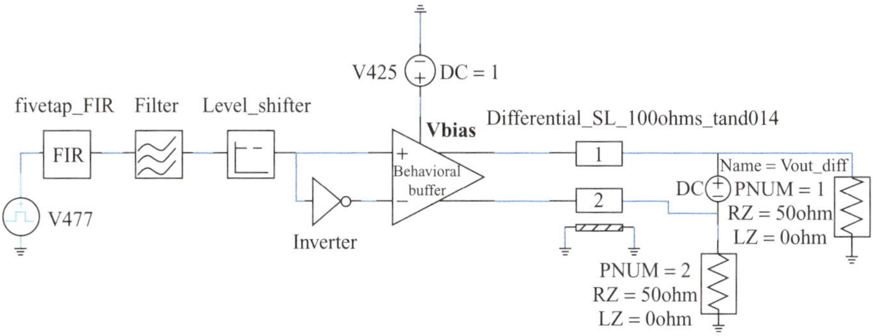

Figure 8.9 Graphical implementation of a high-speed serial driver using Ansoft's circuit simulator Nexxim™.

```
**********************************************************************************************
*                                   Nexxim netlist                                           *
*                              created by Ansoft Designer                                    *
**********************************************************************************************
.option PARHIER='local'
.param _$pre=0
.param _$post1=0
.param _$post2=0
.param _$post3=0

* begin toplevel circuit
.global G_0

.inc E:\Program Files\Ansoft\Designer4\Examples\Nexxia Circuit\\circuit_models.inc
.param UI=1.25e-010
.param trf=1.25e-011
.param pre_lvl=0
.param post1_lvl=0
.param post2_lvl=0
.param post3_lvl=0
.param Trise=1e-011
.param Tfall=1e-011
.param Period=1e-010

.inc "E:\Program Files\Ansoft\Designer4\Examples\Nexxim Circuit\Differential_SL_100ohms_tand014.sp"

.inc "E:\Program Files\Ansoft\Designer4\Examples\Nexxim Circuit\SL_85ohm_7v_6s.sp"

X410 net_9 net_0 "fivetap_FIR" UI='UI' pre='$pre' post1=''_$post1' post2='_$post2' pody3='_$post3'
X411 net_0 net_4 "FILTER" transition_time='trf/UI' crf=1e-012 rtune='10*trf/UI/0.2'
X412 net_4 net_6 "LEVEL_SHIFTER" sumtaps='1 – (pre_lvl + post1_lvl + post2_lvl + post3_lvl)'
X414 net_6 net_47 "INVERTER"
V425 net_10 0 DC=1
V435 net_45 net_46 0 Port1 Port2 0 N=2 L=0.75 TABLEMODEL=Differential_SL_100ohms_tand014
X464 net_6 net_47 net_45 net_46 net_10 "CML_DRIVE" imax=0.02 Rtermp=50 Rtermn=50 ctermn=7.5e-013 ctermp=7.5e-013
V477 net_9 0 PULSE(0 1 0 'Trise' 'Tfall' 'Period/2 - Trise' 'Period')

RPort1 Port1 0 PORTNUM=1 RZ=50 IZ=0
.PORT Port1 0 1 RPort1
RPort2 Port2 0 PORTNUM=2 RZ=50 IZ=0
.PORT Port2 0 2 RPort2

* end toplevel circuit
.end
```

Figure 8.10 Top-level simulation program with integrated circuit emphasis (SPICE) "net list" for a schematic. Note that SPICE is a hierarchical programming language and the complete "net list" calls several subcircuit routines using the "x" call.

The initial conditions of a transient solution are generally taken from the DC solution computed in the previous step unless specifically provided by the user. Options are available to allow a user to input initial conditions; the reader is referred to Kundert[13] for further discussion. The circuit equations must undergo a discreet time approximation to allow for a solution to the nodal equations. This is why a user selects the time step for the circuit simulator to aid it in evaluating the time domain derivative. Numerically integrating the differential equation is commonly referred to as the integration method. Several common numerical integration techniques may be utilized, while the most commonly implemented method is trapezoidal numerical integration.[14] However, many practical circuits have parasitics that are decades apart in value. For instance, a common "net list" may include bypass capacitors that have a value of 0.1 μF but, at the same time, have silicon parasitic capacitances that are approximately 0.1 pF. The difference of six decades is referred to as a *stiff* problem. A stable method such as the backward difference formula, also referred to as Gear's method,[15] and the numerical differentiation formula[16] are well suited for the integration of *stiff* differential equations by using a variable time step.

Once the discretized problem is created, all the nonlinear circuit elements must be linearized around an operating point by means of the NR method similar to that discussed for the DC solution. Then Gaussian elimination or a modified form called LU decomposition is performed to solve for the unknowns. The solutions are checked against the convergence criteria, and, if convergence is met, the solution is saved. The process repeats itself for all time steps until, finally, the complete solution is saved and displayed. It is important for the reader to understand that the intent of creating discretized linear elements is to ensure that traditional linear algebra techniques can be used to solve for the unknowns at multiple time steps.

8.4 EMERGING SPICE SIMULATION METHODS

Simulation methods that take advantage of linear time invariance (LTI) are becoming commonplace to create solutions for high-speed serial applications. These methods combine a single transient step or pulse response to determine the circuit's system response by convolving an input bit pattern with the derivative of the step or pulse response. The assumption of a LTI system is used often in circuit theory because it greatly simplifies the mathematics and allows for quick answers that offer acceptable accuracy. Using superposition within a circuit allows for a large reduction in circuit simulation time. Many nonlinear and time-variant systems are approximated into simpler, special-case LTI systems for this purpose. The limits of these approximations in the case of some important, signal integrity-related circuitry will be discussed along with the tools used to analyze these circuits.

A linear system is one in which the output is a linear function of the input. In the time domain, this means that the signal at each point in time is multiplied by the same constant. In the frequency domain, it means that, for a given input at a certain frequency, no other frequencies can be generated; the output frequencies are the same as the input frequencies, whose amplitudes may be different.

The limitations of a linear system can be illustrated by a simple example of a single amplifier, with a single sinusoidal frequency at the input. If the amplifier is operating linearly, the output will also be sinusoidal, with only a change in the amplitude. Once the amplifier reaches a certain maximum output voltage, the signal will start to "clip" in the time domain, which corresponds to the generation of frequencies at the output that did not exist at the input. Even though an amplifier almost always exhibits this behavior, the nonlinearities can be restricted in such a way as to make them insignificant.

A linear system possesses the property of superposition; in other words, the system possesses both the additive and homogeneity properties; that is, if

$$\begin{aligned} X_1(t) &\Rightarrow Y_1(t) \\ X_2(t) &\Rightarrow Y_2(t) \end{aligned} \tag{8.3}$$

then, by the additive property,

$$X_1(t) + X_2(t) \Rightarrow Y_1(t) + Y_2(t) \tag{8.4}$$

and by the homogeneity property, $aX_1(t) \Rightarrow aY_1(t)$, where $\boldsymbol{a}$ is a constant.[17]

A time-invariant system is one in which the output does not depend on the time, except for a shift or delay in time. In the time domain, the input is exactly recreated at the output, except for a shift in scale or time:

$$X_1(t-\tau) \Rightarrow Y_1(t-\tau), \tag{8.5}$$

where τ is constant.

8.5 FAST CONVOLUTION ANALYSIS

One emerging simulation method is fast convolution analysis. This technique performs a transient SPICE analysis for a step (rising and falling) or a pulse. The derivative of the transient response is then convolved with an input bit pattern to determine the system response. The principle of superposition is used with convolution to create the system response and requires the system to be LTI for accurate analysis (Figure 8.11).

This method provides a great savings in simulation time; a transient solution took 49 min and 30 s, while the fast transient analysis took 13.7 s. Strictly speaking, the entire system must be LTI in order to obtain an accurate analysis. However, Mellitz et al.[11] and Barnes et al.[18] have shown that the assumption of linearity can be relaxed for differential buffers and receivers without a great loss of accuracy (Table 8.1).

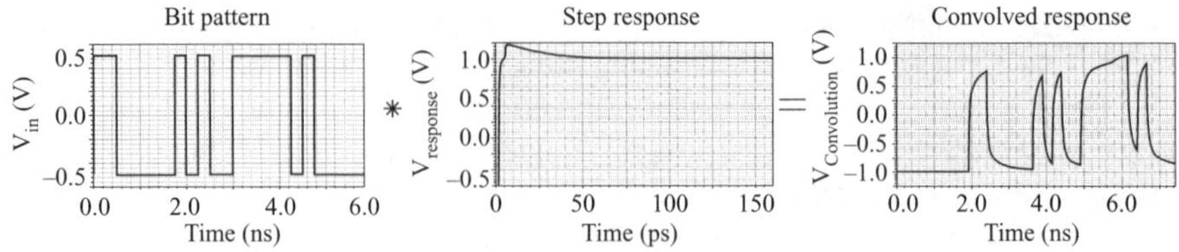

Figure 8.11 Fast transient analysis. The input bit pattern is convolved with the derivative of the step or pulse response of the system yielding the system response.

Table 8.1 Differences between fast convolution and transient solutions for a 12-inch transmission line with via transitions. The output resistances of each driver buffer leg is complementary and varies between $10\,\Omega \le R_{out} \le 90\,\Omega$

Trace characteristics	Transient	Fast transient	% Difference
Simulation time(s)	2970	13.7	198
Bits simulated	98,304	98,304	0
UI (ps)	125	125	0
Vdiff input swing	1	1	0
Eye level zero (mV)	−225	−217	3.86
Eye level one (mV)	205	210	−2.58
Eye amplitude	430	427	0.746
Eye height (mV)	369	369	0.076
Eye signal to noise ratio (SNR)	21.1	21.9	−4.12
Eye opening factor	0.942	0.954	−1.27
Eye width (ps)	108	110	−2.30
Eye P2P jitter (ps)	15.3	14.3	6.78
Eye root mean square (RMS) jitter (ps)	2.87	2.45	15.7
Eye rise time (ps)	46.0	45.85	0.392
Eye fall time (ps)	46.0	45.8	0.440

Another simulation technique uses statistical methods with a step or pulse response to determine the BER of the system. The concept behind statistical analysis is that, in order to generate an eye diagram and bathtub curve for a channel, including the effects of jitter, it is not necessary to run a long transient simulation, implied by the required BER of the channel. By combining the statistics of the bit stream with the variation in transitions caused by jitter, it is possible to generate the information needed with much less computation time.[19]

First, consider a particular point on the step response waveform described above. That point is the sum of all the voltage excursions caused by the various transitions from high to low or low to high at bit interfaces before the time point in question. This is where the statistical concept may be applied. Rather than calculate the voltage excursions for all possible combinations of bits, a probability distribution may be calculated for the sum of all the excursions.

To explain this concept, consider this in terms of what are known as "cursors," the deviations from the ideal waveform. If the channel were perfect, then the response would be identical to the input; the voltage at each point in time would be either +1 V or −1 V, depending on the bit in question.

Referring to Figure 8.7b, compare what the waveform would look like if there were a very large number of 1 s in a row, versus a large number of 0 s, followed by a transition to a 1. In the first case, the waveform would have time to settle to its correct final value of +1 V. In the second case, the waveform would be in transition from −1 V to +1 V because of imperfections in the channel. The difference between those two waveforms is a "cursor," as shown in Figure 8.12.

ISI can now be calculated statistically. Assume that the current bit is a 1. This can happen in two different ways: either the previous bit was a 1, in which case there was no transition between them, or the previous bit was a 0, in which case there was a low-to-high transition between the bits. The probability of a transition is 0.5 if the bits are independent. In the first case, the cursor is 0; in the second case, it has some significant negative value corresponding to the slow rise time of the step response at the end of the channel. Statistically, therefore, the waveform has a 50% chance of having no deviation, and a 50% chance of a negative deviation, equal to the cursor value.

It is clear that this process may be performed both forward and backward, considering each "postcursor" and precursor in turn. The primary difficulty is that, even if the bits are independent, the transitions are not. A high-to-low transition cannot be followed by another high-to-low transition; there must be a low-to-high transition in between. However, this constraint can be handled by the appropriate application of conditional probabilities. The results are probability distribution

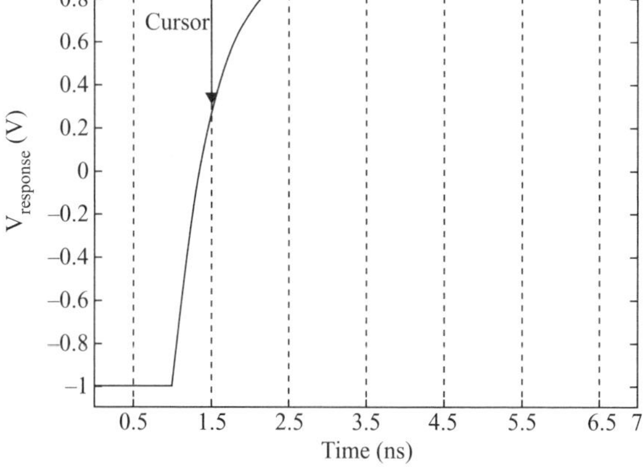

Figure 8.12 Determination of cursors from a rising edge of an RC circuit.

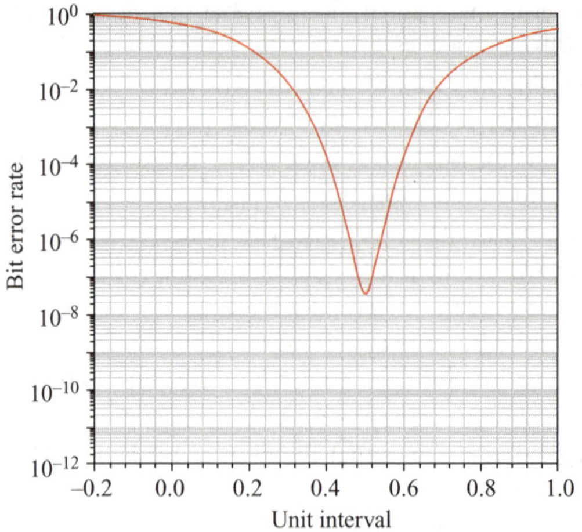

Figure 8.13 Bathtub curve showing bit error rate.

functions for the voltage value of the waveforms of high and low bits (see *Maxwell's Equations*, Appendix A). These functions can be combined into cumulative distribution functions, contours of which give the traditional eye diagram.

Another measure of the signal integrity of a channel is the so-called "bathtub curve" (Figure 8.13). This is a measure of the width of an eye opening at a certain voltage level, usually halfway between the high- and low-voltage levels. The bathtub curve turns out to be particularly easy to generate by using statistical methods: it is simply a slice through the cumulative distribution function determined above.

An advantage of the statistical technique is that the effects of random transmit jitter may be easily incorporated, without resorting to Monte Carlo methods. This is accomplished by the application of the Gaussian probability distribution function to the time location of each cursor. The cursors are, therefore, not just impulses in voltage but are spread out because of the Gaussian distribution in time, as shown in Figure 8.14.

The advantages of edge-based over pulse-based statistical eye calculations are primarily in two areas. First, by separating the rising and falling edges of the pulse, random jitter is truly random because the timing of the edges can be statistically independent. Additionally, effects of duty cycle distortion may be included, which can cause the width of pulse to vary. Another advantage is that the rising edge of a buffer can have different characteristics over that of a falling edge.

8.6 QUASI-STATIC FIELD SOLVERS

Another important type of signal integrity simulation tool is the class of quasistatic field solvers. These types of field solvers restrict the frequency of analysis to the regime of *good* conductors where charge redistribution is fast enough to neglect time

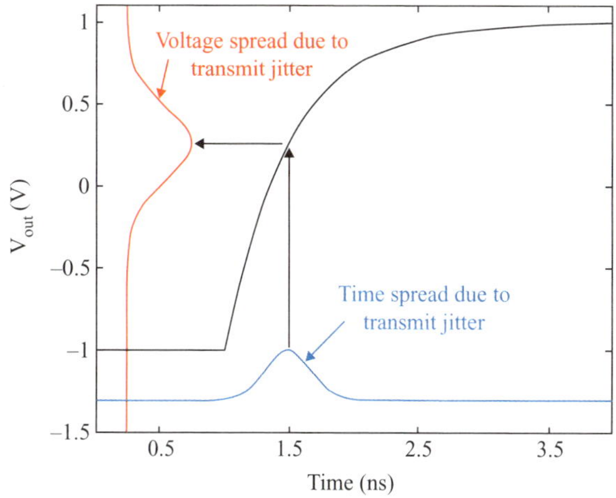

Figure 8.14 Pictorial representation of transmit random jitter smearing for statistical analysis.

derivatives of the electric field intensity so that $\vec{\nabla} \times \vec{E} \approx -\partial\vec{B}/\partial t$ and $\vec{\nabla} \times \vec{H} \approx \vec{J}$. In most of these applications, linear, isotropic media constitutive relations $\vec{D} = \varepsilon\vec{E}$ and $\vec{B} = \mu\vec{H}$ are used Quasistatic solutions are important tools to a signal integrity engineer because, when used properly, solutions are provided that yield 2-D and 3-D field solutions with minimal computer resources. Ansoft's Q3D Extractor™ simulation tool consists of a quasistatic 2-D and 3-D field solver using a graphic user interface to create the geometries to be solved.

In order to create an electromagnetic field solution, Q3D Extractor™ employs a combination of the FEM and the method of moments (MoM). In general, the FEM divides the full problem space into thousands of smaller regions and represents the field in each region (often referred to as an element) with a local function. The MoM divides up the surface of conductors and dielectrics into many triangles to represent the charges and currents on those surfaces. The choice of which method is used depends on the quantity being calculated. DC conductor problems use the FEM to model current flows within the conductors, while inductance and capacitance problems are solved by using the MoM.[20] Note that the simulation tool handles DC conduction problems along with AC conduction problems. Data rates in the multi-gigabit per second range may suffer because of the quasistatic nature of the 3-D field solver.

In addition to the visualization of the quasistatic electromagnetic fields, a SPICE representation of the circuit behavior may be created that can be solved by using circuit simulation techniques (Figure 8.15).

2-D quasi-static field solvers allow visualization of multiple conductors from within a PWB. They provide a means for creation of accurate RLGC matrices that can be used with circuit simulators. These RLGC matrices have the ability to create transmission line models that account for frequency-dependent effects such as dielectric dispersion and surface roughness losses, as covered in previous chapters.

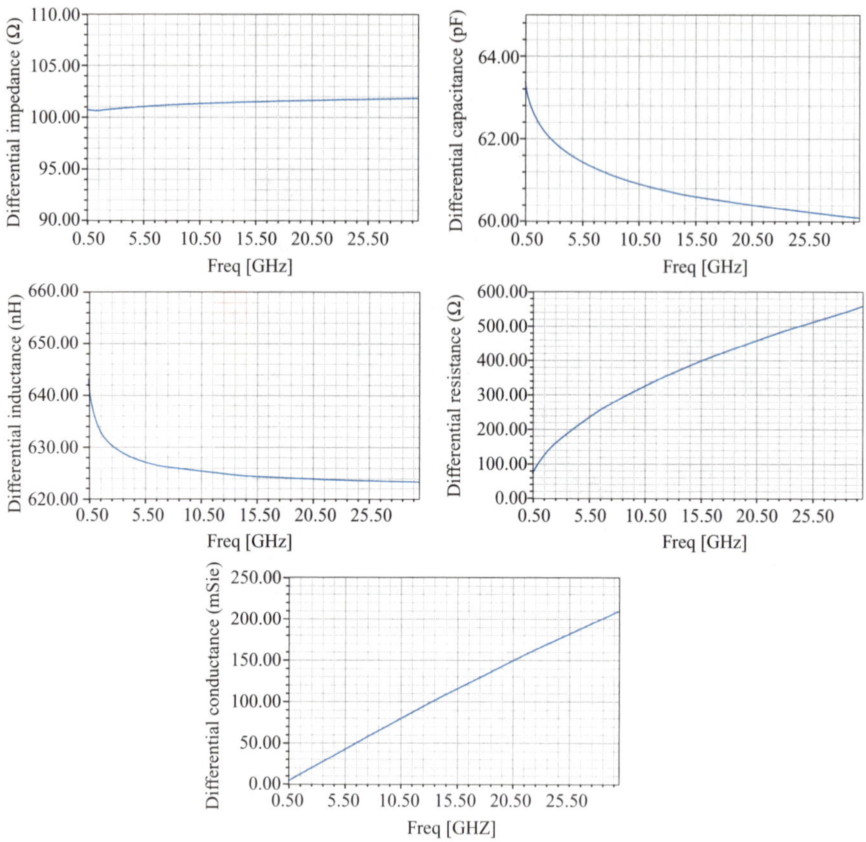

Figure 8.18 Plot of the victim's circuit parameter dependence over frequency.

$$Z_{0diff} = Z_{0odd1} + Z_{0odd2} = \sqrt{\frac{R+j\omega L}{G+j\omega C}}\Bigg|_1 + \sqrt{\frac{R+j\omega L}{G+j\omega C}}\Bigg|_2 \tag{8.6}$$

This model does not account for any surface roughness attenuation. The loss of a real transmission line would thus be greater than that shown in Figure 8.19. A time domain pulse response for the tabular *w*-element described above is shown in Figure 8.19; notice that there are not causality issues at the launch of the waveform (the waveform starts at 0 V and returns to 0 V).

8.7 FULL-WAVE 3-D FEM FIELD SOLVERS

A characteristic of gigabit per second (and beyond) simulations is the modeling of discontinuities caused by 3-D geometries. One way to accurately capture the

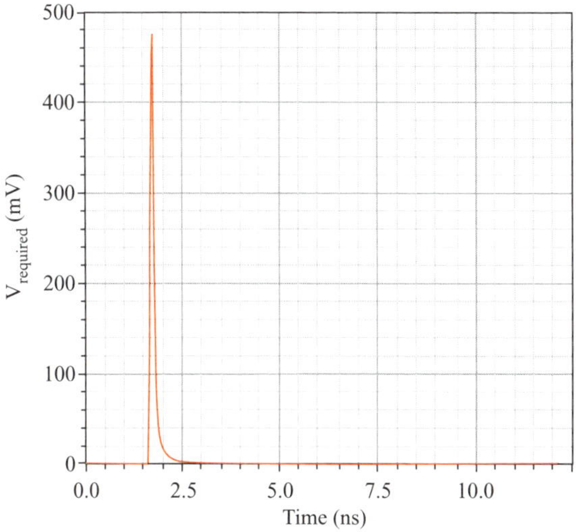

Figure 8.19 10 Gb/s pulse response to the tabular W-element shown in Figures 8.16 through 8.18. The length of the transmission line was set to 10 inches, with a rise time of 15 ps and a fall time of 15 ps.

effects of complicated 3-D structures is through the use of full-wave 3-D field solvers. This section will concentrate on a very specific type of simulator that uses the FEM to create electromagnetic models of arbitrary geometries. For an in-depth introduction to computational electromagnetics and the FEM, the reader is referred to *Computational Electromagnetics for RF and Microwave Engineering*.[24]

Components that are commonly created in 3-D analysis for signal integrity are vias, connectors, and packages. These components are directly affected by the *z*-direction and therefore require more synthesis than can be obtained by using a one-dimensional or 2-D modeling approach. This section will focus on Ansoft's HFSS™. HFSS™ is not limited to solving signal integrity problems because the meshing algorithms may be applied to electromagnetic problems such as antennae designs, RF designs, and filter synthesis (Figure 8.20).

The basic operation is that a tetrahedral mesh that conforms to the arbitrary shape of the object to be modeled is created. Depending on the basis function used, Maxwell's equations are solved at the intersection of the tetrahedron, ensuring that the boundary conditions are satisfied. This means that the accuracy of the results is directly related to the mesh size and number of tetrahedron within the model.

The strength of HFSS™ is that the simulator will create an initial mesh that can be adaptively refined without the user having to guess what type of mesh will provide an accurate solution as shown in Figure 8.20. The adaptive mesh converges based upon the criteria set forth by the user. Generally, the convergence criteria are related to the maximum delta in the s-parameters because of successive mesh solutions.

Simulation of via structures is important in order to account for the interaction of the electromagnetic waves in the PWB. Several different solution types are avail-

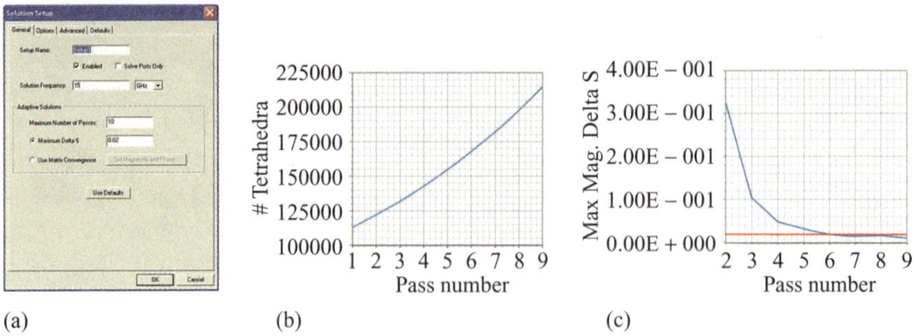

Figure 8.20 (a) Setup of users input for the adaptive mesh refinement process in HFSS^{TM}; (b) plot of the number of tetrahedron increases for each successive mesh refinement; (c) plot of the magnitude of delta s converges for successive mesh refinements.

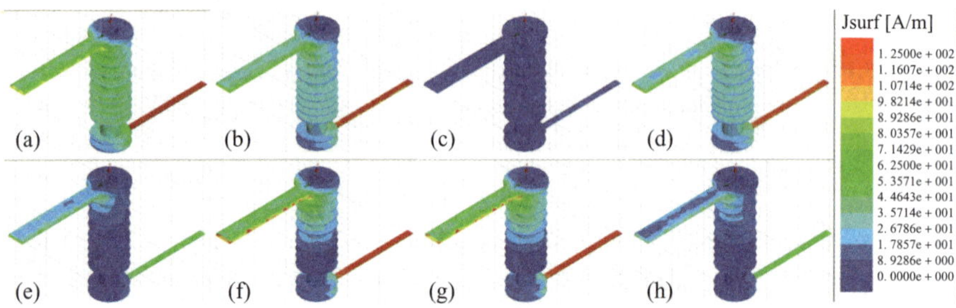

Figure 8.21 (a–d) Surface current on a simple via cross-section at 10 MHz with phase values of 0°, 45°, 90°, and 135°, respectively; (e–h) surface current on a simple via cross section at 1 GHz, with phase values of 0°, 45°, 90°, and 135°, respectively.

able, Driven Modal, Driven Terminal, and Eigenmode Analysis, along with two types of port solutions, Wave Port and Lumped Port; for tutorials on these terms see http://www.ansoft.com/ots/utraining.cfm. In order to see voltages and currents, a driven terminal solution is chosen in Figure 8.21. However, the selection of a port type is important as well. For non-TEM solutions such as a microstrip, a wave port will yield better solutions, assuming the port is set up correctly. However, wave ports are generally limited to an exterior face of the object that drives the need for lumped ports for interior analysis.

In order to determine if the wave port has been correctly set up, a "port only" solution should be run, which will allow the user to look at the electric field lines. The boundary of the wave port can be considered to be a perfect electrical conductor, which forces the any remaining electric fields tangential to the port edge. In order to ensure accurate simulations, the amount of energy forced by the port edges should be minimal (ideally zero). Shown in Figure 8.21 is a simple thru hole via model with 11 PWB layers. Each PWB layer is created by using a frequency-dependent FR-4 dielectric material with $\varepsilon_r = 4.4$ and $\tan\delta = .02$ at 1 GHz.

For Figure 8.21, an analysis was performed from DC to 10 GHz, looking at a simple via model. This model shows the surface current on the via barrel and microstrip sections. Notice the change in the distribution of surface current between 10 MHz and 10 GHz.

Another simple example, shown in Figure 8.22, is a differential via with two grounding vias offset by 20 mils to each side. This via design contains some large cavities within the PWB, which is common when filler material is needed to meet the common PWB thicknesses, for example, 0.225 in.

The traces start as microstrips and end as striplines coming out on layer 3. This leaves a very large via stub that can cause reflections that in turn cause discontinuities, as shown in Figure 8.23. The differential return loss and insertion loss show resonances occurring at 12.57, 27.25, 7.27, and 22.04 GHz, respectively. Notice that, at DC, the return loss is zero, showing that the differential via exhibits an impedance of 100 ohms differential, but, as frequency increases, the matching becomes poor, causing large discontinuities for signals flowing through theses vias because of the via stubs. Resonances in the return loss yield better matching and lower insertion loss, which is favorable; however, the resonances in insertion loss yield additional attenuation of the output signal along with poor impedance matching.

Further investigation of the surface current provides insight to what is occurring within the via. At 1 GHz, the insertion loss is less than −1 dB, with the return loss of around −14 dB and most of the current passes through the signal lines; however,

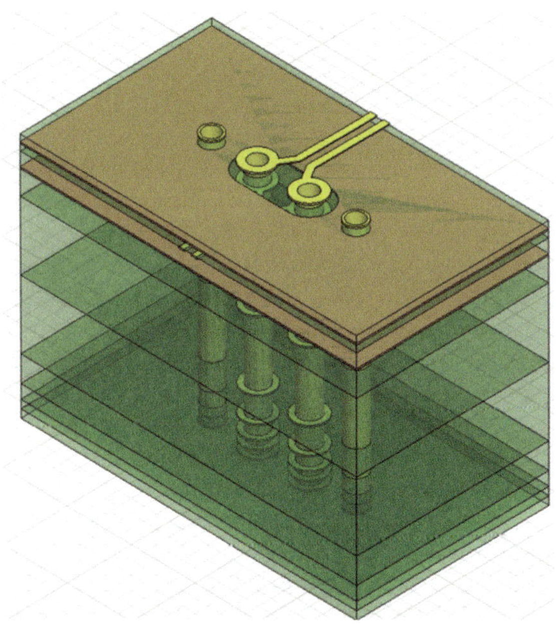

Figure 8.22 Differential via with grounding pins and 173.6 mil via stub.

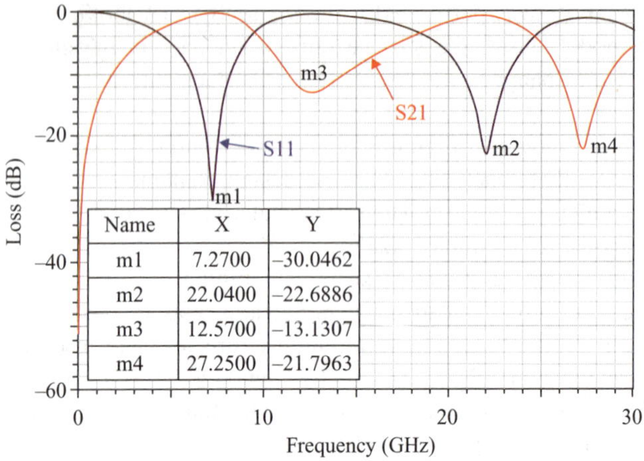

Name	X	Y
m1	7.2700	−30.0462
m2	22.0400	−22.6886
m3	12.5700	−13.1307
m4	27.2500	−21.7963

Figure 8.23 Differential return and insertion loss of a via designed to have a nominal differential impedance of 100 ohms.

at the 7.27-GHz resonance, the through signal is greatly attenuated by current dissipating in the barrel of the via stub. These effects are shown in Figure 8.24.

A topic that was not discussed here is the tools that allow importing of mechanical CAD drawings into the electrical modeling tools. This reduces the amount of work the electrical engineer has to do; however, modifications to the models will then have to be reflected into the mechanical drawings.

8.8 CONCLUSIONS

Chapter Conclusions

Simulation tools permit signal integrity engineers to design modern-day computers, personal electronics such as peak distortion analyses, cell phones, video game consoles, and even radar system electronics through the use of solutions to Maxwell's equations that cannot be handled with analytic theory. The theory developed in Chapters 1–7 has been applied to several examples with complex boundary conditions to exhibit some numerical tools that can be used to analyze high-speed electronic systems. Real-world systems have hundreds of buses that contain tens of thousands of nets in high-speed electronics. Without numerical solution techniques, engineers could not analyze the electrical currents and field intensities produced by high-speed electronic designs, and complex circuits as we know them today would not exist.

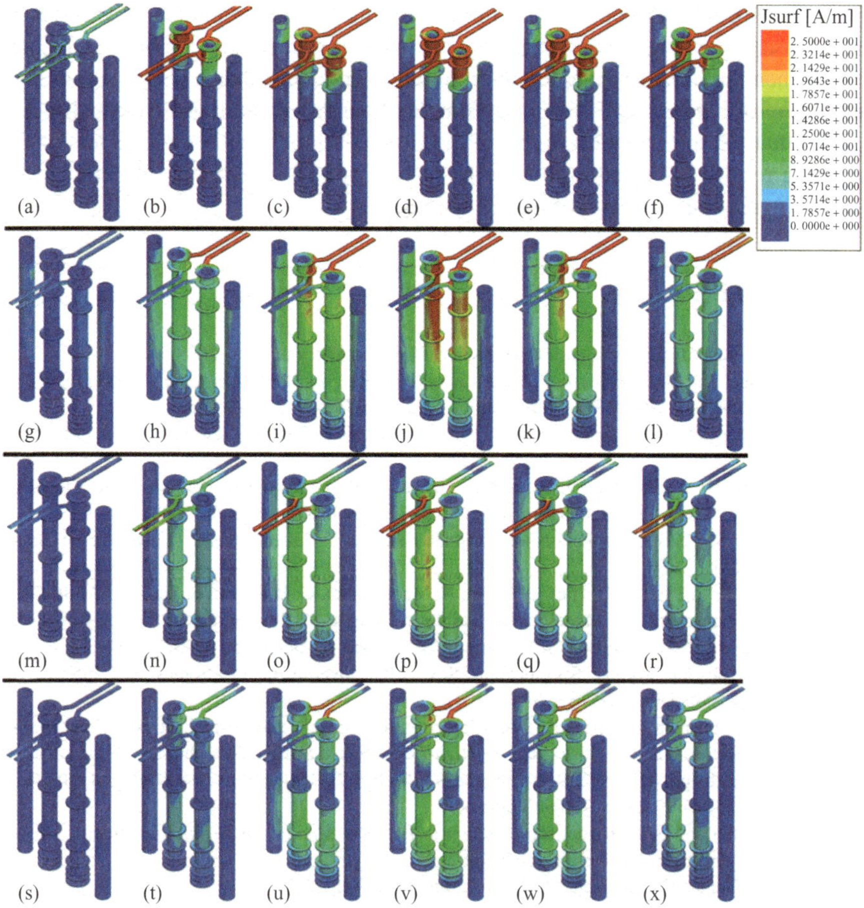

Figure 8.24 Differential via surface current plots: (a–f) Excitation at 1 GHz with phases of 0°, 22.5°, 45°, 90°, 135°, and 157.5°, respectively; (g–l) excitation at 7.27 GHz with phases of 0°, 22.5°, 45°, 90°, 135°, and 157.5°, respectively; (m–r) excitation at 12.57 GHz with phases of 0°, 22.5°, 45°, 90°, 135°, and 157.5°, respectively; (s–x) excitation at 22.04 GHz with phases of 0°, 22.5°, 45°, 90°, 135°, and 157.5°, respectively.

Book Conclusions

This book is not a how-to-use-simulations book and not a book for dummies. It assumes that the engineer has a good mathematical background and that he or she is able to learn complex subjects. The book is also not a rule-of-thumb book; rather, it provides the foundation about how those rules rest on a foundation described by *Maxwell's Equations*. It aims to provide a forefront engineer or physicists a means to evaluate and test emerging software that may make many (sometimes hidden) assumptions by comparing results to a solid theory based on analytical solutions

Bibliography

Abramowitz, Milton, and Irene A Stegun. *Handbook of Mathematical Functions.* National Bureau of Standards Applied Mathematics Series, 55, Superintendent of Documents, U.S. Washington, DC: Government Printing Office, 1964.

Arfken, George. *Mathematical Methods for Phyicists.* 3rd ed. New York: Academic Press, 1985.

Balanis, Constantine, A. *Advanced Engineering Electromagnetics.* Hoboken, NJ: John Wiley & Sons, 1989.

Chen, Yinchao, Qunsheng Cao, and Raj Mittra. *Multiresolution Time Domain Scheme for Electromagnetic Engineering.* Hoboken, NJ: John Wiley & Sons, 2005.

Cheng, David K. *Fundamentals of Engineering Electromagnetics.* Upper Saddle River, NJ: Prentice Hall, 1993.

Clark, Ronald C. *Einstein: The Life and Times.* New York: Harry N. Abrams, 1984.

Collins, Royal Eugene. *Mathematical Methods for Physicists and Engineers.* New York: Reinhold Book Corporation, 1968.

Davidson, David B. *Computational Electromagnetics for RF and Microwave Engineering.* Cambridge: Cambridge University Press, 2005.

DeBye, P. *Polar Molecules.* New York: Dover Publications, 1929.

Dwight, Herbert Bristol. *Tables of Integrals and Other Mathematical Data.* New York: Macmillan, 1961.

Edminister, Joseph A. *Electromagnetics.* 2nd ed. Schaum's Outline series. McGraw-Hill, 1993.

Feynman, Richard P., Robert B. Leighton, and Matthew Sands. *The Feynman Lectures on Physics*, vol. 2: *Electromagnetism and Matter.* Reading, MA: Addison-Wesley, 1964.

Feynman, Richard P., Robert B. Leighton, and Matthew Sands. *The Feynman Lectures on Physics*, vol. 3: *Quantum Mechanics.* Reading, MA: Addison-Wesley, 1965.

Ghosh, S. N. *Electromagnetic Theory and Wave Propagation.* 2nd ed. New Delhi: CRC Press, Narosa Publishing House, 1998.

Gradshteyn, S. I., and I. M. Ryzhik. *Table of Integrals, Series, and Products.* 4th ed. Translated by Yu. V. Geronimus and M. Yu. Tseytlin from Russian. New York: Academic Press, 1965.

Hall, Stephen H., and Howard L. Heck. *Advanced Signal Integrity for High-Speed Digital Designs.* Hoboken, NJ: John Wiley & Sons, 2009.

Hall, Stephen H., Garrett W. Hall, and James A. McCall. *High-Speed Digital System Design: A Handbook of Interconnect Theory and Design Practices.* Hoboken, NJ: John Wiley Interscience, 2000.

Halliday, David, Robert Resmick, and Jearl Walker. *Fundamentals of Physics.* 6th ed. New York: John Wiley & Sons, 2001.

Harrington, Roger F. *Time-Harmonic Electromagnetic Fields.* Hoboken, NJ: Wiley Interscience, John Wiley & Sons, 2001.

Hayt, William H., Jr. and John A. Buck. *Engineering Electromagnetics.* 7th ed. New York: McGraw-Hill, 2006.

Huray, Paul G. *Maxwell's Equations.* Hoboken, NJ: John Wiley & Sons, 2009.

Iles, George. *Flame Electricity and the Camera, Man's Progress from the First Kindling of Fire to the Wireless Telegraph and the Photography of Color*. New York: Doubleday & McClure, 1900.

Isaacson, Walter. *Einstein: His Life and Universe*. New York: Simon and Schuster, 2007.

Jackson, John David. *Classical Electrodynamics*. 3rd ed. Hoboken, NJ: John Wiley & Sons, 1999.

Johnson, Howard, and Martin Graham. *High-Speed Signal Propagation: Advanced Black Magic*, Upper Saddle River, NJ: Prentice Hall, 2003.

Kittel, Charles. *Introduction to Solid State Physics*. 7th ed. Hoboken, NJ: John Wiley & Sons, 1996.

Kittel, Charles. *Quantum Theory of Solids*. Hoboken, NJ: John Wiley & Sons, Inc., 1963.

Laudau, L. D., and E. M. Lifshitz. *The Classical Theory of Fields*. 4th revised English ed. Translated by Morton Hamermesh from Russian. London: Pergamon Press, 1975.

Landau, L. D., E. M. Lifschitz, and L. P Pitaevskii. *Electrodynamics of Continuous Media*. 2nd ed. Translated by J.B. Sykes, J.S. Bell, and M.J. Kearsley from the second edition of Elektrodinamika sploshnykh sred, Izdated'stvo 'Nauka', Moscow, 1982. Amsterdam, The Netherlands: Elsevier, Butterworth-Heinemann, 2006.

Mathews, Jon, and R. L. Walker. *Mathematical Methods of Physics*. New York: W. A. Benjamin, 1965.

Maxwell, James Clerk. "A Comparison of the Electric Units and on the Electromagnetic Theory of Light." *Philosophical Transactions, Royal Society of London* 158 (1868). Reprinted in Maxwell, Scientific Papers.

Maxwell, James Clerk. *A Treatise on Electricity & Magnetism*. Vol. 1. Unabridged, slightly altered, republication of the third edition, published by the Clarendon Press in 1891. New York: Dover, 1954.

Maxwell, James Clerk. *A Treatise on Electricity & Magnetism*. Vol. 2. Reprint of the original edition. Merchant Books, 2007.

Maxwell, James Clerk. *The Dynamical Theory of the Electromagnetic Field*. Reprint of the original. Edited by Thomas F. Torrance. Eugene, OR: Wipf and Stock, 1996.

Ohanian, Hans C. *Gravitation and Spacetime*. New York: W. W. Norton, 1976.

Orlando, Terry P., and Kevin A. Delin. *Foundations of Applied Superconductivity*. Reading, MA: Addison-Wesley, 1991.

Paul, Clayton R. *Electromagnetics for Engineers. with Applications to Digital Systems and Electromagnetic Interference*. Hoboken, NJ: John Wiley & Sons, 2004.

Pozar, David M. *Microwave Engineering*. 3rd ed. Hoboken, NJ: John Wiley & Sons, 2005.

Schelkunoff, S. A. *Electromagnetic Waves*. New York: D. Van Nostrand, 1943.

Schlesinger, Mordechay, and Milan Paunovic. *Modern Electroplating*. Electrochemical Society Series. 4th ed. New York: John Wiley Interscience, 2000.

Scott, William T. *The Physics of Electricity and Magnetism*. New York: John Wiley & Sons, 1962.

Spiegel, Murray R. *Schaum's Outline Series. Complex Variables: With an Introduction to Conformal Mapping and Its Applications*. McGraw-Hill, 1999.

Standage, Tom. *The Victorian Internet; The Remarkable Story of the Telegraph and the Nineteenth Century's On-Line Pioneers*. New York: Walker, 1998.

Tyndall, John Tyndall. *Fragments of Science*. 6th ed. Vols. I and II. New York: Appleton, 1897.

Wills, A. P. *Vector Analysis with an Introduction to Tensor Analysis*. New York: Dover, 1958.

Index

The Foundations of Signal Integrity, by Paul G. Huray